Aquatic Microbial Communities

Garland Reference Library
of Science and Technology
Volume 15

Aquatic Microbial Communities

edited by
John Cairns, Jr.

Garland Publishing, Inc.
New York & London
1977

Library of Congress Cataloging in Publication Data

Main entry under title:

Aquatic microbial communities.

(Garland reference library of science and technology; v. 15)
  Includes bibliographies.
   1.   Aquatic microbiology—Congresses.
I.   Cairns, John, 1923-   II.   Series.

QR105.A69   576'.19'2   76-52711
ISBN 0-8240-9860-9

Printed in the United States of America

CONTRIBUTORS
Current Affiliations

Harold L. Allen, *University of Toledo*

Stuart S. Bamforth, *Newcomb College of Tulane University*

Duncan C. Blanchard, *State University of New York at Albany*

John Cairns, Jr., *Virginia Polytechnic Institute and State University*

G. Dennis Cooke, *Kent State University*

Louis H. DiSalvo, *Naval Biomedical Research Laboratory, Oakland, California*

Thomas Fenchel, *University of Aarhus, Denmark*

Robert W. Gorden, *South Colorado State College*

Barry T. Hargrave, *Bedford Institute of Oceanography, Nova Scotia*

J. E. Hobbie, *Marine Ecology Lab, Woodshole, Massachusetts*

Jaraslov Hrbáček, *Hydrobiologicka Laborator, Czechoslovakia*

Bassett Maguire, Jr., *The University of Texas at Austin*

C. Nalewajko, *University of Toronto--Scarborough College*

Scott W. Nixon, *University of Rhode Island-Narragansett Campus*

Howard T. Odum, *University of Florida*

Bruce C. Parker, *Virginia Polytechnic Institute and State University*

Robert A. Paterson, *Virginia Polytechnic Institute and State University*

Ruth Patrick, *The Academy of Natural Sciences, Philadelphia, Pennsylvania*

Georgina A. Phillips, *Bedford Institute of Oceanography, Nova Scotia*

P. Rublee, *North Carolina State University*

CONTRIBUTORS

George W. Salt, *University of California at Davis*

George W. Saunders, *U. S. Energy Research and Development Administration*

J. K. G. Silvey, *North Texas State University*

Frieda B. Taub, *University of Washington*

Mary Ann Wachtel, *Saint Louis University*

Richard G. Wiegert, *University of Georgia*

J. T. Wyatt, *U. S. Environmental Hygiene Agency*

William H. Yongue, Jr., *Virginia Polytechnic Institute and State University*

CONTENTS

TABLE OF CONTENTS

PREFACE

The predecessor to this book, entitled *The Structure and Function of Freshwater Microbial Communities*, was based on the program of the 1969 Annual Meeting of the American Microscopical Society held in conjunction with the American Institute of Biological Sciences at the University of Vermont. That volume was published in September, 1971, by the Virginia Polytechnic Institute and State University Research Publications Division. All authors were required to attend the symposium and present the papers personally. Since we were not able to offer travel funds, a number of colleagues in the United States and elsewhere could not be invited to attend. This disadvantage was substantially offset by the exchange of information that occurred during the symposium and was summarized in the concluding remarks given by Professor David G. Frey. This volume represents a natural outgrowth of the earlier volume in that it is more inclusive both in geographic distribution of contributors and in the area of coverage. Unfortunately, again travel funds were not available and as a consequence there was no interaction between the participants, which might have produced some interesting insights. However, all participants had the advantage of reading the proceedings of the 1969 symposium. A majority of the original authors have agreed to contribute to this volume and most have substantially revised their contributions and others have modified them slightly. Unfortunately, some of the original authors and several new contributors could not meet the press deadline and requested that their names be dropped.

PREFACE

I am indebted to Darla Davis Donald who corresponded with reviewers, edited copy, and took care of many other details. I am also indebted to the following who helped with the review of parts of this book: J. O. Corliss, University of Maryland; J. W. Winchester, Florida State University; A. F. Carlucci, University of California at San Diego; C. Nalewajko, University of Toronto; D. Botkin, Yale University; and B. Parker, G. Simmons, K. L. Dickson, N. Krieg, A. Hendricks, A. Buikema, and D. McLean of Virginia Polytechnic Institute and State University. The difficult and painstaking job of preparing this book in camera-ready copy was done by Mary Etta Quesenberry. Her diligence and patience are greatly appreciated.

Chapter 1

ADAPTATIONS FOR PHOTOREGENERATIVE CYCLING

Howard T. Odum, Scott W. Nixon, and Louis H. DiSalvo

Department of Environmental Science
University of Florida
Gainesville, Florida

CONTENTS

1

CONTENTS

ABSTRACT

Ecological systems requiring productive photosynthesis and regenerative respiration to maintain smooth flowing mineral cycles may develop under circumstances in
which the regenerative half of the cycle tends to be limiting.

In such stable, less stressed systems as coral reefs,
complex animals and microbial communities accomplish
effective mineral recycling through their specializations
and diversity. Even the regeneration of limestone may be
managed by microbial sediment communities located within
the reef formations and under the adaptive management of
burrowing animals. There are some conceptual similarities
between the reef renewal and urban renewal of our cities.

However, for stressed and other systems in which organic matter and structure accumulate, recycling may be
boosted by adaptations that inject special energy into
catabolism and disordering, thus reducing bottlenecks in
the cycles of regeneration. Aquatic systems of this type
develop in such stressed conditions as brines, temperature
extremes, and pollution that interfere with complex animal
and microbial work. These systems may boost cycling by
diverting solar energy to help break down structure and
storage through photorespiration, drying with fire, and
melting of permafrost. Shift from successional state to
climax state may often require the addition of energies
to the regenerative half of the systems and special adaptation for these processes. The general prevalence of
photoregeneration along with photosynthesis in daylight
cancels many environmental exchanges and interferes with

efforts to measure either process by gas-exchange techniques. Net gains develop only to the extent that the two
processes are momentarily lagging or out of phase. There
is oxygen consumption and carbon dioxide utilization at
sunrise but since the processes giving rise to each are
coupled they are difficult to measure independently. Com-
partmental systems analysis allows separation of the two
processes and the simulation of electrical models provides
some reversed transients of the same class as those observed in gas exchanges of oxygen and carbon dioxide.

Carotenoid pigments may be important in photoregenerative functions. If so, the ratio of yellow to green
plant pigments may measure the relative use of solar
energy in the two processes. When photosynthesis is
limiting, chlorophyll increases; when respiration is potentially limiting, other pigments may increase to accelerate regeneration. The pigment ratio may indicate the
relative switching of energy to the two half systems of
P or R. The pigment ratio increases as climax approaches
only when the initial succession started with a regime of
net production and initial conditions of high nutrients
and low organic stocks. Some photorespiratory capability
may characterize all ecosystems. The photorespiratory
use of solar energy by brine ecosystems contrasts with
the coral reef's channeling of more organic matter into
regenerative food chains, both providing viable formulas
for their environmental circumstance.

INTRODUCTION

Many ecological systems resemble balanced aquaria
with a photosynthetic, productive process (P) balanced
by a regenerative, respiratory process (R) between which
many minerals and work services cycle. Even those systems
that have important import and export flows have P-R

modules in varying quantity. The behavior of systems of
this class has been much studied with modeling and with
measurements. The essence is in two storage units and
one multiplicative feedback loop. Figures 1.1 and 1.2
summarize energy flows, mineral cycles and the essence
of kinetics as represented in the electrical models (Odum
et al. 1963a; Milsum 1966). Figure 1.3 is a legend for
the energy network symbols, each one of which represents
a cluster of associated functions and differential equations.

To be well adapted to available conditions and energies is to process energies competitively with effective
cycling, without blockages, accumulations, or shortages.
This commentary concerns the special adaptations for effective cycling that develop under different environmental
conditions. For many systems the model in Figure 1.1 is
much too simple to predict diurnal patterns of exchange.

DISCUSSION

Regeneration by Microbe-Animal Cities

Many ecosystems, such as coral reefs, contain intri-
cate regenerative systems of animals and microbes that
the animals in part manage. With well-know great diversities, high degrees of behavioral programming, and much
specialization by both animals and microbes, the regeneration is clean, with little organic matter accumulating
or remaining uncycled. In many coral reefs in the Pacific
skeletons are formed and dispersed so as to maintain a
reef platform and structure at a steady state near the
extreme low water spring tide level. Documentation of
the great diversity of microbial processes in these reefs
was provided by DiSalvo (1969). Shown in Figure 1.4 are
evidences of chitin-digesting bacteria that may serve as
a zipper action in releasing the cementing materials of

5

A. Energy

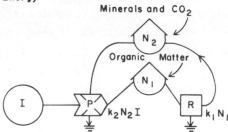

B. Mineral Cycle

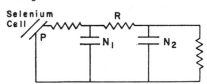

C. Flow of equations in block diagram

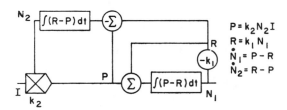

$$P = k_2 N_2 I$$
$$R = k_1 N_1$$
$$\dot{N}_1 = P - R$$
$$\dot{N}_2 = R - P$$

D. Passive analog

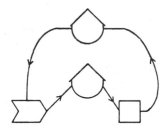

FIG. 1.1. Characteristic basic system with photosynthesis and regenerative respiration balanced with respect to matter. (A) energy; (B) mineral cycle; (C) block diagram of integral equation; (D) passive electrical analog.

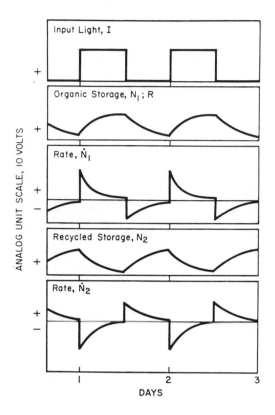

FIG. 1.2. Transient responses of the basic P-R system to on and off light periods with an analog computer circuit wired to represent the block diagram of Fig. 1.1C, using coefficients for a terrestrial microcosm (Odum *et al.* 1970).

the calcareous structure so that the nutrients may be reused and the structures rebuilt to the needs of the new components. Synthesis and replacement of small heads as a group constitutes a steady state of rapid growth and decomposition in the reef as a whole.

That both solid mineral and soft biological structures are so effectively maintained in continual construction and renewal is a triumph of organization and specialization. DiSalvo (1969) described a large quantity of

7

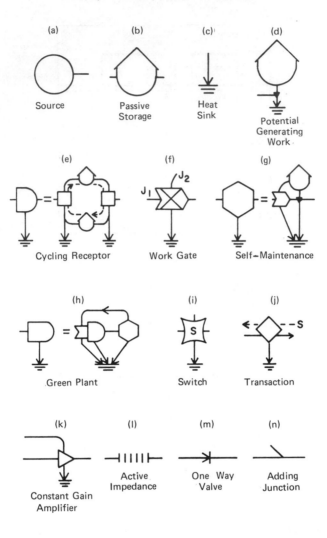

FIG. 1.3. Energy network symbols from Odum and Pigeon (1970).

highly metabolic ooze within the interstices of the super-
ficial reef framework (regenerative spaces) with higher
respiratory rates where current levels are higher. The
energy diagram in Figure 1.5 is compartmentalized to show
one of the overall features in regenerative metabolism

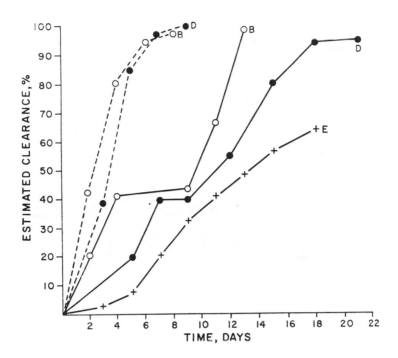

FIG. 1.4. Clearance of particulate chitin in agar by coral reef microbial regenerators obtained from within skeletal spaces, providing evidence for the existence of bacteria capable of breaking down the organic (chitinous) matrix of corals; Kaneohe Bay, Hawaii. B, outer bay; D, midbay; E, inner bay. ——, nonenriched chitin agar; - - -, enriched chitin agar.

and oxygen resulting from symbiotic interactions of infaunal animals that can pump water. The model provides acceleration of pumping by animals when oxygen levels are low inside to maintain a homeostatic balance by re-aerating the inside of the head and reducing the numbers of particulate oxygen consumers through increment filter feeding. The microbial communities are served and managed in this and other ways by the macroscopic motions of the animals. The large microbial diversity provides nutritional specialization for effective mineralization and a high protein supplement to animals within the head. In other

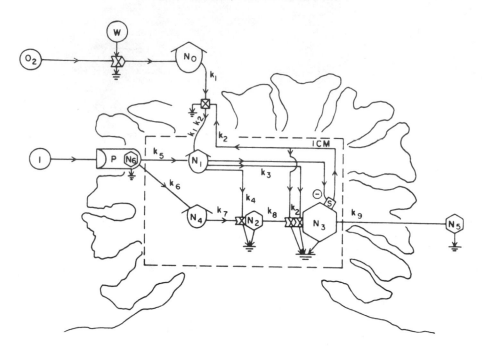

FIG. 1.5. Energy diagram of some compartments of a coral head with animals, microorganisms, and work actions: I, sun's energy; W, wave action; S, sensory mechanism controlling pumping rates; N, oxygen in water; N_1, oxygen in water inside head; N_2, microbial populations; N_3, suspension feeders; N_4, organic substances of microbial ooze; N_5, carnivores, omnivores; N_6, plant populations; k_1-k_9, transfer coefficients.

systems, such as the Texas bays, migratory populations that enter with the tides keep respiration in phase with photosynthesis (Odum 1967).

Under the stable temperature, salinity, and light regimes found on many reefs, the animal-microbial combinations can effectively mineralize without accumulations and deficits. The nutrients released pass to the coral heads on the outside of the live and dead head clusters, where photosynthesis operates in the zooxanthellae. The

resulting production is used locally to support the corals
and the multitude of other animals. Even here, however,
there are auxiliary pigments (Margalef 1959) that may
have regenerative significance by influencing metabolism.

Responses to Limiting Photosynthesis

The adaptive responses of photosynthetic systems to
limits in their input requirements, including nutrients
and light, are well known. With any deficiency in required nutrients, the chlorophyll is decreased, leaves
become chlorotic, and algal cultures turn yellow. Emerson
and Lewis (1943), for example, described deficiencies of
iron reducing chlorophyll, while the ratio of chlorophyll
to photosynthesis was maintained. With limitations in
light chlorophyll is increased, as often described for
shade adaptation. Wassink *et al*. (1956), for example,
reported shade adaptation in a deciduous tree; Gessner
(1937) showed shade adaptation in land and water plants;
and Steeman-Nielsen (1962) showed shade adaptation in
plankton.

The relationship of varying input power to constant
biochemical load of downstream work processes is shown
in Figure 1.6. Here the Michaelis-Menton kinetics of a
cyclic process interacting with an input, in this case
light, provides a varying load ratio of input to output.
For maximum power transfer each step involving an energy
storage against a potential generating force must have
a 50% energy transfer with 50% passing into dispersed
heat, generating entropy in the environment (Odum and
Pinkerton 1955). The system can help to regulate its
input-output loading in the maximum power range (50%
efficiency for each energy storage step) in bright light
regimes by reducing its quantity of cycling receptor
material, in this case chlorophyll, to that required to

11

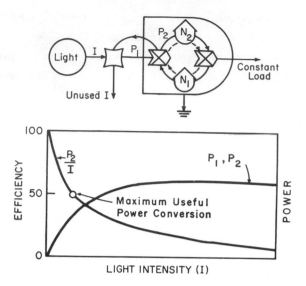

FIG. 1.6. Efficiency of Michaelis-Menton cyclic modules under constant output load and varying input.

absorb the needed input energy load. This allows excess to pass to the next receiver units (next layer of photosynthetic tissue). Each subsequent chlorophyll-bearing unit in turn may adjust its input light absorbed to equal its output load or adjust its load. The consequence of these component mechanisms is overall use of more light at optimum efficiency for maximum power by the ecological system. The predicted theoretical relation of optimum pigment to light input for constant output is given in Figure 1.7 along with an experimental example of response of *Chlorella* given by Phillips and Myers (1954). These graphs show the shade adaptation increase of chlorophyll, which tends to maintain constant output load.

The adaptive chlorophyll system serving as a gear increases with decreased input or increased output. In rich nutritive solution specialists at rapid growth are

12

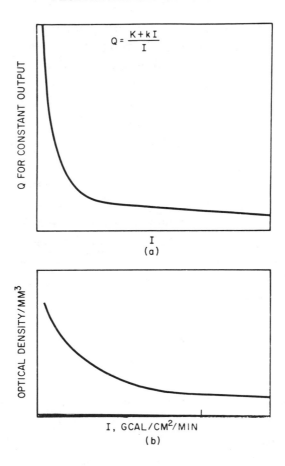

FIG. 1.7. Chlorophyll adaptations to light for maintaining steady output load. (a) theoretical from manipulation of Michaelis-Menton equation; (b) observed by Phillips and Myers (1954) for *Chlorella*.

not limited downstream through such mechanisms as nutrient regeneration and may operate with high chlorophyll. If, however, there is very bright light, they may be relatively limited by the high input-output ratio and reduce chlorophyll. The chlorophyll seems to be servoregulated to adjust the load ratio. When a system has inadequate light

13

it turns chlorophyll up; if its nutrients are limiting, however, the shortage may be solved by fertilization from outside or by some more effective method of recycling, the energy for which may also have to come from outside.

One of the adaptations of specialized algae, such as deep-water reds, to low light intensities is a provision for a second pigment system that contributes to photosynthesis. French and Fork (1961) studied the metabolic transients that result from the interplay of two-pigment systems, developing analog circuits to relate models to transient consequences that were compared with observed data. In a similar way, the arguments that follow on photoregeneration concern analog analysis of transients involving the balance of two-pigment systems, except that one pigment is suggested as photoregenerative.

Problems of Systems Limited by
Slow Respiration

Unlike the coral reef example, many ecological systems have inadequate associations of microbes and animals to fully recycle their organic storages. Various kinds of stresses, such as high, low, and variable temperature; high, low, and variable salinity; poisons; water shortages; and high organic input, make the adaptations of complex animal-microbial organizational patterns relatively expensive in energy costs. Many stress conditions thus lead to reduced diversities of animals and microorganisms. In these environments organic substances accumulate. High organic levels are a part of brines, arctic communities, and toxic situations in pollution. See, for example, the hyperbolic graph of decreasing organic matter with increasing species diversity (Odum 1967). What mechanisms tend to develop for mineral cycling when the energies of the stored matter are not adequate to operate an effective mineral recycling system?

14

PHOTOREGENERATIVE CYCLING

Theory of Photoregeneration

A solution to inadequate recycling is to inject special energies into the regeneration process. If the cycle of P and R is blocked at the R process, it may be economical to divert light energy from photosynthesis to help respiration in an amount required to balance the two processes. Photoregeneration is shown in its simplest form in Figure 1.8. The basic P-R model of Figure 1.1 has been modified by adding one photoregeneration pathway, a multiplicative work gate. Such a mechanism requires some energy absorbing material that is coupled to the regenerative process. Also, since decomposition is in the direction of disorder, less energy may be required than in the initial photosynthesis. When we look at various ecosystems that have organic accumulations, we find pigments and other light-receiving operations that are serving as photoregenerators or may be proposed to have this role pending more experimental testing. In Table 1.1 are listed some systems that seem to fit the hypothesis.

Inverted Transients

In Figure 1.9, from Nixon (1969), are diurnal curves of dissolved oxygen in a brine ecosystem. It was discovered that the normal pattern of day and night oxygen production and consumption was missing. Often there were inverted periods, with the oxygen decreasing after lights were on and increasing briefly when lights were off. The pattern was like that of a cactus (Ting and Dugger 1968). The pH-CO_2 was also reversed on many days. Such curves have been found in aquatic systems before, including some from the list in Table 1.1. The inverted pattern suggested photorespiration, a process well known in physiological work but not appreciated for its adaptive significance to mineral cycling and ecosystem adaptation. The shape of

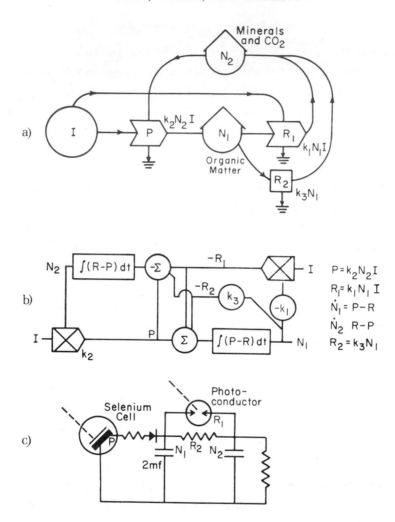

FIG. 1.8. Energy diagram of basic photosynthesis and respiration with photorespiration pathway added. (a) energy diagram; (b) flow of equations in block diagram; (c) passive analog circuit.

the graphs with time suggests time lags in ordinary RC (resistor-capacitor) passive electrical circuits. It was apparent that the behavior might be accounted for by fairly simple time lags involving two storages, representing

Table 1.1

Some Systems Which May Contain Photoregeneration

System	Reference in which data suggest photoregeneration
Mutant *Chlorella* culture lacking chlorophyll	Kowallik (1967)
Brines	Nixon (1969)
Waste stabilization lagoons	Eppley and Macias R (1963)
Shallow marine bay bottom	Bunt (1969)
Chromogenic bacteria in lakes	Henrici (1939), Odum (1967)
Blue-green algae	Odum and Hoskin (1957), Sollins (1969)
Snow algae	Fogg (1967)
Old *Chlorella* cultures	Fogg (1965)
Old blue-green algal cultures	Kingsbury (1956), Manny (1969)
Phytoplankton	Margalef (1959)
Desert vegetation with succulents	Ting and Dugger (1968)

organic and inorganic pools, and a photoconductor pathway, as well as by more elaborate models with more stages of the real regeneration and production represented separately.

Passive Model

In Figure 1.8 is a two-compartment model with a photoconductive element representing the mechanism for the reception of the light stimulus to photorespiration (Nixon and Odum 1970). This may be regarded as an equivalent circuit expressing kinetics in electrical language. It

was also constructed in hardware (Fig. 1.10) and some
transient responses of voltage were recorded following on
and off light input. Time constants were selected by
varying the resistances. As shown in Figure 1.11, a
transient lag was observed with some resistance settings.
In the manner of the theory, changes accumulating upstream
from a high-resisting part of the circuit were flushed
into downstream compartments by the photoconductor when
light was turned on. Conceptually therefore the observed
form of oxygen and carbon dioxide release was compatible
with the theory of photoregeneration as an adaptive recycling mechanism.

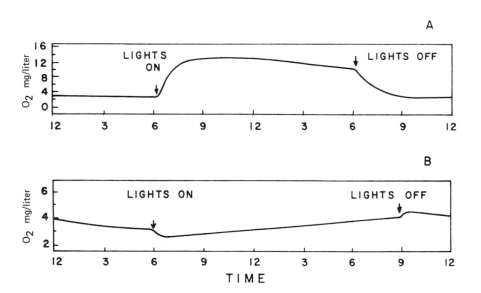

FIG. 1.9. (A) Continuous diurnal curve of dissolved oxygen in a 6%
salt microcosm with a healthy blue-green algal mat and moderate
nutrient levels. Patterns were normal with a square-wave light
input from 6 a.m. to 6 p.m. (B) Continuous diurnal curve of dissolved
oxygen in a 13% salt microcosm with a yellowed blue-green algal mat
and very low nutrient levels. Abnormal patterns were apparent with
a square-wave light input from 6 a.m. to 9 p.m.

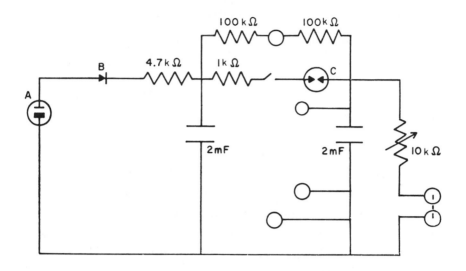

FIG. 1.10. Passive electrical analog model with two storage compartments A, silicon photoconductor; B, diode; C, cadmium sulfate photoconductor.

Operational Analog Model

Another procedure for analog simulation involves the standard operational amplifiers, which were used to model a similar network. In Figures 1.12 and 1.13 a somewhat more complex theory of photoregeneration is modeled. First an energy diagram is drawn (Nixon 1969) that is of the same class of network as the model in Figure 1.10 but derives suggestion from plant physiological work on "photoheterotrophy," such as that by Bunt (1969). The energy diagram automatically defines differential equations that go with each module as also shown in Figure 1.12. Each of these equations is block diagrammed and for each block an electrical analog component is wired (Fig. 1.13). Some experimental manipulation of coefficients (conductivities per unit storage equal to reciprocal of the time constant) provided the lagging transients like that in some observed gas-exchange curves. An example is given in Figure 1.14.

Characteristics of the Photoregenerative Model

The exercises in modeling and simulation show enough
features of the observed systems to help in the interpretation of the more complex real systems that are of this
class of phenomena. Where both P and R are increased by
the light, only the difference in effect on outside nutrient and gas reservoirs is observed. Metabolic quantities
are observed to vary with short transients affected by
the history of previous storages. When photoregeneration
is high, estimates of photosynthetic exchange are almost
meaningless as a measure of the underlying real processes.
Daytime cycles are very rapid, turning off at night with
the light. Stressed systems, which have been thought to
be low in primary productivity, may actually be as productive as other more easily measured types. To regard

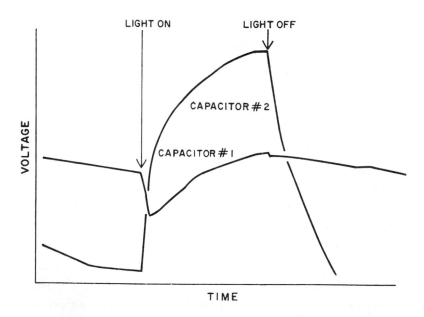

FIG. 1.11. Transients from passive analog (Fig. 1.10) which resemble
observed oxygen transients (Fig. 1.9).

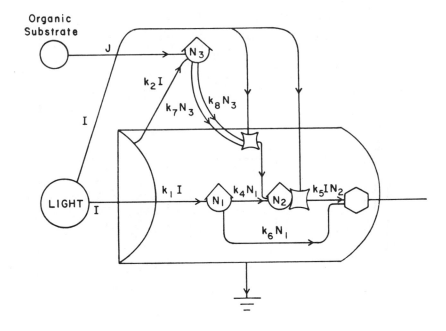

FIG. 1.12. Energy diagram of photoheterotrophy from physiological suggestions (Bunt 1969).

cactus deserts, tropical seas, brines, etc., as low in basic productivity seems erroneous, although we do not know how erroneous until the two processes of P and R can be separated and a realistic measure of gross production obtained. The simulation procedure is one way. Direct measurement of primary photon reception under field conditions is another. Measurements with enzyme inhibitors is another explored by Nixon (1969).

Physiological Mechanisms, Carotenoid Pigments

Kowallik (1967), studying a chlorophyll-free mutant of *Chlorella*, showed data suggesting carotenoids as the receptors of the light energy for photorespiration with

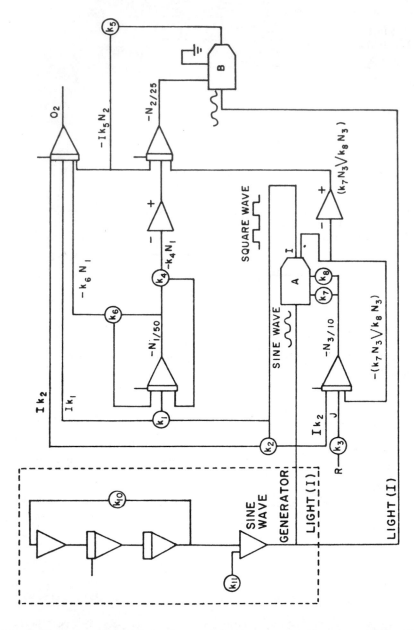

FIG. 1.13. Operational analog model for energy diagram of Fig. 1.10.

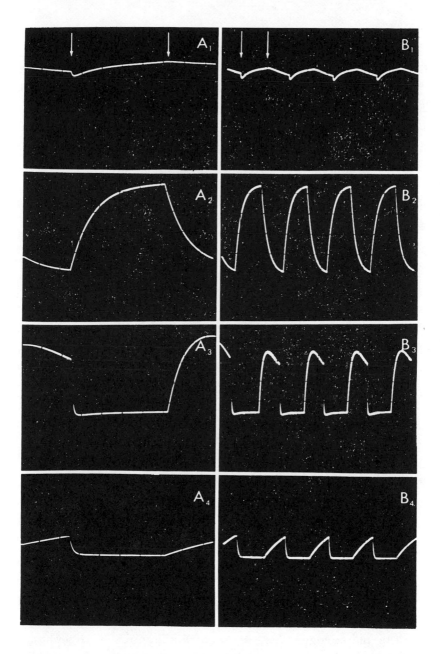

FIG. 1.14. Transients observed with active model of photorespiration. Series A shows the values for 1 day, while series B extends the scan over 4 days to show the steady-state behavior of the model. Arrows indicate when the light goes on (left) and off (right). A_1 and B_1, dissolved oxygen; A_2 and B_2, stored organics (N_1); A_3 and B_3, stored organics (N_2); A_4 and B_4, dissolved organics (N_3). The drop in N_2 during the dark is an artifact resulting from inaccuracy in the comparator at values near zero.

a hyperbolic limiting factor curve of output with intensity of blue light. This curve is consistent with the multiplier relationship in our model (Figs. 1.8 and 1.12). He cites the earlier discovery of similar phenomena by Emerson and Lewis (1943). The saturating intensity for photorespiration with *Chlorella* in blue light is one-tenth that for photosynthesis, as might be expected for a process working toward decomposition and regeneration. Light was also effective under anaerobic conditions in accelerating fermentation.

Carotenoid-Chlorophyll Ratio

The system maintaining chlorophyll has long been known to increase it when needs for photosynthesis are optimal and to decrease it when some other input is limiting. In short the chlorophyll system serves as an input load adjuster to the output chain, which is possibly consistent with other flows. When chlorophyll declines, carotenoids either increase or by comparison are left in relative predominance. There is a yellowing of plant systems when conditions for rapid net photosynthesis disappear. Margalef (1965) proposed the ratio of absorption of pigment extracts at 430 mμm to absorption at 665 mμm as a measure of ecosystem maturity in the sense of succession as well as evolutionary achievement of possible adaptations. He regarded the yellow carotenoids and other pigments as a symptom of increasing complexity of pigments and diversity of maintenance. Kingsbury (1956) and Manny (1969) correlate the ratio with nitrogen shortage.

The photoregeneration theory suggests a somewhat different interpretation. If chlorophyll is adjusting and activating when needed to keep input light load adapted to photosynthetic work chains downstream, perhaps in the same way carotenoids and other pigments of ecosystems

24

(plants, animals, and microorganisms) are adjusting to keep the photoregenerative energies adjusted to maintain respiration and regenerative rates. By this theory, as organic stocks accumulate with danger of becoming blockages, photoregeneration is increased either within plant species that have the mechanisms, by substitution of species, or in other microbial specialists that serve the need. The pigment ratio is thus proposed as a measure of the relative accumulation of organic and inorganic pools and of the switch-controlling diversion of light energy to the one pool that is overstocked, limiting flow downstream.

When systems are high in organic matter and photorespiration is high, the pigment ratio increases; when systems are high in inorganic materials, the ratio goes down. By this view an increasing Margalef ratio is a measure of succession and climax only for the common case where the system starts with low stocks of organics and high concentrations of inorganic nutrients. Species may specialize for this situation simply by omission of photorespiratory systems, and it is these species that are preadapted for our fast net production demands, such as in algal culture, agriculture, and forestry. Zelitch (1967), for example, found photorespiration small in corn. When a steady state develops with high inorganic input and high organic output a low ratio is climax.

When systems start the other way, with high organic stocks and low inorganic stocks, the ratio is predicted to be lower at climax. When input of a steady state is high in organics and low in inorganics the ratio will be high at climax. We were able to switch a flowing stream microcosm containing a blue green algal mat (Odum and Hoskin 1957) from green to yellow or back at will by control of the inorganic nutrition as shown in Figure 1.15.

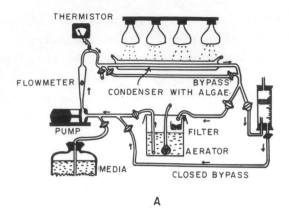

A

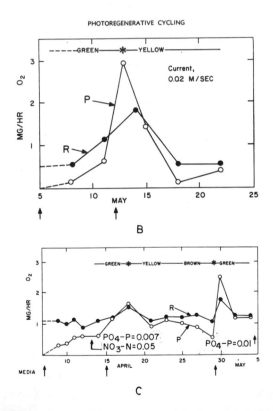

C

FIG. 1.15. Response of blue-green algal mat color and metabolism to nutrient addition (vertical arrows) in an artificial stream microcosm (Odum and Hoskin 1957). (A) circulating stream microcosm; (b) low-current sequence; (c) high-current sequence.

In one series protozoan and flatworm populations developed, making a more effective mineral cycling system. In this series, the color remained green and the orange did not develop. Some pigment ratios are given in Table 1.2.

Further confirmation of the role of photorespiration to drive recycle in ecosystems where organic matter is high and nutrients low was provided by studies of *Halobacteria* by Stoeckenius (In press). In studies of photochemical response of isolated membranes of *Halobacteria* containing bacteriorhodopsin, he found the membrane acting as a light-driven proton pump. In the energy diagram in Figure 1.16 the action of light in generating positive protons is shown as means for oxidizing organic matter; the negative charges generated by light at the same time are neutralized by reaction with oxygen.

Table 1.2

Some High Carotenoid: Chlorophyll Ratios

System	D430/D665[a]	Source
Brines	5–8	Nixon (1969)
Coral reef in Puerto Rico. *Acropora sp.*	5.7	Odum and Cintrón (1970)
Phytoplankton off the coast of Spain	2–10	Margalef (1965)
Climax blue-green algal mat in a flowing water microcosm	8	Odum and Hoskin (1957)
Lichen growing on granite rock. *Cladonia* sp.	10.8	Lugo (1969)

[a]Optical densities of acetone extracts.

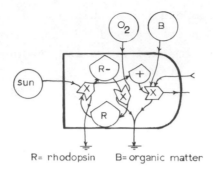

R= rhodopsin   B= organic matter

FIG. 1.16. Energy circuit model of photooxidation with rhodopsin (See text for explanation).

Examples of Systems with Photoregenerative Tendencies

While almost all ecosystems except specialized early successional pioneer associations may have the photoregenerative mechanisms, there are some systems where evidence is already available. In other situations this theory seems to explain some phenomena that had questionable adaptive significance before. Included in Table 1.1 are many systems in which light energy may be diverted to accelerate regenerative flux, helping to maintain overall metabolism at high levels.

Blue-Green Algal Mats

The blue-green mat systems in shallow marine waters in Texas and in culture tanks like the stream microcosms are green when nutrients are added and yellow when nutrients are low, but total metabolism and chlorophyll values remain high even when the predominant color at the surface is yellow (Odum *et al.* 1958, 1963b, c). Other stressed systems with yellowed blue-greens are the hot springs studied by Brock and Brock (1969), the brines described

by Nixon (1969), and the cultures of Kingsbury (1956).
Sollins (1969) found delayed transients of oxygen in odd
blue-green mats in the laboratory. Pink bacteria accompanied blue-green mats in Texas lagoons receiving oil-rich
bleedwater wastes (Odum *et al.* 1963b).

Chlamydomonas in Anaerobic Waste
Stabilization Lagoons

A well-documented case of photoassimilation is given
by Eppley and Macias (1963) where a species of *Chlamydomonas* in waste ponds in the Mohave desert used pigments
to absorb and photorespire acetate produced by anaerobic
bacteria. Many kinds of high-organic pollutions have
colored and specialized algae that may turn out to operate
similarly.

Many algae, such as *Chlorella* and *Haematococcus*, in
old or nutrient-limited cultures develop yellow pigments
along with excesses of fats and other organics and their
cells may increase in size. Fogg (1965) summarized these
characteristics and used a colored plate to dramatize the
deep green versus red phases of growing and stationary
populations. These characteristics make sense mechanisti-
cally and adaptively as a switch in regimes for recycling
under conditions when there is not a complex ecological
community to perform the service of renewal.

A Diatom in Organic Marine Sediments

Bunt (1969) isolated and studied a diatom from organic sediments of Biscayne Bay, which was found to assimilate a variety of organic compounds at different rates
in the light and in the dark. Metabolism of the assimilated organics, however, proceeded only in the light, a
process Bunt termed "photoheterotrophy."

Corals

DiSalvo (1969) presented spectrophotometric curves of coral pigments from Kaneohe Bay, Hawaii. Inner bay stations stressed with terrestrial runoff and sewage pollution had higher carotenoid to chlorophyll ratios (3.4) compared to outer bay stations (2.4) and midbay stations (2.4). Corals with their symbiotic zooxanthellae do have abundant carotenoid pigments and adequately recycle their own photosynthetic production.

Brines

Medium and high brines have colored pigments in all their populations, including *Artemia*, halophilic bacteria, the alga *Dunaliella salina*, and the black pigmented ciliate *Fabrea* (Nixon 1969). High brines are high in dissolved organic matter, diversity of microbes is small, and photorespiratory energy may be essential to cycling. Brines are green with blue-green algae and *D. viridis* at an earlier lower salinity, high-nutrient stage. Nixon (1969) provides systems analyses of these systems.

Chromogenic Bacteria in Normal Waters

In the surface layers of normal types of waters, general plating of bacteria in low-nutrient agar in the dark or light yields chromogens, brightly colored bacterial colonies of many kinds, which develop slowly on the agar. These are much more numerous from lighted surface waters than from dark environments. Henrici (1939) described the pattern in lakes of Wisconsin; Odum (1967) described this pattern in Silver Springs, and others have found this in the sea. DiSalvo (1969) found more pigmented bacteria near the shore in Hawaii, where organic matter was more concentrated than in Eniwetok. Are these

pigmented nonphotosynthetic forms pigmented for photoregenerative purposes? If so, then photorespiration may be
a normal part of most aquatic ecosystems. Many workers
have mentioned occasional unpublished experiences with
dark-light bottle work in rich organic waters in which
light bottles have consumed more oxygen than the dark
controls. The extent to which this is prevalent is the
extent of an error in measuring photosynthesis by known
methods.

Sitka Spruce and Permafrost

Although not photorespiration, an adaptation of dwarf
Sitka Spruce in Fairbanks, Alaska, may be another case of
photoregeneration. There the spruce boughs are short, not
spreading widely as they do farther south. The ground
between trees is left exposed to summer light. If the
limiting property is regeneration of minerals hindered by
soil ice, the ground heating provides a shunt of light to
regenerative melting.

Fire Climax

Monk (1966) in dry areas of Florida suggests ever-
green leaves as an adaptation for cycling through continuous fall, drip, and other characteristics. Some, but
not all, dry systems develop storages of organic matter
that accumulate until they are mineralized by fire. The
adaptation of plant structure to permit this and to provide driable and inflammable substrates provides a means
for diverting the potential energies of the sun's heating
into drying so that the substrates may support fire regeneration, as in the Southeastern United States.

Succulents

The well-known metabolic day-night inversion of gas exchange in such succulents as cactus seems to be another case of delayed transients. Carbon dioxide is fixed and oxygen released at night. For example, Ting and Dugger (1968) describe intermediates in the biochemical processing so that light fixed during the day is stored in chemical potential energy of organic acids (malic acid) without full processing until night. Malic storages decay faster in light. Recognized already as a water conservation mechanism, the possibilities of an accelerating photorespiration role may be considered when outside nutrients and water are limiting and only the internal cycle is available. Cactus chlorophyll is very high when water is abundant and much less when conditions are dry outside.

Autumn Leaves

In the temperate forest in autumn and in tropical forests at the time of the dry season, the leaves turn red and yellow as chlorophyll is removed and other pigments remain. Where light is brighter, the color seems brighter and develops sooner. Are these pigments serving to accelerate reduction, catabolism, and translocation of valuable nutrients prior to leaf fall? In bright light where photosynthesis has accumulated more photosynthates, a program of reduction may be greater.

Many shade plants, such as aquatic plants or tropical forest bromeliads, turn red when placed in the sun. With their load of utilization of photosynthate fixed by genetic characteristics, excess light may unbalance the ratio of photosynthesis to respiration. Decreasing chlorophyll and increasing other pigments may keep the system's internal cycles operable and more nearly balanced.

Red Algae on Snow

On the surface of old snow there may be growths of red algae. In this stressed environment, the pigments may serve to accelerate the recycling by utilization of light. At least there is heating through thermal absorption. Are there more specific energy pathways of photorespiration? Fogg (1967) found 4500 cells/cc with predominance of carotenoid pigments and low rates of apparent photosynthesis.

New Yellow Leaves in Climax Forests

The new growing leaves of rain forests and temperate deciduous forests emerge yellow with pigments prior to addition of chlorophyll. At this stage respiration is in excess of photosynthesis, and organic substrates are being translocated in from other parts of the tree for growth. Is it possible that photorespiration is a predominant process at this stage, the extra energy of the sun coupled to help with the growing processes?

Lignin Derivatives in Waters

Heavily colored swamp waters and derivatives from paper mills have aromatic compound fractions of lignin breakdown that are highly colored, absorbing bright lights in tropical waters, such as the Amazon. Many such waters, including the Santa Fe River in Florida, have very high respiration, their oxygen level holding half saturation with a balance between oxygen diffusing in and consumption of oxygen by the aquatic system. Sometimes experiments on oxygen change in light and dark bottles show greater

respiration in the light bottles than in the dark ones.
If through temperature elevation or other means, some
photochemical boost to consumption is involved either
directly or through bacterial actions, dark swamp waters
may qualify for another case of photoregeneration.

Primitive or Specialized Roles

Since photorespiration seems so widely distributed
in many kinds of communities that seem primitive, is it
possible that this was a more prevalent process in the
past? How many of our fossil plants, and even animals,
derived aid from adaptive coupling of light to consumptive
and growth processes? Was the origin of life a coupling
of two photic processes, photopolymerization and photo-
dissociation?

Implication for Urban Man

If the many special adaptations for diverting energies
into regenerative recycling and reorganization of natural
systems is a part of effective persistence and survival,
some parallel may be seen in man's urban culture, which
is in danger of becoming clogged with accumulations, wastes,
old structures, and other manufactured units of our system.
The instinctive tendencies of some youth to be destructive
may be serving now as in the past as part of a probing for
the emerging self-design process, stabilizing man's system
by developing specialist modules for improving the cycling
and reordering of our cities and other institutions.

The decaying cores of many of our largest urban
centers stand as mute evidence of the need for continual
regeneration and regrowth. Man can choose to channel his
energies in this direction through uncontrolled fires,
riots, and continual decays or through the best efforts
of controlled urban renewal and social concern.

34

ACKNOWLEDGMENTS

Dr. A. Moore, Lawrence Burns, and Robert Kelley aided with simulation and scaling of the operational analogs.

REFERENCES

Brock, T. D., and Brock, M. L. 1969. Effect of light intensity on photosynthesis by thermal algae adapted to natural and reduced sunlight. *Limnol. Oceanog. 14*, 334–341.

Bunt, J. S. 1969. Observations on photoheterotrophy in a marine diatom. *J. Phycol. 5*, 37–42.

Caperon, J. 1967. Population growth in micro-organisms limited by food supply. *Ecology 48*, 709–889.

DiSalvo, L. H. 1969. Regenerative functions and microbial ecology of coral reefs. Ph.D. dissertation, University of North Carolina, Chapel Hill.

Emerson, R., and Lewis, C. M. 1943. The dependence of the quantum yield of *Chlorella* photosynthesis on wave length of light. *Am. J. Bot. 30*, 165–178.

Eppley, R. W., and Macias, F. M. 1963. Role of the alga *Chlamydomonas mundana* in anaerobic waste stabilization lagoons. *Limnol. Oceanog. 8*, 411–416.

Fogg, G. E. 1965. *Algal Culture and Phytoplankton Ecology*. University of Wisconsin Press, Madison and Milwaukee. 126 pp.

Fogg, G. E. 1967. Observation on the snow algae of the South Orkney Islands. *Phil. Trans. Roy. Soc.* Series B, *252*, 279–287.

French, C. S., and Fork, D. C. 1961. Computer solutions for photosynthesis rates from a two-pigment model. *Biophys. J. 1*, 669–681.

Gessner, F. 1937. Untersuchungen uber Assimilation und Atmung submerser Wasserpflanzen. *Jb. wiss. Bot. 85*, 267–328.

Henrici, A. T. 1939. Problems of lake biology. *Am. Assoc. Advan. Sci. Sympos. 10*, 39–64.

Kingsbury, J. M. 1956. On pigment changes and growth in the blue green alga, *Plectonema nostocorum* Bornet ex Gomont. *Biol. Bull. 110*, 310–319.

Kowallik, W. 1967. Chlorophyll-independent photochemistry in algae. In *Energy Conversion by the Photosynthetic Apparatus*. Brookhaven Symposia in Biology, No. 19, pp. 467–477.

Lugo, A. E. 1969. Energy, water, and carbon budgets of a granite outcrop community. Ph.D. dissertation, University of North Carolina, Chapel Hill.

Manny, B. A. 1969. The relationship between organic nitrogen and the carotenoid to chlorophyll-a ratio in five freshwater phytoplankton species. *Limnol. Oceanog. 14*, 69-79.

Margalef, R. 1959. Pigmentos assimiladores extraidos de los colonias de celenteros de los arrecifes de coral y su significado. *Ecol. Invest. Pesq. 15*, 81-101.

Margalef, R. 1965. Ecological correlation and the relationship between primary productivity and community structure. *Mem. Ist. Ital. Idrobiol. 18*, 335-364.

Milsum. 1966. *Biological Control Systems Analysis.* McGraw Hill Book Co., New York, 466 pp.

Monk, C. D. 1966. An ecological significance of evergreenness. *Ecology 47*, b. 504-505.

Nixon, S. 1969. Characteristics of some hypersaline ecosystems. Ph.D. dissertation, University of North Carolina, Chapel Hill.

Nixon, S., and Odum, H. T. 1970. A model for photoregeneration in brines. *ESE Notes 7* (i), 1-3.

Nixon, S., and Odum, H. T. 1957. Trophic structure and productivity of Silver Springs, Fla. *Ecol. Monogr. 27*, 55-112.

Odum, H. T. 1967. Biological circuits and the marine systems of Texas. In *Pollution and Marine Ecology*, T. A. Olson and F. J. Burgess, (Eds.). Interscience, New York, pp. 99-157.

Odum, H. T., Beyers, R. J., and Armstrong, N. E. 1963a. Consequences of small storage capacity in nonnoplankton pertinent to measurement of primary production in tropical waters. *J. Mar. Res. 21*(3), 191-198.

Odum, H. T., and Cintrón, G. 1970. Forest chlorophyll and radiation. In *A Tropical Rain Forest*, H. T. Odum and R. F. Pigeon, (Eds.). Div. of Techn. Information, U. S. Atomic Energy Commission.

Odum, H. T., and Hoskin, C. M. 1957. Metabolism of a laboratory stream microcosm. *Publ. Inst. Mar. Sci. 4* (2), 115-133.

Odum, H. T., Lugo, A., and Burns, L. 1970. Metabolism of forest floor microcosms. In *A Tropical Rain Forest*, H. T. Odum and R. F. Pigeon, (Eds.). Div. of Techn. Information, U. S. Atomic Energy Commission.

Odum, H. T., McConnell, W., and Abbott, W. 1958. The chlorophyll "a" of communities. *Publ. Inst. Mar. Sci., Univ. Texas*, 5, 65-96.

Odum, H. T., and Pigeon, R. F. 1970. *A Tropical Rain Forest*. Div. of Techn. Information and Education, Oak Ridge, Tenn.

Odum, H. T., and Pinkerton, R. C. 1955. Time's speed regulator, the optimum efficiency for maximum power output in physical and biological systems. *Am. Sci.* 43, 331-343.

Odum, H. T., Cuzon, R., Beyers, R. J., and Allbaugh, C. 1963b. Diurnal metabolism, total phosphorus, Ohle anaomaly, and zooplankton diversity of abnormal marine ecosystems of Texas. *Publ. Inst. Mar. Sci. 9*, 404-453.

Odum, H. T., Siler, W. L., Beyers, R. J., and Armstrong, N. 1963c. Experiments with engineering of marine ecosystems. *Publ. Inst. Mar. Sci. 9*, 373-403.

Phillips, J. N., and Myers, J. 1954. Measurement of algal growth under controlled steady state conditions. *Plant Physiol. 29*, 148-152; 152-161.

Sollins, P. 1969. Measurements and simulation of oxygen flows and storages in a laboratory blue-green algal mat ecosystem. M.A. thesis, University of North Carolina, Chapel Hill.

Steeman-Nielsen, E. 1962. Inactivation of the photochemical mechanism in photosynthesis as a means to protect the cells against too high light intensities. *Physiol. Plant. 15*, 161-171.

Stoeckenius, W. In press. Bacteriorhodopsin, a biological signal and energy transducer. *Conference Internationale de la Physique Theorique a la Biologie*, Institute de la Vie, Paris. (Abstract).

Ting, I. P., and Dugger, W. M. 1968. Non-autotrophic carbon dioxide metabolism in cacti. *Bot. Cag. 129*, 9-15.

Wassink, H. C., Richardson, S. D., and Preters, G. A. 1956. Photometric adaptation to light intensity in leaves of *Acer pseudoptatanus*. *Acta Bot. Neerlandica,* *5*, 247-256.

Zelitch, I. 1967. Water and CO_2 transport in the photosynthetic process. In *Harvesting the Sun in Photosynthesis in Plant Life*, A. S. Pietre, F. A. Greer, and T. J. Army (Eds.). Academic Press, New York, pp. 231-248.

Chapter 2

MODELS, PREDICTIONS, AND EXPERIMENTATION
IN ECOLOGY

George W. Salt

Department of Zoology
University of California
Davis, California

CONTENTS

ABSTRACT

There is a pressing current need in ecology for the ability to make accurate predictions. Such predictions result from deductions from verified models. Models used in ecology occupy a gradient ranging from abstract models at one end to descriptive models at the other. Simulation models lie midway between these two extremes. Each type has certain unique properties and potentially for generating predictions, but for all of them experimentation provides the essential test of either the predictions, the assumptions. or the structural interrelations between the elements of the model.

INTRODUCTION

Ecologists are increasingly finding themselves called upon to participate in decisions that may result in irreversible modifications of the habitat. The tons of cement that are poured in the formation of a dam are not easily removed if the decision to build the dam proves to have been unwise. Similar remarks apply to the filling of bays, the construction of refineries, and the clear-cutting of forests. In all these instances, if the ecologist is called upon the participate in the formation of the decision, he or she is asked in effect to make predictions.

The search for predictive ability is not new. In the past comprehension of a system or a phenomenon was expected to result in an ability to foretell the behavior of the system under changed conditions. What are new are the demands being placed on ecologists for highly quantified forecasts. The engineers, economists, and cost accountants who participate in the decision making or advising process with the ecologist cast their remarks in precise terms, often providing estimates quantified to one or more decimal places. If the ecologist can only phrase his or her judgments in general terms, such as more than or less, or provide only a broad view of the outcome of the manipulation, then his or her opinions are likely to carry little weight, particularly if they are at variance with or in opposition to those of the other counselors. Consequently, it is possible to argue that one of the pressing needs in contemporary ecology is the ability to make accurate, quantitative predictions.

Unless one proposes to study each individual system in advance before manipulating it, the obvious method of generating predictions is by the application of deductive logic to an accepted generalization. Scientists have a variety of tools at their disposal for the establishment of generalizations—observation, the postulation of one or more hypotheses and their subsequent testing by experimentation, and the formulation of models. The construction of models is particularly appealing because models are inexpensive when compared to the cost of a lengthy and elaborate program of investigation and experimentation. Furthermore, because quantitative modeling is relatively new, it has acquired a sanctity that, to some degree, insulates it from criticism. The object of this article is to examine the interaction of observation, experimentation, and the formulation of models of various kinds and to determine, if possible, how that interaction can best be structured for the production of quantified predictions.

As Patten (1971) has pointed out, a model can only reflect part of the performance of the system, and the type of model constructed will determine, or be determined by, the kind of predictions desired. Whatever the kind of model, it is only acceptable as a general statement after it has been verified, either by comparison with the system that it is supposed to mirror or by experimentation. An unverified model is a hypothesis, not a generalization (Kowal 1971).

Models have been described as having three competing properties: generality, precision, and reality (Levins 1966). The term "generality" means that the model successfully describes the responses of a wide array of species living in differing habitats and assemblages. The word "precision" can be ambiguous in that different persons use it to mean different things. More suitably, it refers to the closeness with which the output of the model fits the data on which it is based. A model that generates an equation for a curve that passes exactly through the data points on a graph has high precision. Sometimes, unfortunately, the same word is used to indicate the accuracy with which the model can predict the future performance of the system or population that it is supposed to mimic. For this characteristic, "predictive ability" is probably better. The term "accuracy" might also be used just as well. The degree of realism in a model is determined by the number of characteristics of the parent system that are included in the model. All models are to some degree unrealistic, as one can hardly describe the parent object completely. The degree of unreality is determined by the number of characteristics that are excluded and the number of simplifying assumptions that are made.

It is accepted that all three properties cannot be maximized at the same time in the same model. This is another way of saying that models may be highly abstract and apply to a large set of populations or be specific and precise and apply to a small set. If the model is internally consistent, that is, if it works at all, it will generate predictions. Whether these predictions are

45

valuable insights into the performance of the system or
pure rubbish depends on the accuracy of the model. The
generation of predictions, in itself, tells one little
about the character of the model.

If one examines the models constructed by ecologists,
one finds that they range the gamut from highly abstract
to very specific. One also finds that the sequence of
steps followed in the construction of these various kinds
of models is not the same nor is the role of experimentation in them identical.

ABSTRACT MODELS

Abstract model building has been eloquently defended
by Levins (1966). Its chief benefits are a high degree
of generality, for which precision and reality are sacrificed. Such models permit the generation of a large number of predictions. Models of this kind have been constructed by a number of persons, of whom Levins (1968),
MacArthur and Wilson (1967), Rosenzweig and MacArthur
(1963), May (1973), and Smith (1969) can be cited as examples.

The construction of abstract models begins with the
observation of nature, either by the modeler or others.
Various simplifying assumptions about the characteristics
of the organisms in question are made and the performance
of these idealized organisms or systems is formulated in
mathematical terms. Manipulations of the expressions
then provide predictions of how the organisms perform under
changing circumstances. Up to this point, no experimentation has taken place.

Experimentation in abstract model building follows
the announcement of the predictions resulting from the
model. Usually these experiments are performed by some-
one other than the formulator of the model. It is an

interesting peculiarity of abstract models that although
they are formulated in general terms based on highly simplified assumptions, because of the deductive mathematical
manipulations used in their formulation, they generate
quantifiable predictions. MacArthur's (1957) celebrated
broken stick model is an example. MacArthur concluded
from observations cited by Lack (1954) that bird populations generally were at or near equilibrium states. He
then offered three proposed models of niche structure to
account for species abundances. In a subsequent paper
(MacArthur 1960) he provided deductions from the model
and a comparison of expected values with observed values,
collected by others. Efforts toward verification of the
model continued for several years. Ultimately Hairston's
(1959) tests demonstrated that the proposed model was un-
satisfactory and it was rejected.

The model of island biogeography proposed by MacArthur
and Wilson (1967) is also a well-known abstract model.
It has been tested in experiments by Wilson and Simberloff
(1969) and in a succession of experiments by Cairns and
his associates (Cairns et al. 1969; Cairns and Ruthven
1970; Cairns et al. 1973). To date, the model appears
to be at least partially verified, although a number of
unsuspected phenomena have emerged from the experiments
designed to test the original model, as might be expected.

DESCRIPTIVE MODELS

At the opposite end of the spectrum are highly specific descriptive models. Those assembled by King and
Paulik (1967), Holling (1963, 1966a, b), Griffiths and
Holling (1969), and Hughes and Gilbert (1968) are examples.
Here, each component of the behavior of an individual organism or a population of a single species is subjected
to close measurement and quantification. A mathematical

expression descriptive of the component is then fit to
the data. Finally, a model is constructed by assembling
all the components into one integrated whole, usually in
a computer program.

Because the predictions of each equation are checked
against the performance of the subject organism or population, when the entire model is assembled the likelihood
of closeness of fit between it and the original population
is high. Observation of the performance of a different
population and the degree to which the model mimics that
performance constitute the verification. Models of this
style have a high degree of precision and are high in
reality. However, an individual model is usually low in
generality. Broad generalizations can result from descriptive models through a comparative study of a number
of them, all of which are descriptive of similar activities. At present we do not have a sufficient array of
such models for comparative study.

We may examine, as an example of a descriptive model,
that of Gilbert and Gutierrez (1973) for the aphid
Masonaphis maxima (Mason), its plant food, and the parasite
that attacks it. From detailed field observation on the
growth of the aphid population and its plant host in the
field, they derived mathematical expressions for various
components of population growth and change in age structure. These field observations were supplemented by
laboratory measurements of such parameters as individual
rates. When the entire model was assembled, it was found
to reflect quite accurately population growth and age-class
change of the aphids. Note, however, that no verification
occurred, as the model was never tested against new data
collected after the model was constructed.

There are several features in this process worthy of
note. First, although unannounced, the authors must have
had a rough model of the performance of the population in

their heads when they began. This unspecified model undoubtedly dictated the choice of components to be included
in the finished model. Second, no experimentation, in the
strict sense, took place throughout the construction process, although if one considers quantified laboratory
measurement to be experimentation, it did. However, true
experimentation, in the sense of hypothesis testing, took
place not on the aphids but on the completed model. The
potential for experimentation of this kind is one of the
chief virtues of descriptive models (Watt 1966).

The verification of a descriptive model is itself an
experiment. The model generates predictions about the
performance of the population in the future (hypotheses).
Comparison of the performance of the population with that
predicted provides evidence for the acceptance or rejec-
tion of the model. If the model is verified, it then
can be considered a generalization and deductions may be
made from it concerning the performance of the population
under different circumstances.

During the assembling of the model, experiments may
be needed to determine how the components should be fit
together. With the various components in hand, the modeler
must decide whether the effects of two or more components
are interactive or not. He or she may base his or her
decision on information derived from three sources: intuitive appreciation of the biology of the organisms being
studied, experimental data on other species reported by
another investigator, or the results of experiments designed specifically to test for interaction effects within
the system being modeled.

Intuitive decisions are as reliable as the perception
of the observer. The difficulty is that there is no way
to determine whether one possesses talent in this activity,
particularly as nearly every biologist assumes that he or
she does, whether his or her fellows agree or not.

Moreover, if one sets up a program of experimentation to test one's own intuition, bias in the results is a foregone conclusion. One must conclude, therefore, that to rely on intuition is to risk making serious errors in construction.

The reliability or applicability of experimental results derived from another species is probably in inverse proportion to the phylogenetic distance between the species involved. For example, Hassell (1971) experimented on the interaction in insects between the host density and the parasite density as determinants of the rate of parasitization. The results showed that these two factors interacted. Experiments on the effects of predator density and prey density in protozoa, *Didinium* and *Paramecium* (Salt 1974), revealed that these two controls were non-interactive but were instead alternative controls of the rate of capture by the predator. The use of the results on insects in the formulation of a model of protozoa would have introduced an error in the structure of the model and a consequent loss of predictive power.

Curds and Cockburn (1968) constructed a model to reflect the rate of consumption of *Klebsiella aerogenes* by *Tetrahymena pyriformis*. In it, they assumed that the densities of both the protozoan and the bacterium interacted in determining the rate of removal of the bacterium. This instance serves as a useful example of the difficulties of basing model structure on intuitive decisions. The results of the study on feeding rate in a protozoan cited above make clear that an alternate relationship between these two variables is possible. This suggestion is further strengthened by certain anomalies in the data of Curds and Cockburn (their Fig. 5), which they observed but decided to ignore. Those data become explicable if one postulates alternative, noninteracting controls.

However, the predicted values from their equations do not differ radically from those observed. In consequence, one can neither argue against their decision nor be totally convinced that it has been the correct one. An experimental design that permitted the use of an analysis of variance would have settled the matter for all concerned.

Experimentation may also serve in the construction of descriptive models by revealing the presence of unsuspected feedback controls between elements of the system or population being modeled. Metabolites from one organism may affect others with which the organism does not otherwise interact. An observer would be unlikely to detect this relationship in the absence of direct experimental test for its existence. For example, in the food chain consisting of dead leaves or grasses, *Aerobacter aerogenes, Paramecium aurelia,* and *Didinium nasutum*, one would expect that there would be interactions that controlled the operation of the system between any two adjacent elements in the chain, as well as internal interactions between the individuals within each population. One would not be likely to postulate interactions between members of nonadjacent elements. Yet Beers (1927) demonstrated experimentally that fresh plant extracts were inhibitory to the *Didinium* and that the bacteria, not *Paramecium*, provided the stimulation for excystment of *Didinium*. Neither of these controls within the system would have been a reasonable assumption in the absence of direct experimental evidence.

Thus, one may conclude that in spite of the seeming lack of necessity for experimentation in the formulation of descriptive models, in fact experimentation is required. The results of appropriate experiments must dictate the structure of the model by revealing whether components should be linked in additive or multiplicative fashion

and whether unsuspected feedbacks occur between nonadjacent
elements in the system.

SIMULATION MODELS

Midway between abstract and descriptive models are
explanatory models. To these one may properly apply the
term "simulation," in the dictionary sense of "having the
appearance but not the reality" of a system. In the formulation of a simulation model, the investigator carries
out detailed quantified studies of a population, food
chain, or food web. Armed with the insights provided by
this study, the investigator formulates a series of assumptions about the internal performance of the system. These
assumptions are cast into mathematical expressions and
assembled into a provisional model. The model is then
checked by comparing its output with the prior observations
on the system.

There are a number of such simulation models. Jost
et al. (1973) found that Monod's (1949) descriptive model
of the growth of bacterial populations did not fit the
performance of multiple species interactions in a chemostat. They then postulated the existence of multiple
states within the bacterial populations and formulated a
simulation model based on this and other assumptions
(Frederickson *et al.* 1973). The output of this model
mirrored their experimental results better than the Monod
model. Similarly, Rapport (1971) studied the phenomenon
of food preference in *Stentor coeruleous*. Following his
studies, he assembled a simulation model designed to represent the performance of the protozoan as well as that
of the metazoan predators that Murdoch (1969) considered
in his studies on switching. Williams' (1971) model of
algal population dynamics is an elegant example of a
simulation model based on 12 postulated general properties

of microorganisms and 6 basic assumptions. Royama (1971) has also formulated a model for food gathering by predators.

Verification of a simulation model proceeds in the same fashion as that for descriptive models, that is, through a comparison of predicted performance with new observations on the population or system being modeled.

The point of application of experimentation in simulation modeling is obviously at the assumptions. These are essentially hypotheses. The investigator is saying, "If the organisms are behaving in such and such a way, then their performance should be as follows." Usually the assumptions are formulated by a highly informed observer, and their probability of being correct is consequently high. However, they represent essentially intuitive guesses and possess all the previously listed hazards of such judgments. Experimental tests of the assumptions supply two values. One, they insure that the model is not producing the right results for the wrong reasons through some fluke (Dayton 1973), thus, verifying the explanatory character of the model. Two, if the assumptions are tested on a species different from that studied by the author of the model and are found to apply, then it is reasonable to assume that the model applies to the second species as well, thus helping to establish the generality of the model. Without the experimental testing, the model remains a more or less intriguing example of the ability of the investigator to make informed guesses about the performance of the system which he or she has studied.

CONCLUSION

This analysis of the characteristics of modeling and experimentation provides a convenient format for analyzing the structure of different kinds of models and the values

of each. It obviates making value judgments as to which
is better or worse or more or less worthwhile. The limitations of each and their suitability for different purposes become more apparent. Abstract models can be constructed as fast as the insight and creativity of the
modeler permit. They do not require elaborate data collection or observation. Moreover, they generate large
numbers of predictions, most of which, because of their
limited data base and the fallibility of human beings,
must prove to be wrong. Experimentation tests the state-
ments (hypotheses) provided by abstract models. A series
of confirmatory experiments on a variety of populations
or systems will establish the model as a generalization,
or a type ideal for generating predictions about many
systems.

Simulation models require more time and effort, both
in data collection and in their formulation. Experiments
on the assumptions of a verified simulation model serve
to establish the generality of the assumptions and the
usefulness of the model for predicting the performance of
other populations or systems.

Descriptive models are the most laborious and time
consuming to construct. Furthermore, comparative study
of a group of such models, while undoubtedly the most
powerful method of discovering widely applicable generalizations, is not yet possible. Experimentation, in the
classic sense, is used to determine the internal structure
of the system being modeled. Computer experimentation
provides a means of experimentation without manipulation
of the system itself. Thus, it can generate predictions
about the system on which it is based. In the present
state of knowledge, descriptive models are the only ones
capable of providing the kind of reliable predictions required for intelligent decision making on environmental
problems.

MODELS, PREDICTIONS AND EXPERIMENTATION

We return now to the initial contention that a need exists for predictive ability in ecology. If this need is to be satisfied by deductions from verified generalizations or models, then the foregoing discussion makes it clear that regardless of the type of model formulated, its predictive ability must be determined to a considerable degree by the amount of experimentation that has gone into its formulation and verification. Therefore, one may argue that a need for experimentation exists in ecology related to the need for predictive ability. If this conclusion is true, then the current level of experimentation in ecology as compared to observation and description is far too low. A portion of the collective energy that now goes into these latter two types of study can profitably be devoted to experimentation.

REFERENCES

Beers, C. D. 1927. *J. Morph. Physiol., 43,* 499.

Cairns, J., Jr., Dahlberg, M. L., Dickson, K. L., Smith, N., and Waller, W. T. 1969. *Amer. Nat. 103,* 439.

Cairns, J., Jr., and Ruthven, J. A. 1970. *Trans. Amer. Microsc. Soc. 89,* 100.

Cairns, J., Jr., Yongue, W. H., Jr., and Boatin, H., Jr. 1973. *Revista Biol. 9,* 35.

Curds, C. R., and Cockburn, A. 1968. *J. Gen Microbiol. 54,* 343.

Dayton, P. K. 1973. *Amer. Nat. 107,* 662.

Frederickson, A. G., Jost, J. L., Tsuchiya, H. M., and Ping-Hwa Hse. 1973. *J. Theoret. Biol. 38,* 487.

Gilbert, N., and Gutierrez, A. P. 1973. *J. Anim. Ecol. 42,* 323.

Griffiths, K. J., and Holling, C. S. 1969. *Can. Entomol. 101,* 785.

Hairston, N. G. 1959. *Ecology 40,* 404.

Hassell, M. P. 1971. *J. Anim. Ecol. 40,* 473.

Holling, C. S. 1963. *Mem. Entomol. Soc. Can.* No. 32, 22.

Holling, C. S. 1966a. *Mem. Entomol. Soc. Can.* No. 48, 1.

Holling, C. S. 1966b. In *Systems Analysis in Ecology,* K. E. F. Watt (Ed.) Academic, New York, pp. 195-214.

Hughes, R. D., and Gilbert, N. 1968. *J. Anim. Ecol. 37,* 553.

Jost, J. L., Drake, J. F., Frederickson, A. G., and Tsuchiya, H. M. 1973. *J. Bacteriol. 113,* 834.

King, C. E., and Paulik, G. T. 1967. *J. Theoret. Biol. 16,* 251.

Kowal, N. E. 1971. In *Systems Analysis and Simulation in Ecology,* Vol. I, B. C. Patten (Ed.) Academic, New York, pp. 123-196.

Lack, D.   1954.   *The Natural Regulation of Animal Numbers*, Oxford University Press, Oxford, p. 343.

Levins, R.   1966.   *Amer. Sci. 54,* 421.

Levins, R.   1968.   Evolution in changing environments. *Monogr. Pop. Biol. 2,* Princeton University, Princeton, N. J., p. 120.

MacArthur, R. H.   1957.   *Proc. Natl. Acad. Sci. 43,* 293.

MacArthur, R. H.   1960.   *Amer. Nat. 94,* 25.

MacArthur, R. H., and Wilson, E. O.   1967.   The Theory of Island Biogeography. *Monogr. Pop. Biol. 1,* Princeton University, Princeton, N. J., p. 203.

May, R. M.   1973.   Stability and Complexity in Model Ecosystems. *Monogr. Pop. Biol. 6,* Princeton University, Princeton, N. J., p. 235.

Monod, J.   1949.   *Ann. Rev. Microbiol. 3,* 371.

Murdoch, W. W.   1969.   *Ecol. Monogr. 39,* 335.

Patten, B. C.   1971.   In *Systems Analysis and Simulation in Ecology,* Vol. I, B. C. Patten (Ed.). Academic, New York, pp. 3-121.

Rapport, D. J.   1971.   *Amer. Nat. 105,* 575.

Rosenzweig, M. L. and MacArthur, R. H.   1963.   *Amer. Nat. 97,* 209.

Royama, T.   1971.   *Res. Pop. Ecol. Suppl. 1,* 91.

Salt, G. W.   1974.   *Ecology 55,* 434.

Simberloff, D. S., and Wilson, E. O.   1969.   *Ecology 50,* 278.

Smith, F. E.   1969.   In *Eutrophication: Causes, Consequences, Correctives.* Natl. Acad. Sci., Washington, D. C., pp. 631-645.

Watt, K. E. F.   1966.   In *Systems Analysis in Ecology,* K. E. F. Watt (Ed.). Academic, New York, pp. 1-14.

Williams, F. M.   1971.   In *Systems Analysis and Simulation in Ecology,* Vol. I, B. C. Patten (Ed.). Academic, New York, pp. 197-267.

SALT

Wilson, E. O., and Simberloff, D. S. 1969. *Ecology 50,*
    267.

Chapter 3

EXPERIMENTAL AQUATIC LABORATORY
ECOSYSTEMS AND COMMUNITIES

G. Dennis Cooke

Biological Sciences
Center for Urban Regionalism and Environmental Systems
Kent State University
Kent, Ohio

CONTENTS

INTRODUCTION

Theory regarding structure and function of communities and ecosystems has come largely from summation and
integration of separate investigations on segments of
these systems, and many questions therefore remain to be
asked about the development, activity, and arrangement of
these levels of organization. Research, until recently,
has been largely descriptive, in part because experiments
in nature are very difficult to perform. Ecosystems are
usually very large and the investigator must study a species, trophic level, or process and not the system itself.
This has resulted in theory developed almost completely
by summary of our knowledge of parts, and not by observation and experimentation on the systems themselves.

One approach to the problem of experimentation with
these levels of organization has been the use of aquatic
laboratory ecosystems and communities, or small-scale
field enclosures. These systems have ranged from hollow
bamboo stems (Kurihara 1960), tissue culture flasks (Coler
and Gunner 1970), "Winogradski Columns" (Castenholz 1972),
to large outdoor ponds (Hall *et al.* 1970). There are two
basic types of experimental laboratory systems. One, the
closed beaker or flask type, is considered to be an ecosystem (as defined by E. P. Odum 1971) and is often called
a microcosm or microecosystem. The other is the open
laboratory community, containing autotrophic, heterotrophic, and abiotic components, which is part of a larger
outside ecosystem. In both, the assemblage of organisms
is natural and is usually obtained from nature as a unit
and then allowed to reorganize and reassemble under laboratory conditions in several to many replicate units so

61

that the investigator has a series of controllable, repeatable, manipulatable, miniature ecosystems or communities. In some studies (e.g., McIntire 1968a), the laboratory model is designed to duplicate as nearly as possible an actual community. In all cases the systems are
simplifications, often abstractions, and are usually not
meant to be like any other ecosystem except in properties
fundamental to all ecosystems (Taub 1971). Laboratory
ecosystems and communities differ fundamentally from such
systems as resuspended BOD bottles or polyethylene enclosures in that these latter systems cannot be repeated.

The object of this chapter is to describe some basic
properties of laboratory ecosystems and communities, particularly with regard to the development of theory related
to ecological succession, to suggest some additional areas
of study, and to indicate how our understanding of laboratory systems may result in ecosystems of direct use to
man.

DISCUSSION

The Microcosm Method

The use of aquaria or large tanks for research and
teaching purposes is well known. In most instances, these
are uncontrolled systems, without replicate units, and
successive sampling by the investigator progressively
alters them. Often, the medium is not defined and the
identification of the vast bulk of organisms is unknown.
Beyers (1962, 1963a) was among the first to employ many
small replicate laboratory ecosystems, and his technique
is representative of what has become known as the "microcosm method."

Beyers' basic laboratory ecosystem was developed from
a composite sample of a sewage oxidation pond near Austin,
Texas. The material from the pond was placed under a

pondlike laboratory regime of temperature and light and
then allowed to go through succession. A large "stock"
system was thus created.

Two different types of media are used by Beyers. The
basic medium is sterile one-half strength Taub and Dollar
(1964) No. 36 solution. When 5 mg thiamine/liter plus 35
mg of fixed nitrogen/liter are added to this base medium,
the microecosystems undergo an autotrophic succession when
an aliquot of the stock system is added to each sterile
flask. If 0.05% proteose-peptone is added instead to the
base medium, the new ecosystems experience a bacterial
bloom or a heterotrophic succession, which then is followed by the autotrophic succession. In either case, the new
ecosystems are carefully cross-seeded to minimize divergence. Depending upon the objectives of the experiment,
physical conditions, such as light and temperature or the
composition of the medium, can be varied. The standard
light source used by Beyers has been Gro-lux (Sylvania),
with a cycle of 12 hr light-12 hr dark. From a few to
hundreds of replicate microecosystems are thus established
by the investigator under the desired conditions of medium
and physical environment.

Standard ecological techniques are employed for the
analysis of the structural and functional features of the
experimental ecosystems. Community photosynthesis and
respiration are usually measured by the pH method (Beyers
et al. 1963) and at intervals during the experiment systems
are sacrificed for measures of populations, biomass, nutrients, chlorophyll, and so forth. One of the most valuable
features of the microcosm method is that the measurements
do not affect the overall results, since sufficient replicates are established to allow the investigator to sacri-
fice several ecosystems at frequent intervals during the
study.

The extent of replicability with regard to basic
characteristics such as population density and rates of
succession has not been adequately established in many
studies. Some divergence, even in carefully cross-seeded
replicate systems, has been noted, and the variation to
be tolerated in experimental ecosystems remains to be
established. A larger variance than customarily found
in experimental work may have to be accepted, since small
differences at the outset of the experiment may be magni-
fied as succession proceeds. There is good evidence to
show that replicate microcosms do not differ significantly
with respect to levels of community metabolism (Abbott
1966) and the replicate systems seem to go through succession to maturity at the same rate (Cooke 1967). Further work with regard to replicability is needed.

Beyers' basic microecosystem is gnotobiotic. The
species have been listed by Gorden *et al.* (1969). The
ecosystem has three producer, five consumer, and eleven
bacterial species. His microsystem is probably the most
complex gnotobiotic ecosystem known to science. Taub
(1969) has also developed a gnotobiotic ecosystem consisting of one producer species, one consumer species,
and three decomposer species; and Nixon (1969) has described a gnotobiotic brine microecosystem.

The Ecology of Laboratory Ecosystems

Diurnal Metabolism

The metabolic activity of laboratory microecosystems,
like large units of nature, has a distinct diurnal pattern
(Beyers 1963a,b, 1965). Figure 3.1, based on the average
of 100 diurnal metabolism measurements from 12 benthic
microcosms, is typical of the daily variation in metabolism
of a microcosm under 12 hr of constant illumination and
12 hr of dark. Just after the lights are turned off, the

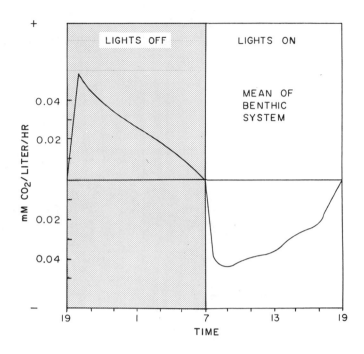

FIG. 3.1. Diurnal rates of carbon dioxide uptake and release during net photosynthesis and nighttime respiration in 12 benthic fresh-water microcosms. This curve is the mean of 100 curves. From Beyers (1963b). Reproduced with permission of the author.

ecosystems exhibit a burst of respiratory activity, which then declines over the night. At lights on, there is a burst of photosynthesis, which also declines during the remainder of the period. An identical type of metabolism curve was found in a wide variety of microecosystems studied by Beyers (Fig. 3.2). The simplest ecosystems, such as the $51^{\circ}C$ and the brine systems, exhibit the largest amplitude of variation, while more complex ecosystems, such as the pond type, seem to have less diurnal variation.

Other investigators have reported identical patterns in laboratory microcosms (H. T. Odum *et al*. 1963b; Butler

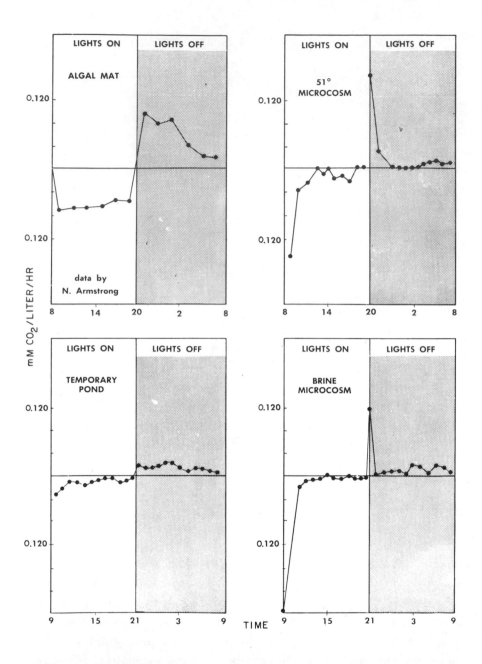

FIG. 3.2. Diurnal rates of carbon dioxide uptake and release during net photosynthesis and nighttime respiration in a marine algal mat microcosm, a temporary pond microcosm, and a brine microcosm. From Beyers (1963b). Reproduced with permission of the author.

1964). Abbott (1966) noted that in addition to the day-
night pattern there was also a midday depression in photosynthesis, which he interpreted to be an endogenous rhy-
them. Similar diurnal variations in metabolism have not
been noted in laboratory stream communities (McIntire and
Phinney 1965), although they have been observed in natural
streams (Owens 1965).

A complete explanation for the shape of the diurnal
curve is apparently not possible at this time, although
Beyers (1963a, 1965) has offered several hypotheses.
Initially, it was thought that the burst of respiration
at the onset of dark was caused by the stimulatory effect
of the dissolved oxygen that had built up during the day.
When dissolved oxygen was held constant by bubbling the
microcosm, however, the burst of respiration, as measured
by pH changes, was still present. Beyers (1965) also
suggested that a pool of highly labile organic molecules
will be available in greater quantities early in the
evening and will stimulate respiratory activity. However,
as Abbott (1966) has pointed out, there seems to be little
evidence for such an accumulation, except for large organic molecules entering storage.

Beyers assumed that under constant illumination gross
community photosynthesis was also constant and that respiration increased during the lighted period, thus accounting for the high net community photosynthesis early
in the day and a steadily decreasing rate as the day progressed. Since gross photosynthesis and day respiration
occur simultaneously, it is difficult to determine whether
one or both behave in this way during the illumination
period. However, assuming the hypothesis to be correct,
an electrical analog circuit model was constructed to
simulate the metabolism in a closed laboratory ecosystem
(H. T. Odum *et al.* 1963a). The computer curves were very
similar to the observed pattern of metabolism in the

microcosms, thus lending indirect evidence to support the hypothesis that under constant illumination gross community photosynthesis remains steady while day respiration increases.

The effect of a lengthened photoperiod (24 hr of light followed by 24 hr of dark) on the metabolism of steady-state microecosystems was also investigated (Beyers 1963a). There was no net community photosynthesis during the second half of the light period, nor any measurable respiration during the second half of the dark period. Beyers interpreted the reduction in photosynthesis to be caused by limitation by nutrient availability. The reduction of respiration suggested a conservative mechanism in ecosystems of considerable survival value. During prolonged periods of reduced light (bad weather, for example) the respiratory machinery of ecosystems may be reduced to the minimum rate, conserving structure and allowing the system to rapidly return to a usual metabolic rate when light levels become normal. Apparently in balanced, steady-state ecosystems, photosynthesis and respiration are closely coupled.

To summarize, laboratory microecosystems have a distinct diurnal pattern of community metabolism in which rates of photosynthesis or respiration are not constant from hour to hour, even under constant illumination. Natural ecosystems exhibit a similar pattern. Thus estimates of daily community metabolism rates from short-term observations should be avoided. Apparently more complex ecosystems (in terms of numbers of species) have reduced amplitude of this diurnal curve.

Ecological Succession in Microecosystems

Introduction

The principles of succession form the basis of a unifying theory of ecology (Margalef 1963a,b, 1968;

EXPERIMENTAL ECOSYSTEMS AND COMMUNITIES

E. P. Odum 1969, 1971). The relationships among such ecosystem attributes as species diversity, biomass, and productivity have been formulated into an overall qualitative
model of succession (E. P. Odum 1969). We are just beginning to quantitatively examine these principles, particularly in relation to human activities in changing the
landscape or introducing materials into these levels of
organizations, and only very recently have we made sustained efforts to apply these principles to the development of ecosystems of use to man, such as a bioregenerative
life support system for space travel. The study of ecological succession in laboratory microecosystems has contributed substantially to our understanding of succession
and has provided a means for experimentally examining
some of man's impacts on ecosystems and his attempts to
construct them.

Succession is a complex process of community development that involves a directional, predictable change in
community structure and function that results from the
modification of the physical environment by the community
and that culminates in as stable an ecosystem as is biologically possible on the site in question (E. P. Odum
1963). Succession in the closed microecosystem is unnat-
ural in the sense that there is no immigration or emigra-
tion, only an input of light and an exchange of gases with
the atmosphere. Therefore, the species composition does
not change except in terms of relative abundance and the
number of species in encystment stages. This property
of laboratory ecosystems may be particularly advantageous
for future work since one may compare ecosystems of varying diversity. Successional processes and the overall
pattern of succession in the flask or beaker type of laboratory ecosystem are considered to be representative of
a large-scale unit of nature in which the magnitude of
exchange with adjacent systems is negligible when compared
to the activities within the system.

The Pattern of Succession in Laboratory Microcosms

As in larger natural ecosystems, various groups of species gain ascendancy in terms of activity and abundance during succession and then are replaced or become less active. Cooke (1967) observed that plankton-feeding *Daphnia* were abundant only during the early "bloom" stage of succession and then were not active in the systems, whereas ostracods, which are detritus feeders, became more numerous as succession proceeded. Of the 30 genera of algae in the ecosystem only about 10 or 12 were abundant at any one time, and at maturity there was a stable species composition. Other investigators have made similar observations (Beyers 1963a; Taub 1969; Nixon 1969).

Gorden (1967) and Gorden *et al.* (1969) demonstrated that not all species in microecosystems were active at the same time and that some species altered conditions so as to favor other species. Gorden found that all of the bacterial species in his microecosystem (Beyers' basic system) were capable of stimulating the growth of *Chlorella*, the metabolically most important autotroph, by releasing thiamine and carbon dioxide. Figure 3.3 shows the great enhancement of *Chlorella* population growth when bacteria were present in comparison to growth in various media without bacteria but with thiamine or thiamine and sterile air (CO_2 enrichment). Filtrates containing dissolved organic material from a *Chlorella* culture and a bacterial culture were capable of supporting bacterial growth. Thus, taxonomically diverse species interacted to bring about succession in this microecosystem.

The changes in microecosystem metabolism and biomass were studied by Cooke (1967) during an autotrophic succession and by Gorden (1967) and Gorden *et al.* (1969) for heterotrophic succession. These events in Cooke's study

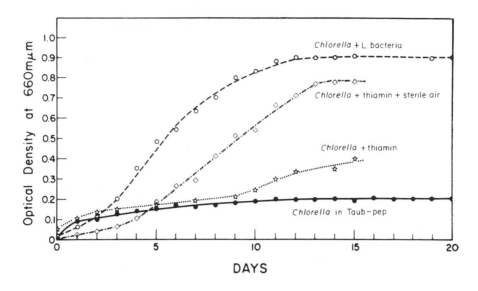

FIG. 3.3. Optical density measurements of the growth of *Chlorella* in Taub-pep, Taub-pep thiamin, Taub-pep plus thiamine and sterile air, and Taub-pep plus L bacterium. From Gordon *et al*. (1969). Reproduced with permission of the authors.

are summarized in Figures 3.4 and 3.5. During autotrophic succession there was an initial burst of photosynthetic activity, in excess of community respiration, so that bio-mass accumulated. Metabolism increased to a maximum and then decreased until photosynthesis equaled respiration (P/R ratio of about 1) and there was no further increase in biomass. During the first five days of Gorden's hetero-trophic succession, respiration exceeded photosyntheses (PR < 1) because of the bacterial bloom; then an autotro-phic succession, like that shown in Figures 3.4 and 3.5, took place. A mature ecosystem has been defined as one in which levels of biomass and metabolism are stable (Cooke 1967).

71

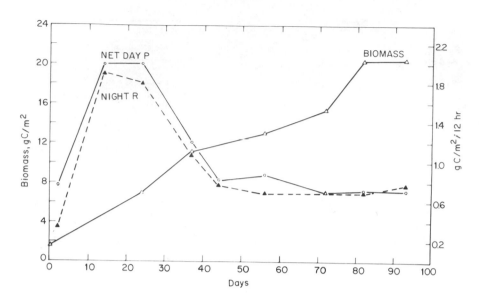

FIG. 3.4. Pattern of net community photosynthesis, night respiration, and biomass changes during autotrophic succession of laboratory microecosystems. From Cooke (1967).

Early in succession, during the algal bloom, the ratio of gross community photosynthesis to biomass (P/B) was high (Fig. 3.5). That is, a small amount of biomass was very productive. As succession proceeded, without any addition of nutrients, this type of efficiency, called ecosystem production efficiency, dropped rapidly so that at more mature stages there was a large relatively unproductive amount of organic matter.

The ratio of biomass to gross community photosynthesis (B/P) is a measure of system maintenance efficiency (Fig. 3.5). Early in succession the ratio was low, but at maturity the cost to the system of maintaining large structure was small; photosynthesis per unit of biomass was low.

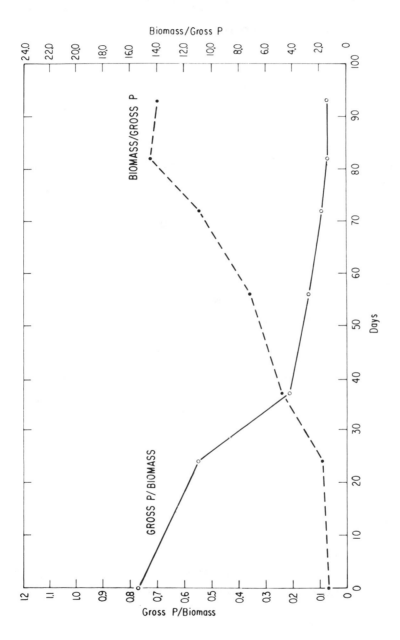

FIG. 3.5. Changes in ratios between gross community photosynthesis and biomass (production efficiency) and biomass and gross community photosynthesis (maintenance efficiency) during autotrophic succession of laboratory microecosystems. From Cooke (1967).

As in large natural systems, microecosystems become increasingly stratified and more heterogeneous during succession, and the importance of detritus as a source of energy and a location for many of the species increases. The algal bloom is replaced by large patches of detritus in which many of the bacteria and algae are intimately associated. Presumably the ostracods, the top consumer in Beyers' and Cooke's ecosystems, grazed on clumps of these organisms in the detritus.

Another change in laboratory ecosystems during succession is the shift in color from bright green in early stages to a yellow-brown in mature stages, often with small patches of green where ostracods are grazing and releasing nutrients (Cooke 1967; Gorden 1967; Taub 1969). This gradual change in color accompanies the increase in stratification and the formation of detritus. Margalef (1965) has suggested that this shift in the ratio of pigments (carotenoids to chlorophylls), which is based on the premise that the commonly observed successional pattern is largely caused by the gradual depletion of free plant nutrients, can be used as an index to the nutritional status of the autotrophic component of an ecosystem and thus to the age of the system.

Whether the Margalef index can be used as a sensitive measure of nutritional status of autotrophs and of the stage of succession requires much additional work with both field and laboratory systems. Numerous studies (e.g., Wilhm and Long 1969; Mathis 1972) suggest caution in the use of this index until further experimental work is available to quantify the relationships among pigment ratios, nutrients, and community metabolism.

Other biochemicals, in addition to pigments, may be expected to increase as aquatic ecosystems age, so that biochemical diversity may be an important measure of succession in aquatic ecosystems. Some preliminary

observations on the relationship between types of fatty
acids and algal species composition and diversity in laboratory stream communities have been reported by McIntire
et al. (1969), and an increase in phosphatases in microecosystems during succession has been shown (Cooke and
Heath, pers. comm.). A quantitative study of the transformation of elements, particularly nitrogen and phosphorus, and the corresponding changes in plant pigments and
other biochemicals during succession is needed.

Work with radioactive tracers (Whittaker 1961; Confer
1972) and with adaptively produced phosphatases (Cooke and
Heath, pers. comm.) in laboratory ecosystems lends further
observations to support the hypothesis that nutrient (phosphorus in these cases) limitation and recycling are key
factors in regulating the pattern of succession.

Recent work in the laboratory of Dr. Robert T. Heath
and me has shown that the *Chlorella* sp. in Beyers' gnotobiotic ecosystem adaptively produces acid and alkaline
phosphatases as a response to limiting phosphorus conditions in the medium. We isolated the *Chlorella* and grew
it in phosphorus-free and phosphorus-enriched Bold's basal
medium and found that it produced phosphatases under pho-
sphorus-limiting conditions. Then we aseptically inoculated aliquots of the Beyers' ecosystem into sterile flasks
containing 100 ml of half-strength Taub and Dollar (1964)
medium, supplemented by 5 mg/liter thiamine and 17 mg/liter
fixed nitrogen. The initial phosphorus level (as ^{31}P)
was 300 μg PO_4-P/liter, a value typical of enriched aquatic
communities. Levels of orthophosphate and filterable un-
reactive phosphorus (FUP) were determined by the EPA method
(*Methods for Chemical Analysis of Water and Wastes*, United
States Environmental Protection Agency [USEPA], 1971),
biomass of entire ecosystems was obtained by transferring
the ecosystems, with appropriate techniques for removing
side wall organic matter, to tared crucibles, and

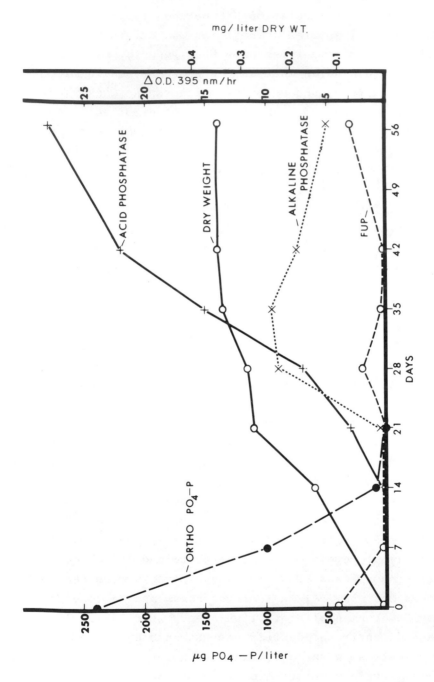

FIG. 3.6. Levels of phosphorus fractions, dry weight, and phosphatase activity during autotrophic succession.

phosphatases were measured as described by Fitzgerald and Nelson (1966). Figure 3.6 illustrates the results. As expected, orthophosphate dropped rapidly to undetectable levels by day 21. FUP remained at low levels throughout succession. Although community metabolism was not measured, an apparent P/R of 1 was reached at day 35 (as indicated by a leveling of biomass accumulation) and the amount of enzyme in the ecosystem began to rise at this stage of succession. Apparently during the early stages of succession phosphate is accumulated rapidly and biomass increase is slower. At maturity, there is no change in biomass and levels of phosphatases are high, suggesting that the final level of biomass and community metabolism is related to rates of recycling as well as to the initial amount of nutrients at the start of succession.

Whittaker (1961) added radioactive phosphorus to aquaria-type microcosms and examined the rate of uptake and the recycling of the element during succession. Phosphorus was taken up rapidly into plankton and then more slowly accumulated into other fractions of the ecosystem (Fig. 3.7). With low initial levels of phosphorus (oligotrophic aquaria), biomass developed toward a steady state after eight weeks of succession, with most of the added phosphorus in seston, side wall and bottom algae, and sediment. Very little was free in the water. In eutrophic aquaria steady state was not reached during the term of the experiment and, as pointed out by H. T. Odum and Hoskin (1957) for their enriched lotic ecosystem, succession was more rapid than in nutrient-poor systems. Whittaker also studied an artifical outdoor pond and observed a successional pattern of phosphorus transformations similar to that of the aquaria.

The utility of microecosystems in the study of the ecosystem level of organization is supported by the fact that succession in the beaker-type microecosystem closely

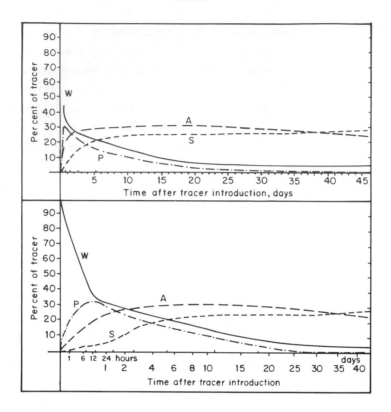

FIG. 3.7. Experiment 1. Distribution of ^{32}P in oligotrophic aquaria (Columbia river water) in relation to time after tracer introduction, with time on linear scale above, square root scale below. Vertical ticks above the baseline for time indicates times of sample removal. Aquarium fractions are: water (W, solid line), algae attached to aquarium surface (A, longer dashes), plankton or seston (P, dot-and-dash), and sediment (S, shorter dashes). From Whittaker (1961). Reproduced with permission of the author.

resembles both the long-term pattern of succession and many short-term seasonal cycles in large natural ecosystems. The trends in photosynthesis, respiration, and biomass during succession of a forest ecosystem and a microcosm are similar (E. P. Odum 1969). As in the microcosms, forest succession leads to an ecosystem where rates of photosynthesis and respiration are nearly equal (Ovington 1962), and the early high rate of photosynthesis in an abandoned agriculture area gradually decreases (E. P. Odum 1960), indicating that as succession toward a mature forest proceeds, the P/B ratio declines. In a stable marine environment in which nutrients have been depleted, the rate of metabolism is regulated by the rate of nutrient turnover (Ketchum *et al.* 1958).

Confer (1972) has criticized the relevance of these studies on "closed" microecosystems and nutrient levels to conclusions regarding succession and eutrophication, such as those developed by E. P. Odum (1969). In Confer's view, lakes with high import of nutrients proceed toward mature eutrophy; the conclusion that eutrophic lakes are immature, as suggested by the resemblance of their properties to those of developmental stage of microcosms, is wrong.

What is often forgotten is that eutrophication is a process that occurs at the ecosystem level of organization, and that lake communities that are becoming eutrophic are responding to events on the terrestrial portion of the watershed ecosystem. With regard to developmental trends in biomass, nutrients, and metabolism, the beaker-type system closely resembles a large-scale unit of nature; these systems are not intended to represent lake communities.

Input-output aquaria, like lakes on watersheds that have had nutrient diversion, should become more mature and oligotrophic if nutrient input becomes less. Such responses have been documented for deep, basically unproductive lakes (Stockner and Benson 1967; Edmondson 1968), which have returned to communities with the attributes of mature oligotrophy as the terrestrial influence became like that of a more mature ecosystem with good nutrient conservation and closed nutrient cycles. It would be very

useful if a replicable and controllable laboratory ecosystem with both terrestrial and aquatic components could
be developed so that experimental and holistic approaches
to questions of eutrophication, ecosystem maturity, and
watershed management strategies could be emphasized.
Confer's (1972) input-output aquaria represent a significant step in this direction.

Regulation in Mature Microecosystems

The relationship between the rate of nutrient recycling and the metabolism of autotrophs at various stages
of succession or the close coupling of photosynthesis and
respiration, as demonstrated in Beyers' (1963a) photoperiod
experiments, are examples of internal regulation in ecosystems. Another aspect of regulation in ecosystems is
related to the hypothesis that mature multispecies ecosystems are more resistant to external perturbations than
are less mature or complex systems (Cooke *et al.* 1968).
This hypothesis, difficult to examine in large natural
ecosystems, can be studied in laboratory ecosystems. The
results of some preliminary studies on this hypothesis are
presented here.

Beyers (reported in E. P. Odum and Beyers 1965; Cooke
et al. 1968) compared the effect of an acute dose of irradiation upon laboratory ecosystems and on a unialgal
culture. This experiment provides some evidence for the
built-in homeostasis of a mature balanced ecosystem. A
mature microecosystem was divided into two parts; one-half
served as a control and the other half was irradiated at
10^6 rads in an acute dose. There were no visible effects
from the irradiation except for the elimination of the
ostracods. However, when the control and experimental
portions of the microecosystem were used as stocks to
start new successions just after irradiation, and at weekly

intervals thereafter, the results became apparent (Fig. 3.8). At first the accumulation of biomass in new micro-cosms started from the irradiated portion was slow, when compared to controls. The effect decreased with time after irradiation, indicating the system's ability for self-repair. The longer the period after irradiation, the greater the recovery. The dominant primary producer in Beyers' microecosystem (*Chlorella* sp.) showed no effects of irradiation up to 2×10^6 rads and then took 40 days to die at that dose. In contrast, Posner and Sparrow

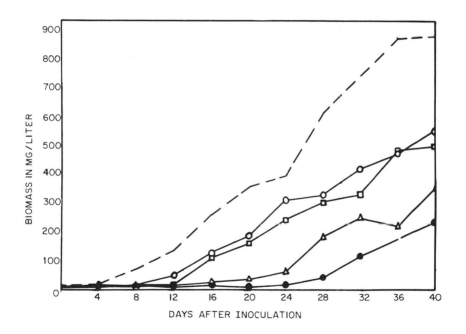

FIG. 3.8. Course of biomass increase with time in an autotrophic succession in a laboratory microecosystem irradiated at 10^6 rad. Successions were initiated by inoculating samples of the irradiated mature microecosystems at 1 (●), 8 (△), 15 (□) and 22 (○) days after irradiation. Control curve (- - -) is from nonirradiated microeco-systems. From Cooke *et al.* (1968).

(1964) reported that 90% of the individuals in a pure culture of *Chlorella* died after a dose of only 23,000 rads. The multispecies mature ecosystem seems to confer radiation protection to its parts.

The internal regulatory activities of ecosystems have also been studied by Copeland (1965). A steady-state laboratory ecosystem, which had been maintained at 1500 footcandles, was switched to a 230 footcandle environment. Metabolism dropped rapidly and a heterotrophic succession took place, followed later by an autotrophic one. Turtle grass, the dominant producer at 1500 footcandles, was replaced by a blue-green alga as metabolism of the reduced light ecosystem gradually returned, through the autotrophic succession, to the level of metabolism maintained by control turtle grass ecosystems at 1500 footcandles. Return to the stable, balanced original level of community metabolism was thus accomplished by a change in the dominant producer. Ecosystems apparently also have a strong tendency to regulate to a balance between rates of photosynthesis and respiration.

Summary

Recent research on the flask- or beaker-type laboratory ecosystems has resulted in some important structural and functional data about the pattern of succession and the development of organization in ecosystems. Studies on these microcosms also demonstrated certain regulatory properties of ecosystems that would have been very difficult to examine under experimental conditions in nature.

The basic trend in ecological succession, as demonstrated in microecosystems, is a shift from a structurally and metabolically unstable ecosystem to one of great homeostasis. The early high efficiency of production, an efficiency related to nutrient availability, shifts to a high maintenance efficiency at maturity, an efficiency

apparently related to a tie up of nutrients in biomass
and a stable nutrient turnover rate. Further study will
indicate whether measures of biochemical diversity, such
as plant pigment ratios, will serve as indices of the status of nutrients during succession in aquatic ecosystems.
During succession, the population density and activity of
the component species change, and some species alter conditions in such a way as to favor the development of other
species. Although not studied in microecosystems as yet,
it is well known that some protists, including *Chlorella*,
release inhibiting substances as well. Thus internal activities of these ecosystems alter conditions, to the extent permitted by energy input and other physical factors,
to bring about an evolution of an ecosystem from a situation of early metabolic and structural instability to one
of greater homeostasis. Mature ecosystems, as shown in
the study of balanced laboratory microcosms, are apparently
far more resistant to perturbations than are individual
components of the system. The successional behavior of
these laboratory ecosystems closely resembles that of
large natural ecosystems, thus strengthening the contention that future study of microecosystems will prove to
be of great value in understanding this level of biological organization.

Microcosms as Life Support Systems

The contrast between developmental and mature stages
is very apparent in certain more practical approaches to
ecosystems. Saunders (1968) has reviewed the progress in
the development of a bioregenerative life support system
for space travel. For long-term space flight (hundreds
or thousands of days), a closed, completely regenerative
life support system will be needed that will metabolize
byproducts of astronaut metabolism and will provide them
with food, water, gases, and environmental diversity and

protection in much the same way that these are provided
in man's ecosystem on earth. This life support system
will properly be called a microecosystem, of which man is
a component species.

One type of proposed microecosystem for space travel
is to consist of only two to a few species (for example,
Chlorella-man or *Hydrogenomonas*-man) plus supportive
plumbing and backup systems in place of natural recycling
and homeostatic mechanisms. Such an ecosystem would have
the ecological characteristics of an immature system and
would require a large amount of energetically expensive
external controls in order to prevent succession and to
maintain long-term stability. Such an ecosystem would
have the advantage of high production efficiency (high
P/B) and low biomass, thus reducing payload weight. Any
failure of external homeostatic devices or a change in
the activity of the support organism would be fatal to
man since the system would then undergo ecological succession and might no longer support the astronaut. The
probability of long-term reliability of a mechanical system
with low biological diversity and the properties of an
early successional stage is remote.

The alternative to the low diversity microcosm for
life support in space is a multispecies mature ecosystem
with its many built-in homeostatic mechanisms (H. T. Odum
1963; Cooke *et al*. 1968; Cooke 1971). As shown in studies
of laboratory ecosystems, developmental stages go through
a process of organization that results in an ecologically
mature system that has long-term stability. Mature stages
of ecosystems also seem to be more resistant to external
perturbations, such as acute doses of radiation. Finally,
mature stages, in contrast to developmental stages in
microecosystems, are metabolically stable (Fig. 3.4), and
the development of an AQ/RQ imbalance between astronauts
and autotrophs would be very unlikely because in

multispecies systems the RQs of the several heterotroph
species (including astronauts) would balance the AQs of
the several autotroph species, and a temporary imbalance
with one species would be compensated by other species.

Several disadvantages of the multispecies microecosystem for life support in space have been shown by studies
on laboratory ecosystems. The diurnal variation in ecosystem metabolism (Fig. 3.1) could lead to deleterious
effects in any type of support system. Beyers' (1963b)
data (Fig. 3.2) indicate, however, that the amplitude of
the daily variation is reduced with increasing diversity
of the system. Also, the mature ecosystem has a low P/B
ratio, and such a system must be large (and heavy), when
compared to the algae-man system, to support the astronauts.

The development of a reliable microecosystem for
space travel will depend, in part, upon studies of small
laboratory microcosms and upon the application of the
principles of ecology, particularly those dealing with
ecological succession and the contrast between the properties of developmental and mature stages. The result
of such studies may also be of immense benefit to all men
since what must be designed for space travel is the minimum
ecosystem consistent with long-term survival of man. This
is the problem that we face with our biosphere, and it is
imperative to understand whether it is possible to have
long-term stability of ecosystems along with the requirements for high yield. That is, can man expect long-term
stability of the biosphere if more and more of its area
becomes high yield and low diversity, with properties of
developmental stages?

Laboratory Streams

Introduction

Laboratory streams are not considered to be ecosystems
in the sense of E. P. Odum (1971) but are usually "open"

communities containing living and nonliving fractions of
a stream, estuary, or other flowing systems, which have
been allowed to reassemble and reorganize in the laboratory
under conditions of water flow and other desired physical
and chemical regimes. The earlier laboratory streams
(e.g., H. T. Odum and Hoskin 1957; Kevern and Ball 1965)
recirculated the same water over the community. More
realistic lotic communities were later developed, often
by diverting natural stream flow into a holding tank and
thence through the laboratory streams (e.g., McIntire
1966a; Patrick 1968).

The most realistic duplication of a natural stream
is the set of six replicate streams constructed by C. D.
McIntire and his associates at Oregon State University
(McIntire *et al.* 1964; Davis and Warren 1965; Fig. 3.9).
Water from a nearby stream is pumped into a wooden storage
tank, through a filter to the laboratory streams, and then
out an effluent pipe. Water chemistry, temperature, turbidity, organic load, and other factors vary with seasonal
changes in the outside stream. The investigator controls
such factors as light intensity, photoperiod, and current
velocity. Small trays with rocks and rubble from the outside stream are placed in the troughs of the laboratory
streams, and organisms in the incoming water seed the
streams. Each tray can be used as a sample of a small
area of the stream and can be removed to a photosynthesis-
respiration chamber (Fig. 3.10) for brief measurements of
community metabolism under carefully controlled conditions.
McIntire's model stream closely resembles a section of a
natural stream. Rates of photosynthesis, efficiency of
light fixation, amount of chlorophyll a per square meter,
species composition, and the light intensity, photoperiod,
and current velocities are very similar to values that
have been regularly reported for a variety of natural
streams (McIntire and Phinney 1965). The replicability

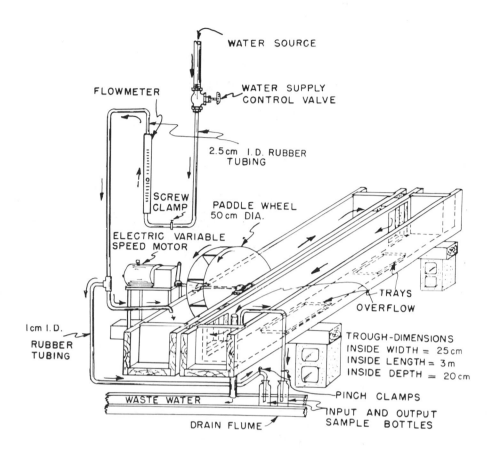

FIG. 3.9. Diagram of one of the six laboratory streams, showing the paddle wheel for circulating the water between the two inter-connected troughs and the exchange water system. From McIntire *et al.* (1964). Reproduced with permission of the authors.

of the six streams, at least with respect to rates of community metabolism, has been demonstrated (McIntire *et al.* 1964) and the pattern of total community energy flow closely duplicates that reported for natural streams (Davis and Warren 1965).

87

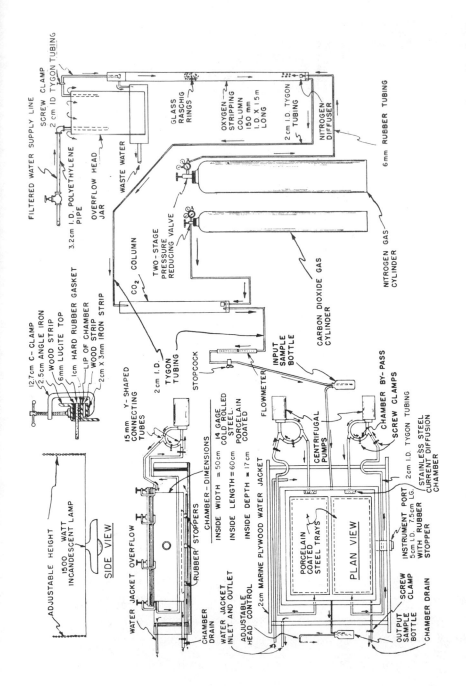

FIG. 3.10. Diagram of the photosynthesis-respiration chamber, showing the chamber with its circulating and exchange water systems, the water jacket for temperature control, the nutrient and gas concentration control system, and the light source. From McIntire et al. (1964). Reproduced with permission of the authors.

EXPERIMENTAL ECOSYSTEMS AND COMMUNITIES

Succession in Laboratory Streams

In natural streams, two types of ecological succession
are observed (H. T. Odum 1956): (1) longitudinal upstream-
downstream succession, and (2) the short-term response of
a stream section to variations in import and to seasonal
changes in such physical factors as light, temperature,
and current velocity. Streams have a gradation of maturity from source to mouth but stream "climax," in the
sense described for beaker-type ecosystems, must be interpreted in relation to a terrestrial-aquatic type of climax
ecosystem (Margalef 1960). A terrestrial-aquatic model
stream ecosystem has not yet been studied in detail and
present laboratory streams have insufficient length to
exhibit an upstream-downstream effect.

Studies of ecological succession in laboratory streams
have been concentrated almost exclusively on short-term
succession. Seasonal changes in import and local physicochemical conditions produce a pronounced response in the
structure and metabolism of any given stream section, and
if these conditions remain fairly constant, the stream
will come to a steady state relative to the conditions.
This is short-term ecological succession in streams. One
of the difficulties in the analysis of this type of succession in nature is to ascertain which factors are most
important in determining a particular level of structure
and metabolism of a stream section. Laboratory streams,
where all factors but one or a combination may be held
constant, have been very useful in providing data about
this problem and have been used as the basis for a simulation model (McIntire 1973).

Two different approaches to the study of seasonal
succession in McIntire's streams have been made. In one
set (McIntire and Phinney 1965) two streams, a light-
adapted one and a shape-adapted one, were studied. Length

of the daily photoperiod was altered every two weeks to correspond to seasonal changes, and seasonal variations in the structure and metabolism of these two streams were then examined. In the other study (McIntire 1968a,b), six streams were set up, three at 150 footcandles and three at 700 footcandles. Three current velocities, 0, 14, and 35 cm/sec, were established within each light regime.

An analysis of the response of benthic algal species to the six current and light regimes was made. Current velocity and low light had a statistically significant positive effect on most of the diatom species (79 of the 89 species in the streams were diatoms), but a few diatoms were most successful at high light and low current. McIntire's (1968a) data indicate that at any given season there is an algal community with a characteristic species composition for that season. An example, the seasonal responses of populations of three algal species *Nitzschia linearis* (Bacillariophyta), *Phormidium retzii* (Cyanophyta), and *Tribonema minor* (Chrysophyta) in the six different streams, is shown in Figure 3.11. The data of Figure 3.11 illustrate the value of controlled laboratory communities in autecological analyses. The algal populations grow under conditions much like that found in natural streams, in contrast to the artificially of unialgal cultures, and the investigator can hold all but one or two factors constant.

Changes in community metabolism were shown to be related to the previous history of illumination of the community and to water temperature, velocity, and amount of nutrients. Stream sections that developed and were exposed to higher light intensity had a higher rate of photosynthesis. The enhancement of the photosynthetic rate with an increase in light was linear up to about 1000 footcandles in communities developed at 500 footcandles

90

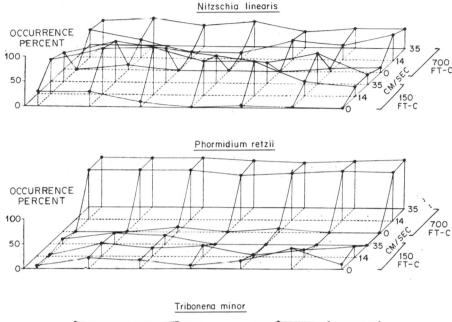

Nitzschia linearis

Phormidium retzii

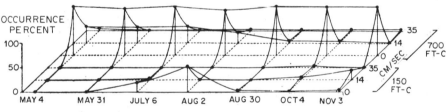

Tribonena minor

FIG. 3.11. Observed occurrence of *Nitzschia linearis, Phormidium retzii,* and *Tribonema minor* in the six laboratory streams on the seven sampling dates during the experiment. Occurrence is expressed as the percentage of microscope fields in which the organisms were observed. From McIntire (1968a). Reproduced with permission of the author.

(Phinney and McIntire 1965; McIntire and Phinney 1965; McIntire 1968a), and shade-adapted communities reached light saturation at values only slightly less than light-adapted communities.

The rate of current velocity is frequently believed to be a very important factor in the rate of community metabolism because rapid water flow apparently produces a steeper diffusion gradient to absorbing cells and thus facilitates exchange between cells and water (Whitford and Schumacher 1961). McIntire (1966a,b; 1968a) has shown that the metabolism of stream communities is enhanced by high current velocity as well as temperature and light. His data indicate that communities that develop in rapid current exhibit a greater metabolic response to changes in current velocity than do communities that develop in slow current.

The level of biomass at any given time in the season was related not only to those factors that regulate community metabolism but also to the rate of export from the stream section. In laboratory streams, peaks of export were often found just after a biomass development peak. As precipitation increased in the fall, the silt load entering the laboratory stream increased and apparently had a scouring effect. Communities that developed at low light had many diatoms and were not as vulnerable to export, while light-adapted communities had many filamentous algae that were easily dislodged. Both slow-current and fast-current communities often had nearly identical biomass levels, even though export from the fast-current community was much higher. Fast-current streams had a higher rate of net community photosynthesis.

The amount of chlorophyll a per square meter of stream community and the ratio of caroteniods to chlorophyll a have been shown to be reliable indices to the conditions to which the stream has been exposed (McIntire and Phinney 1965; McIntire 1968a). The greatest concentrations of chlorophyll a were found in communities that developed at high light and rapid current. The Margalef pigment index (Margalef 1965) was highest in streams developed at high

light and no current (pondlike), indicating that nutrient
depletion can be expected in stream communities exposed
to near stagnant conditions. During the summer, the index
remained fairly stable in all streams but increased during
the onset of fall, when temperature and nutrient supply
were decreased. McIntire's data indicate this index to
be a rather sensitive one to changes in the nutrition of
algal communities in streams.

To summarize, the rate at which a community was established through ecological succession in any particular
section of a stream, as shown by studies of laboratory
streams, was related to current velocity, light intensity,
previous illumination history, and such factors as silt
load. The rate of photosynthesis during a season was
often light limited and was greater in sections with high
current and high export. Temperature was usually not a
factor on a day-to-day basis because it remained rather
steady over a season, except in slow-current areas, and
was effective only at high light intensities (above 1000
footcandles). The type of community that developed, in
terms of species, was related to current and light. It
appeared that the Margalef pigment index was sensitive to
changes in the plant nutrient levels in streams. McIntire
has summarized some of his studies with laboratory streams
(McIntire 1969) and has developed a simulation model of
periphyton dynamics from the data (McIntire 1973).

The study of laboratory streams, where the effects
of single factors on community activity may be analyzed
and where the rate and type of succession over an entire
year under various conditions of illumination and current
speed can be examined, has provided some valuable data
that will be very useful in interpreting the overall pattern of longitudinal succession and organization of a
natural stream. The development of a terrestrial-aquatic

laboratory stream, with upstream-downstream succession, would be very valuable for future investigations.

CONCLUSIONS

The preceding discussions have provided evidence that laboratory microecosystems and complex communities have the properties of larger natural systems. This has been demonstrated in laboratory studies of diurnal community metabolism, ecological succession, and community regulation, as well as in experimental investigations on the ecology of one or more small groups of species.

The use of "laboratory strains" of organisms is well known in biology. These have been developed as a means of testing hypotheses and for suggesting theories through experimentation on a controlled and defined, but complex, unit of biological organization. For example, *Drosophila, Neurospora, Escherichia coli,* and *Tetrahymena* have been thoroughly studied by geneticists and cell biologists. Similarly, animal physiologists have used the white rat, embryologists the amphibian, and population ecologists the flour beetle and *Daphnia*. The laboratory microcosm may well become an ecological "white rat" for the study of the ecosystem level of biological organization. As illustrated by laboratory streams and by the recent use of experimental laboratory systems to examine the effect of simulated tidal cycles on marine benthic diatom communities (McIntire 1969; McIntire and Wulff 1969), autecological studies, as well as investigations about more complex levels of organization, may be conducted in the laboratory without resorting to the artificiality of unispecific cultures.

In addition to the study of the basic principles of ecology, the laboratory community or ecosystem may also become a very valuable diagnostic tool. For example,

94

many new organic compounds are developed each year, and
most of these compounds are released to the environment
before we understand their possible effects. The usual
procedure, if testing is done, is to test one or more
vertebrate species (game species or species with many genes
in common with man). Unfortunately, such tests are relatively meaningless since such pollutants as pesticides
are applied, in practice, to the community and ecosystem
levels of organization and not to these organism or population levels. Laboratory ecosystems make excellent devices for the examination of the short- and long-range
effects of potential pollutants on the actual level of
biological organization to which they are applied. Such
studies can be carried out before the potential pollutants
are released. Some examples (Hueck and Adena 1968; Cairns
1969) illustrate the effect of some common pollutants on
the activity of laboratory ecosystems and communities,
and Södergren (1973) constructed a highly simplified and
artifically compartmentalized laboratory ecosystem in
which he was able to test the effects of certain pesticides
on three trophic levels. By introducing the material into
one compartment and then, after an appropriate interval,
feeding some of the living material that had been exposed
to the pesticide to species of another trophic level in
another compartment, he was able to ascertain whether
these pesticides were degraded through a food chain. Cooper
and Copeland (1973) used laboratory ecosystems to study
the response of regions of an estuary, defined in the ecosystems by rate and retention of freshwater flow, to
drought, degree of dependence on input of fresh water, and
combinations of industrial effluent and fresh water. Their
work provided useful evidence on the impact of reducing
river flow and increasing industrial effluents on fisheries
and tourism of Galveston Bay estuary. Their approach seems
to be the type that is required to provide the scientific

basis for environmental planning and illustrates the utility of laboratory ecosystems for environmental impact work.

A logical extension of the laboratory microcosm method is to construct miniature, replicable outdoor ecosystems and communities. For example, Whitworth and Lane (1969) used a series of wading pools as controlled ecosystems in order to assess the long-term effect of a group of toxicants on ponds, and Goodyear *et al.* (1972) used similar pools to experimentally examine the relation between level of pond fertilization and fish productivity. As H. T. Odum *et al.* (1963b) point out, the construction of ecosystems is actually an experiment in ecological engineering that may result in some new systems of value to man.

Finally, the microcosm approach is very useful in teaching, since the student is continuously presented with a holistic view of ecology and at the same time is provided with a system amenable to laboratory and field experimentation (Bovbjerg and Glynn 1960).

A holistic theory of ecology will not be developed, of course, by a study of laboratory ecosystems alone. Observations of nature are also required, as are studies of all levels of organizations. However, experimentation is needed in ecology and laboratory ecosystems and communities can play a significant role in filling that need. There are several interesting questions in ecology that may be answered through the microcosm method. For example, what is the relationship between species and pattern diversity and the structural and metabolic stability of an ecosystem? Do ecosystems die?

ACKNOWLEDGMENTS

Preparation of this chapter was supported by a Summer Research Fellowship, by the Center for Urban Regionalism and Environmental Systems, and by the Department of

Biological Sciences at Kent State University. I thank
Drs. C. D. McIntire, R. W. Gorden, and R. T. Heath for
their constructive criticisms of the chapter. To Drs.
R. J. Beyers, R. W. Gorden, C. D. McIntire, and R. H.
Whittaker I owe a particular debt of gratitude for their
permission to reproduce some of their figures for use in
this chapter.

REFERENCES

Abbott, W. 1966. Microcosm studies on estuarine waters. I. The replicability of microcosms. *J. Water Pollut. Control Fed. 38*, 258-270.

Beyers, R. J. 1962. The metabolism of twelve aquatic laboratory microecosystems. Ph.D. dissertation. The University of Texas, Austin.

Beyers, R. J. 1963a. The metabolism of twelve laboratory microecosystems. *Ecol. Monogr. 33*, 281-306.

Beyers, R. J. 1963b. A characteristic diurnal metabolic pattern in balanced microcosms. *Publ. Inst. Mar. Sci., Texas 9*, 19-27.

Beyers, R. J. 1965. The pattern of photosynthesis and respiration in laboratory microecosystems. *Mem. Inst. Ital. Idrobiol., 18 Suppl.*, 61-74.

Beyers, R. J., Larimer, J. L., Odum, H. T., Parker, R. B., and Armstrong, N. E. 1963. Directions for the determination of changes in carbon dioxide concentration from changes in pH. *Publ. Inst. Mar. Sci., Texas 9*, 454-489.

Bovbjerg, R. V., and Glynn, P. W. 1960. A class exercise on a marine microcosm. *Ecology 41*, 229-232.

Butler, J. L. 1964. Interaction of effects by environmental factors on primary productivity in ponds and microecosystems. Ph.D. dissertation, Oklahoma State University, Stillwater.

Cairns, J. Jr. 1969. Rate of species diversity restoration following stress in freshwater protozoan communities. *Univ. Kansas Sci. Bull 48*, 209-224.

Castenholz, R. W. 1972. Hot spring microbial communities recreated in modified Winogradski columns. *Limnol. Oceanog. 17*, 767-772.

Coler, R. A., and Gunner, H. B. 1970. Laboratory enclosure of an ecosystem response to a sustained stress. *Appl. Microbiol. 19*, 1009-1012.

Confer, J. L. 1972. Interrelations among plankton, attached algae, and the phosphorus cycle in artificial open systems. *Ecol. Monogr. 42*, 1-23.

Cooke, G. D. 1967. The pattern of autotrophic succession in laboratory microcosms. *Bioscience 17*, 717-721.

Cooke, G. D. 1971. Ecology of Space Travel. In *Fundamentals of Ecology* (3rd ed.) E. P. Odum (Ed.). W. B. Saunders Co., Philadelphia, pp. 498-509.

Cooke, G. D., Beyers, R. J., and Odum, E. P. 1968. The case for the multi-species ecological system, with special reference to succession and stability. In *Bioregenerative Systems* J. F. Saunders (Ed.). NASA, SP-165, pp. 129-139.

Cooper, D. C., and Copeland, B. J. 1973. Responses of continuous-series estuarine microecosystems to point-source input variations. *Ecol. Monogr. 43*, 213-236.

Copeland, B. J. 1965. Evidence for regulation of community metabolism in a marine ecosystem. *Ecology 46*, 563-564.

Davis, G. E., and Warren, C. E. 1965. Trophic relations of a sculpin in laboratory stream communities. *J. Wildl. Manage. 24*, 846-871.

Edmondson, W. T. 1968. Water-quality management and lake eutrophication: The Lake Washington case. In *Water Resources Management and Public Policy* T. H. Campbell and R. O. Sylvester (Eds.). University of Washington Press, Seattle, pp. 139-178.

Fitzgerald, G. P., and Nelson, T. C. 1966. Extractive and enzymatic analyses for limiting or surplus phosphorus in algae. *J. Phycol. 2*, 32-37.

Goodyear, C. P., Boyd, C. E., and Beyers, R. J. 1972. Relationships between primary productivity and mosquito fish (*Gambusia affinis*) production in large microcosms. *Limnol. Oceanog. 17*, 445-450.

Gorden, R. W. 1967. Heterotrophic bacteria and succession in a simple laboratory aquatic microcosm. Ph.D. dissertation, The University of Georgia, Athens.

Gorden, R. W., Beyers, R. J., Odum, E. P., and Eagon, R. G. 1969. Studies of a simple laboratory microecosystem: Bacterial activities in a heterotrophic succession. *Ecology 50*, 86-100.

Hall, D. J., Cooper, W. E., and Werner, E. E. 1970. An experimental approach to the production dynamics and structure of freshwater animal communities. *Limnol. Oceanog. 15*, 839-928.

Hueck, H. J., and Adena, D. M. M. 1968. Toxicological investigations in an artificial ecosystem. A progress report on cooper toxicity towards algae and *Daphnia*. *Helgolander Wiss. Meersunters 17*, 188-199.

Ketchum, B. H., Ryther, J. H., Yentsch, C. S., and Corwin, N. 1958. Productivity in relation to nutrients. *Rapp. Proc. -Verb. Cons. Int. Explor. Mer. 144*, 132-140.

Kevern, N. R., and Ball, R. C. 1965. Primary productivity and energy relationships in artificial streams. *Limnol. Oceanog. 10*, 74-87.

Kurihara, Y. 1960. Biological analysis of the structure of microcosm with special reference to the relations among biotic and abiotic factors. *Sci. Rep. Tohoku Univ. Ser. IV (Biol.) 26*, 269-296.

Margalef, R. 1960. Ideas for a synthetic approach to the ecology of running water. *Int. Rev. ges. Hydrobiol. 45*, 133-153.

Margalef, R. 1963a. On certain unifying principles in ecology. *Am. Nat. 97*, 357-374.

Margalef, R. 1963b. Succession of marine populations. *Adv. Front. Plant Sci. 2*, 137-188 (India).

Margalef, R. 1965. Ecological correlations and the relationship between primary productivity and community structure. *Mem. Ist. Ital. Idrobiol. 18 Suppl.*, 355-364.

Margalef, R. 1968. Perspectives in Ecological Theory. The University of Chicago Press, Chicago. viii+ 111 pp.

Mathis, B. J. 1972. Chlorophyll and the Margalef pigment ratio in a mountain lake. *Am. Midl. Nat. 88*, 232-235.

McIntire, C. D. 1966a. Some effects of current velocity on periphyton in laboratory streams. *Hydrobiology 27*, 559-570.

McIntire, C. D. 1966b. Some factors affecting respiration of periphyton communities in lotic environments. *Ecology 47*, 918-930.

McIntire, C. D. 1968a. Structural characteristics of benthic algal communities in laboratory streams. *Ecology 49*, 520-537.

McIntire, C. D. 1968b. Physiological-ecological studies of benthic algae in laboratory streams. *J. Water Pollut. Control Fed. 40*, 1940-1952.

McIntire, C. D. 1969. A laboratory approach to the study of the physiological ecology of benthic algal communities. In *Proceedings of the Eutrophication-Biostimulation Assessment Workshop*. E. J. Middlebrooks, T. E. Maloney, C. F. Powers and L. M. Kaack (Eds.). pp. 146-157.

McIntire, C. D. 1973. Periphyton dynamics in laboratory streams: A simulation model and its implications. *Ecol. Monogr. 43*, 399-420.

McIntire, C. D., Garrison, R. L., Phinney, H. K., and Warren, C. E. 1964. Primary production in laboratory streams. *Limnol. Oceanog. 9*, 92-102.

McIntire, C. D., and Phinney, H. K. 1965. Laboratory studies of periphyton production and community metabolism in lotic environments. *Ecol. Monogr. 35*, 237-258.

McIntire, C. D., Tinsley, I. J., and Lowry, R. R. 1969. Fatty acids in lotic periphytons: another measure of community structure. *J. Phycol. 5*, 26-32.

McIntire, C. D., and Wulff, B. L. 1969. A laboratory method for the study of marine benthic diatoms. *Limnol. Oceanog. 14*, 667-678.

Nixon, S. W. 1969. A synthetic microcosm. *Limnol. Oceanog. 14*, 142-145.

Odum, E. P. 1960. Organic production and turnover in old field succession. *Ecology 41*, 34-39.

Odum, E. P. 1963. *Ecology*. Holt, Rinehart, and Winston, New York, vii+ 152 pp.

Odum, E. P. 1969. The strategy of ecosystem development. *Science 164*, 262-270.

Odum, E. P. 1971. *Fundamentals of Ecology* (3rd ed.), W. B. Saunders, Co., Philadelphia, 574 pp.

Odum, E. P., and Beyers, R. J. 1965. *Biodynamics of Microecosystems*. Report to NASA, August 1, 1965, 6 pp.

Odum, H. T. 1956. Primary production in flowing water. *Limnol. Oceanog. 1*, 102-117.

Odum, H. T. 1963. Limits of remote ecosystems containing man. *Am. Biol. Teach. 25*, 429-443.

Odum, H. T., and Hoskin, C. M. 1957. Metabolism of a laboratory stream microcosm. *Publ. Inst. Mar. Sci., Texas 4*, 115-133.

Odum, H. T., Beyers, R. J., and Armstrong, N. E. 1963a. Consequences of small storage capacity in nannoplankton pertinent to measurement of primary production in tropical waters. *J. Mar. Res. 21*, 191-198.

Odum, H. T., Siler, W. L., Beyers, R. J., and Armstrong, N. E. 1963b. Experiments with engineering of marine ecosystems. *Publ. Inst. Mar. Sci., Texas 9*, 373-403.

Ovington, J. D. 1962. Quantitative ecology and the woodland ecosystem concept. *Adv. Ecol. Res. 1*, 103-192.

Owens, M. 1965. Some factors involved in the use of dissolved-oxygen distributions in streams to determine productivity. *Mem. Ist. Ital. Idrobiol. 18 Suppl.*, 209-224.

Patrick, R. 1968. The structure of diatom communities in similar ecological conditions. *Am. Nat. 102*, 173-183.

Phinney, H. K., and McIntire, C. D. 1965. Effect of temperature on metabolism of periphyton communities developed in laboratory streams. *Limnol. Oceanog. 10*, 341-344.

Posner, H. B., and Sparrow, A. H. 1964. Survival of *Chlorella* and *Chlamydomonas* after acute and chronic gamma radiation. *Rad. Bot. 4*, 253-257.

Saunders, J. F. (Ed.). 1968. *Bioregenerative Systems*. NASA Special Publication 165. U.S. Govt. Printing Office, Washington, D.C., viii+ 153 pp.

Södergren, A. 1973. Transport, distribution, and degradation of chlorinated hydrocarbon residues in aquatic model ecosystems. *Oikees 24*, 30-41.

Stockner, J. G., and Benson, W. W. 1967. The succession of diatom assemblages in the recent sediments in Lake Washington. *Limnol. Oceanog. 12*, 513-532.

Taub, F. B. 1969. A biological model of a freshwater community: A gnotobiotic ecosystem. *Limnol. Oceanog.* *14*, 23-34.

Taub, F. B. 1971. A continuous gnotobiotic (species defined) ecosystem. In *The Structure and Function of Freshwater Microbial Communities*, J. Cairns, Jr. (Ed.). Research Division Monograph 3, Virginia Polytechnic Institute and State University, Blacksburg, pp. 101-120.

Taub, F. B., and Dollar, A. M. 1964. A *Chlorella-Daphnia* food-chain study: The design of acompatible chemically-defined culture medium. *Limnol. Oceanog. 9*, 61-74.

Whitford, L. A., and Schumacher, G. J. 1961. Effect of current on mineral uptake and respiration by a freshwater alga. *Limnol. Oceanog. 6*, 423-431.

Whittaker, R. H. 1961. Experiments with radiophosphorus tracer in aquarium microcosms. *Ecol. Monogr. 31*, 157-188.

Whitworth, W. R., and Lane, T. H. 1969. Effects of toxicants on community metabolism in pools. *Limnol. Oceanog. 14*, 53-58.

Wilhm, J. L., and Long, J. 1969. Succession in algal mat communities at three nutrient levels. *Ecology 50*, 645-652.

Chapter 4

A CONTINUOUS GNOTOBIOTIC
(SPECIES-DEFINED) ECOSYSTEM

Frieda B. Taub

College of Fisheries
University of Washington
Seattle, Washington

CONTENTS

ABSTRACT

The responses of nutrient-limited autotrophic communities to change in light intensity and dilution rate were studied by means of a two-stage continuous-culture apparatus. The communities consisted of the single alga *Chlamydomonas reinhardtii*, the protozoan *Tetrahymena vorax*, and the bacteria *Pseudomonas fluorescens* and *Escherichia coli*; 0.5 mM NO_3 was the limiting nutrient. The protozoan population became extinct when its removal rate by dilution exceeded its food-limited growth rate. An algal density of 2 x 10^6 cells/ml permitted the protozoan population to persist with a dilution rate of 0.73/day, while algal densities of 0.3 x 10^6 cells/ml resulted in protozoan extinction at 0.75 or 1.0/day. The reductions in algal density were achieved by increasing flow rate or decreasing light intensity. *Escherichia coli* populations became extinct at high dilution rates but not under low light intensity. *Pseudomonas fluorescens* was less sensitive to the changes. Kinetic relationships between algal cell concentration and protozoan growth rates accurately predicted extinction, but not steady-state populations of the protozoan.

The downstream and yield communities were progressively denser and more stable under environmental changes.

INTRODUCTION

The experiment reported here was designed to show the responses of a chemostat community to changes in light intensity and dilution rate. It represents an advance over our previously reported studies (Taub 1969a; Taub 1969b) in that continuous-culture methods provided for continuous growth during the duration of the study and allowed the estimation not only of standing crop but also of net growth rates and productivity rates during steady-state conditions. The use of a "gnotobiotic" (Dougherty 1959) community permitted enumeration of the populations of all component species and minimized uncontrolled variables.

Interaction between the organisms was assured by the fact that the heterotrophs depended on the autotrophs for their source of organic compounds, and the protozoa could ingest both the algal cells and the bacteria in addition to using extracellular products. The system was used as a model of a real biological community in the sense of Patten (1968); as a model, it is an abstraction and a simplification but not a substitute for a real system. The behavior of the model must not be inferred to apply to real communities unless support for such an inference is supplied. In this study the model is used almost entirely as an analytical tool.

Previous work on microbial ecosystems has been reviewed by Dennis Cooke in this volume.

MATERIALS AND METHODS

The organisms consisted of the alga *Chlamydomonas reinhardtii*, Strain 90, of the Indiana culture collection; the protozoan *Tetrahymena vorax*, Strain V_2S, from G. G. Holtz, Jr.; and the bacteria *Pseudomonas fluorescens*, 13525, and *Escherichia coli*, 14948, both of the American Type Culture Collection. The chemical environment was medium 63 (Taub and Dollar 1968a). This medium is composed of reagent grade inorganic chemicals, with the exception of 0.0284 mM EDTA, and does not support significant growth of the heterotrophic organisms but supplies all of the requirements for autotrophic growth of *Chlamydomonas*. Each continuous-culture apparatus consisted of

(a) a medium reservoir, which delivered aseptic medium by means of a digital pump; (b) a first (upstream) growth flask, the overflow of which was delivered to (c) a second (downstream) growth flask, the overflow from which was delivered to (d) a yield reservoir. The flow into the first flask was determined by a digital metering pump (the medium remaining within closed Tygon tubing). Since the volume of liquid in each flask was maintained constant by an overflow mechanism, the volume delivered to the yield bottle represented the flow into each of the growth flasks. Mixing was achieved both by aeration with a 2% CO_2-enriched air mixture and rocking at 17°, at 20 cycles per min. Loose glass beads on the bottom of each flask were used to dislodge cells that adhered to the sides of the flasks. (The flask is described in detail in Taub and Dollar 1968b, 1965.) Temperature was controlled at $20 \pm 1^{\circ}C$ and lights were supplied for 12 hr daily by combinations of fluorescent and incandescent bulbs in a Sheer growth unit. Light intensity was measured by a GE 213 light meter from which the filter had been removed. Therefore, these values were approximately twice as high as would have been attained had the filter been used. (Removal of the filter in part eliminated the relative insensitivity of the meter to wavelengths less than 5000 A or greater than 7000 A.) Two replicate systems designated green (system 1) and red (system 2) were studied. The only recognized difference between the systems was a greater volume in flask 1 of the green system caused by a crack in the overflow tube; volumes are shown in Figures 4.1 and 4.2.

Chlamydomonas were enumerated by hemacytometer, *Tetrahymena* by the dilution method of Sonneborn (1950), and the bacteria by plating on King's medium B (King *et al.* 1954), and EMB media (Difco 1953). The lowest detectable populations for these methods were 1 x 10^4 *Chlamydomonas*/ml, 5 *Tetrahymena*/ml, and 10^2 bacteria/ml.

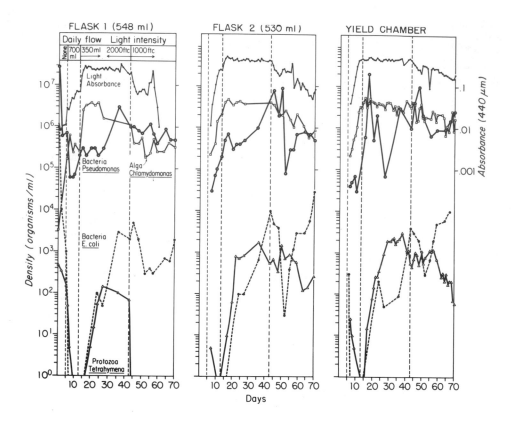

FIG. 4.1. Population densities and light absorbances for the three communities of the green system (system 1). Continuous system (replicate "green").

Pigment measurements were conducted in accordance with the methods of Strickland and Parsons (1968) and the use of a 1-cm cuvet and the PS formula. Analyses of NO_3 and NO_2 were determined by the method of Wood *et al.* (1967) and Bendschneider and Robinson (1952). Oxidizable carbon was determined by a modification of the method of Strickland and Parsons (1968); whole samples instead of particulate matter were analyzed since medium 63 did not interfere as would sea water.

A CONTINUOUS GNOTOBIOTIC ECOSYSTEM

All of the organisms were inoculated into the first flask on day 0 and permitted to grow as batch cultures until day 5. A flow of approximately 750 ml/day was initiated on day 5 and maintained until day 13, at which time the flow was decreased to approximately 400 ml/day. Light intensity was maintained at approximately 2000 foot-candles until day 43, when it was reduced to half that intensity. Optical density was measured daily in each flask by use of the nephelometer tube. Samples were

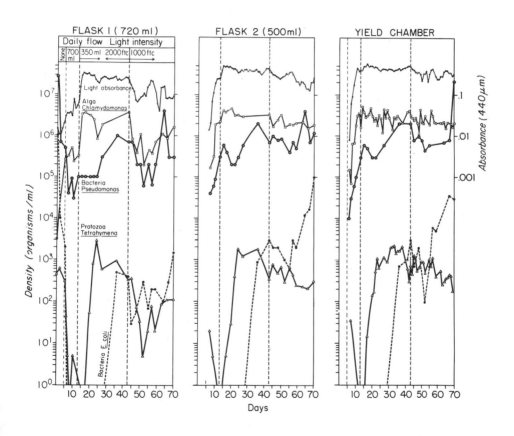

FIG. 4.2. Population densities and light absorbances for the three communities of the red system (system 2). Continuous system (replicate "red").

aseptically drawn from each of the flasks and yield reservoir for species enumeration two or three times weekly. The chemical analyses were largely limited to samples from the yield flask. Yield was removed daily.

RESULTS AND DISCUSSION

Standing Crop Relationships

The light absorbance and densities of the organisms are shown in Figures 4.1 (green replicate) and 4.2 (red replicate). During the first five days of batch growth, only the algal population increased rapidly. Densities and exponential growth rate

$$\mu = 2.3 \ \frac{(\log x_1 - \log x_0)}{t}$$

are shown in Table 4.1. When high flow was initiated (Table 4.2), the algal and *Pseudomonas* populations remained stable, whereas the protozoa and *E. coli* washed out. When the flow was reduced (Table 4.3), algal concentrations increased and assumed a steady state, the *Pseudomonas* populations stayed at the same density for almost two weeks and then increased in both replicates, and the protozoan and *E. coli* populations rose from undetectable levels to low, steady states. When the light intensity was reduced on day 43 (Table 4.4), the algal population decreased to a new steady state; the protozoan population in the green replicate decreased to 2×10^2 and that in the red replicate became undetectable; the *Pseudomonas* population decreased slightly; and the *E. coli* population increased slightly. A comparison of the growth rates during initial batch growth with those during the dilution phases indicates that algal growth was not at its maximum and there was no growth on the part of the other organisms, probably as a result of residual lag stages.

Table 4.1
Growth Characteristics During Batch Growth

	Density		Exponential (ln) growth rate (per day) (μm)
	Initial	Day 5	
System 1 (green)			
O.D.	0.015	0.03	0.14
Alga	3.3×10^3	2.7×10^5	0.88
Protozoan	4.5×10^2	3.1×10^2	-0.08
Pseudomonas	2.8×10^7	4.5×10^5	-0.825
E. coli	1.0×10^6	2×10^3	-1.29
System 2 (red)			
O.D.	0.012	0.028	0.17
Alga	3.3×10^3	1.7×10^5	0.78
Protozoan	4.5×10^2	1.5×10^2	-0.22
Pseudomonas	2.8×10^7	7×10^5	-0.74
E. coli	1×10^6	1×10^3	-1.38

The downstream communities received their population upon initiation of overflow from the upstream community on day 5. In general, all populations were slightly denser and showed less fluctuation than the upstream community did; e.g., in the upstream community, the algal populations fluctuated over a sevenfold range during the three time periods, whereas in the downstream communities the changes were only over a threefold maximum change. The protozoan populations were especially more stable in the downstream communities. Once they became established

Table 4.2

Density, Growth Rates, and Production Rates of Organisms, High Light, High Dilution Rate, Days 7 to 13 Inclusive

System 1

Avg. flow 749 ml/day; avg. dilution rate for flask 1 = 1.040, flask 2 = 1.497; avg. light intensity 2000 ft-c[a]

	Standing crop (cells/ml)		Exponential (ln) growth rate (per day)		Production (cells/day)	
	Flask 1	Flask 2	Flask 1	Flask 2	Flask 1	Flask 2
O.D.	0.05	0.16	1.04	1.06	3.5×10^1	1.2×10^2
Alga	2.9×10^5	1.1×10^6	1.04	1.12	2.1×10^8	8.4×10^8
Protozoan	2.5	6.3	(1.04)	0.90	(1.9×10^3)	4.7×10^3
Pseudomonas	6.3×10^4	1.1×10^5	1.04	0.65	4.7×10^7	8.2×10^7
E. coli	0	(1.0×10^{-5})	0	0	0	0
	Yield		Yield		Yield	
O.D.	0.27		0.48		2.0×10^2	
Alga	1.74×10^6		0.44		1.3×10^9	
Protozoan	7.0		0.11		5.2×10^3	
Pseudomonas	1.2×10^5		0.10		9.1×10^7	
E. coli	(1.0×10^{-5})		0		0	

114

System 2

Avg. flow 759 ml/day; avg. dilution rate for flask 1 = 1.379, flask 2 = 1.431; avg. light intensity 2000 ft-c[a]

	Standing crop (cells/ml)			Exponential (ln) growth rate (per day)			Production (cells/day)		
	Flask 1	Flask 2	Yield	Flask 1	Flask 2	Yield	Flask 1	Flask 2	Yield
O.D.	0.05	0.01	0.02	1.38	0.93	0.60	4.3×10^1	1.2×10^2	2.2×10^2
Alga	3.7×10^5	1.1×10^6	1.7×10^6	1.38	0.95	0.41	2.8×10^8	8.3×10^8	1.3×10^9
Protozoan	(1.25)	(1.25)	8.0	(1.38)	(0)	1.85	(9.5×10^2)	(9.5×10^2)	6.1×10^3
Pseudomonas	9.6×10^4	1.1×10^5	8.4×10^4	1.38	0.19	-0.28	7.3×10^7	8.4×10^7	6.4×10^7
E. coli	0	(1.0×10^{-5})	(1.0×10^{-5})	0	1.43	0	0	(7.6×10^{-3})	(7.6×10^{-3})

[a]Figures in parentheses indicate extinct for practical purposes, average less than detectable limits.

Table 4.3

Density, Growth Rates, and Production Rates of Organisms, High Light, Low
Dilution Rate, Days 24 to 43 Inclusive

System 1

Avg. flow 409 ml/day; avg. dilution rate for flask 1 = 0.568, flask 2 = 0.818;
avg. light intensity 2000 ft-c

	Standing crop (cells/ml)		Exponential (ln) growth rate (per day)		Production (cells/day)	
	Flask 1	Flask 2	Flask 1	Flask 2	Flask 1	Flask 2
O.D.	0.24	0.39	0.57	0.32	9.8×10^1	1.6×10^2
Alga	2.1×10^6	3.2×10^6	0.57	0.29	8.5×10^8	1.3×10^9
Protozoan	1.2×10^3	1.3×10^3	0.57	0.07	4.9×10^5	5.4×10^5
Pseudomonas	5.3×10^5	8.8×10^5	0.57	0.33	2.2×10^8	3.6×10^8
E. coli	2.3×10^2	9.8×10^2	0.57	0.63	9.2×10^4	4.0×10^5
	Yield		Yield		Yield	
O.D.	0.42		0.08		1.7×10^2	
Alga	3.0×10^6		-0.06		1.2×10^9	
Protozoan	1.4×10^3		0.06		5.7×10^5	
Pseudomonas	1.4×10^6		0.43		5.5×10^8	
E. coli	9.4×10^2		-0.04		3.84×10^5	

System 2

Avg. flow 402 ml/day; avg. dilution rate for flask 1 = 0.730, flask 2 = 0.758; avg. light intensity 2000 ft-c

	Standing crop (cells/ml)		Exponential (ln) growth rate (per day)		Production (cells/day)	
	Flask 1	Flask 2	Flask 1	Flask 2	Flask 1	Flask 2
O.D.	0.24	0.43	0.73	0.33	9.7×10^1	1.7×10^2
Alga	2.1×10^6	3.8×10^6	0.73	0.34	8.6×10^9	1.5×10^8
Protozoan	8.4×10^1	9.9×10^2	0.73	0.70	3.4×10^4	4.0×10^5
Pseudomonas	1.0×10^6	2.0×10^6	0.73	0.37	4.0×10^8	7.9×10^8
E. coli	8.4×10^2	2.7×10^3	0.73	0.52	3.4×10^5	1.1×10^6
	Yield		Yield		Yield	
O.D.	0.46		0.06		1.8×10^2	
Alga	3.5×10^6		-0.08		1.4×10^9	
Protozoan	1.7×10^3		0.52		6.7×10^5	
Pseudomonas	1.4×10^6		-0.36		5.5×10^8	
E. coli	1.2×10^3		-0.82		4.8×10^5	

Table 4.4

Density, Growth Rates, and Production Rates of Organisms, Low Light, Low
Dilution Rate, Days 45 to 70 Inclusive

System 1

Avg. flow 402 ml/day; avg. dilution rate for flask 1 = 0.558, flask 2 = 0.804;
avg. light intensity 1220 ft-c

	Standing crop (cells/ml)		Exponential (ln) growth rate (per day)		Production (cells/day)	
	Flask 1	Flask 2	Flask 1	Flask 2	Flask 1	Flask 2
O.D.	0.12	0.26	0.56	0.45	4.6×10^1	1.1×10^2
Alga	7.6×10^5	2.1×10^6	0.56	0.51	3.1×10^8	8.2×10^8
Protozoan	1.9×10^2	4.4×10^2	0.56	0.46	7.6×10^4	1.8×10^5
Pseudomonas	6.2×10^5	1.1×10^6	0.56	0.35	2.5×10^8	4.4×10^8
E. coli	1.1×10^4	1.6×10^4	0.56	0.22	4.6×10^6	6.3×10^6
	Yield		Yield		Yield	
O.D.	0.33		0.22		1.3×10^2	
Alga	2.4×10^6		0.2		9.6×10^8	
Protozoan	6.3×10^2		0.36		2.6×10^5	
Pseudomonas	3.6×10^6		1.18		1.4×10^9	
E. coli	1.2×10^4		-0.29		4.7×10^6	

System 2

Avg. flow 411 ml/day; avg. dilution rate for flask 1 = 0.747, flask 2 = 0.775; avg. light intensity 1220 ft-c

	Standing crop (cells/ml)		Exponential (ln) growth rate (per day)		Production (cells/day)	
	Flask 1	Flask 2	Flask 1	Flask 2	Flask 1	Flask 2
O.D.	0.04	0.16	0.75	0.57	1.7×10^1	6.7×10^1
Alga	3.6×10^5	1.8×10^6	0.75	0.62	1.5×10^8	7.3×10^8
Protozoan	0	5.6×10^2	0.75	0.78	0	2.4×10^5
Pseudomonas	7.4×10^5	2.2×10^6	0.75	0.52	3.0×10^8	9.2×10^8
E. coli	2.4×10^4	5.9×10^3	0.75	-2.34	9.7×10^6	2.4×10^6
		Yield		Yield		Yield
O.D.		0.25		0.41		1.0×10^2
Alga		1.9×10^6		0.08		7.9×10^8
Protozoan		5.8×10^2		0.04		2.4×10^5
Pseudomonas		2.7×10^6		0.20		1.1×10^9
E. coli		3.3×10^3		-0.57		1.4×10^6

during the low dilution period, they reached a steady density of ca. 1 x 10^3 cells/ml. They were little affected
by the drop in light intensity and there was only a slight
but constant downward trend.

The populations in the yield reservoirs were usually
slightly denser and more stable than in the downstream
community.

Growth Rates and Production Estimates

Assumptions and Their Validity

Unlike the data analyses above, the estimation of
growth rates and production rates required various assumptions, outlined below. These have been adapted from the
excellent review of Herbert (1964). Our system conformed
to a single-stream, multistage continuous culture. Culture medium containing the growth-limiting nutrient s at
a concentration s_o (NO_3, 0.5 mM) entered flask 1 at a
flow rate f and culture flowed from flask 1 to flask 2
and finally out of the latter at the same flow rate f.
The dilution rates were respective $D_1 = f/v_1$ and $D_2 = f/v_2$,
where v is the culture volume. The concentrations of cells
and of substrate in the two flasks were, respectively,
x_1, x_2 and s_1, s_2. Growth was assumed to follow the exponential growth equation $dx/dt = \mu x$, where μ = exponential
growth rate.

In the first flask, there was no inflow of cells, and
the equation for cell balance is increase = growth - outflow; $dx_1/dt = u_1 x_1 - D_1 x_1$. Under steady-state conditions,
$dx_1/dt = 0$ and $u_1 = D_1$; i.e., the growth rate in the steady
state was equal to the dilution rate. This condition is
obvious when higher or lower growth rates are considered;
the populations would increase or decrease so that a
steady state would not prevail.

A CONTINUOUS GNOTOBIOTIC ECOSYSTEM

In the second flask, the equation for cell balance is: increase = inflow - outflow + growth: $dx_3/dt = D_2x_1 - D_2x_2 + \mu_2x_2$. At a steady state, $dx_2/dt = 0$ and

$$\mu_2 = D_2 \frac{(x_2 - x_1)}{(x_2)}$$

These relationships assume perfect mixing and representative overflow.

The yield reservoir represents a special problem, since the population started from zero volume each day and accumulated gradually as the reservoir filled until it was replaced and sampled. Dilution did not occur as there was no overflow. On a population basis the exponential growth rate was calculated as:

$$\mu = 2.3 \frac{(\log x_{yield} - \log x_2)}{t}$$

where $t = 1$ day.

It may be argued that, on a physiological basis, individual cells were growing at twice this rate, since cells that entered at minute 1 had the full day to grow, whereas those cells that entered immediately before the replacement had no time to grow.

The relationship between limiting substrate and growth rate, which has been shown to hold in many cases, is described in the Michaelis-Menten relationship:

$$\mu = \mu_{max} \frac{(s)}{K_s + s}$$

where K_s = "saturation constant," numerically equal to substrate concentration at which $\mu = \frac{1}{2} \mu_{max}$. This relationship has been shown by Monod (1942) to apply to bacteria in fermenters and has been demonstrated since by Jannasch (1967) and Wright and Hobbie (1965) to apply to natural populations of bacteria. The relationship has been verified for the alga *Chlorella* by Pearson et al. (1969).

121

Monod (1942) also showed that the yield, Y, where

$$Y = \frac{\text{weight of bacteria formed}}{\text{weight of substrate used}}$$

is a constant (at least to a first approximation).

The doubling rate may be calculated from the exponential (ln) growth rate μ by the relationship, doubling rate = $0.693/\mu$.

The continuous-culture systems that utilize limited nutrient input as an external control on growth rate have been thoroughly reviewed by Novick (1955). He shows that the stability of the system is caused by interaction between the population density and the nutrient concentration: the rising population density lowers the concentration of the required factor until the population growth rate slows. The growth rate decreases until it becomes equal to the washout rate. He also shows that the concentration of the controlling factor in the growth flask, s_1 or s_2, depends only on the growth rate (= flow rate) selected and is independent of the input concentration, s_0. (Notation changed to conform with Herbert's system.)

Production rate for each flask is $P = fx$, the flow rate multiplied by the density. Thus the production in the second flask is the result for the first and second flask, and that in the yield reservoir represents the yield of the total system. The yield produced within the second or yield chambers can be calculated by subtraction of the value of the previous chamber.

Before these assumptions were applied, consideration was given to their probable validity.

Since the densities measured represent population levels after whatever consumption may have occurred, the growth and production rates are minimal, or net, rates. Gross growth and production would include those algal and bacterial cells that were produced and consumed before measurement.

Perfect mixing was probably not achieved, but the error would be small and was ignored. Growth on the chamber walls and tubes was minimized until the end of the experiment but did become dense enough by day 70 so that the experiment was terminated. Adherence of organisms on the walls would result in higher estimates of growth rates than had actually occurred since a larger population than measured would be contributing offspring. The surfaces may have served as a refuge for the protozoa and *E. coli* while they were undetectable in the liquid samples, although this is not necessary since the enumeration methods would not have shown very low populations.

The usual assumption that Y is constant and independent of s is not strictly valid for *Chlamydomonas*. In a nitrate-rich medium, the cells formed can be 2.5 times as high in N content (percent of ash-free dry weight) as cells are in a medium in which the nitrate has been depleted (Taub, unpublished data). Also, the size and dry weight of the *Chlamydomonas* can vary at least twofold, and the *Tetrahymena* can vary more than fourfold (Taub, unpublished data). Therefore, the usual tacit acceptance of cell number as an estimate of substrate removed and biomass produced can be used only very approximately. (At the time of this study we did not have the capacity to measure cell volumes.)

So that the simplified formulas shown above could be used, it was necessary to assume steady-state population densities. Periods of obvious rapid change, such as immediately after environmental changes, were not considered in these calculations, i.e., flow was initiated on day 5, but data were not used until day 7, days 14-24 were eliminated from consideration since the protozoa and *E. coli* populations were rapidly changing, and days 43-44 were eliminated from the calculations. Otherwise, the day to day changes were assumed to be caused by experimental

error, and all measurements within each of the three time
periods were averaged (Tables 4.2-4.4). This method ig-
nores the midperiod change in *Pseudomonas* density during
the second period and the gradual downward trend of the
protozoan density during the third period.

Upstream Community

It will be immediately noted that in the first flask
the exponential growth rates were equal to the dilution
rate so long as the population maintained itself (since
we defined these populations as being in a steady-state
condition). These values indicate which organisms, under
the stated conditions, were capable of growing at the dilution rate. For example, the loss of the protozoan and
E. coli populations indicates that they could not maintain growth rates of 1.04 or 1.38 per day (high flow, high
light, Table 4.2). Their growth rate could be calculated
by a comparison of the washout rate with the dilution rate
(Jannasch 1969). Both of these organisms were able to
grow at a rate of 0.57 and 0.73 per day when dilution was
lower and algal density higher (low flow, high light,
Table 4.3). *Escherichia coli* could also grow at a rate
of 0.75 per day when light intensity was reduced and the
algal concentration decreased (Table 4.4). In contrast,
the protozoan population could grow at 0.56 but not 0.75
under the reduced light and decreased algal cell concentration. Since we have no evidence that the change in
light intensity directly influenced the protozoan population, its limitation is presumably related to the algal
density.

The limitation of protozoan density and growth rate
poses the question of limiting factor(s). Since this
protozoan reaches population densities of 10^5 cells/ml in
rich organic media, no direct density-dependent limitation

is postulated. The possible substrates for protozoan nutrition were (1) algal cells, (2) algal extracellular products, and (3) bacteria. Protozoa were cultured on algal cultures with densities ranging from 8.4 to 3.0 x 10^6 cells/ml, and a μ_{max} of 1.1 and a $K_s = 0.7$ x 10^6 cells/ml were obtained (algal cell concentrations 0.57 and 0.16 x 10^6 cells/ml did not conform to the Michaelis-Menten line). On the assumption either that the algal cells were the sole nutrient source or that extracellular products were proportional to cell concentration, the kinetic relationships defined above might be expected to explain protozoan behavior. The dilution rate, the algal standing crop, and the calculated growth rates are compared in Table 4.5. Extinction did occur in all three cases, as predicted by the kinetics. A steady-state population did occur in the one case where it was predicted. However, in the two cases where increasing populations were predicted, the populations were actually steady state. Theory would predict that the protozoan populations would have increased to the point where they cropped the algal population down to the density where the protozoan growth rate was reduced to the dilution rate. This did not occur. So that it could be determined whether extracellular products were a significant source of nutrient for protozoan growth, an experiment was conducted in which the algal concentration, was varied from nil to ca. 4 x 10^6 cells/ml while extracellular material was kept constant. In the absence of algal cells, the protozoa showed no significant growth, and growth was directly related to cell concentration. These results suggest that the extracellular products either by themselves, or with a limited concentration of algal cells, do not form a significant source of nutrient. The importance of the bacteria as a nutrient source is difficult to evaluate. The protozoan does not require

Table 4.5

Comparison of Actual Protozoan Population Behavior with That Predicted by Kinetics[a]

	Algal concentration ($\times 10^6$ cells/ml)	μ	D	Predicted	Actual	Cells/ml
System 1 (days)						
7-13	0.29	0.31	1.0	Extinction	Extinction	
24-43	2.1	0.82	0.57	Increasing Population	Steady state	1.2×10^3
45-70	0.76	0.56	0.56	Steady state	Steady state	1.9×10^2
System 2 (days)						
7-13	0.37	0.37	1.38	Extinction	Extinction	
24-43	2.1	0.82	0.73	Increasing Population	Steady state	8.4×10^1
45-70	0.36	0.36	0.75	Extinction	Extinction	

[a]

$$\mu = \mu_{max} \frac{s}{K_s + s}$$

where $K = 0.73 \times 10^6$ algal cells/ml and $\mu_{max} = 1.1$.

the presence of bacteria to grow on algae; the kinetic
values were obtained from cultures in which no bacteria
were known to be present. Also the presence of *E. coli*
in the experiments described here does not correlate directly with the success of the protozoa since the *E. coli*
populations increased under reduced lighting, whereas the
protozoan populations diminished greatly in one case and
became extinct in the other (Table 4.4). Nor did the
Pseudomonas densities relate to protozoan populations.
Of course the bacterial generation times could have varied
and thus provided more cells during certain periods.

The absence of significant cropping of the algal population with the expected predator-prey cycles also deserves consideration. As in the previous experiments
(Taub 1969a; Taub 1969b), no significant cropping of algal
populations could be demonstrated as long as light was
supplied. When the protozoan is introduced into algal
cultures and incubated in the dark for 4 days, 50-80% of
the algal cells disappear, and chlorophyll a and ^{14}C uptake ability are significantly decreased (Taub, unpublished). By variation of the concentration of NO_3, the
μ_{max} of 2.0 and K_s of 0.16 mM were obtained for the *Chlamydomonas* under the high-light conditions (12-hr light-dark
cycle). Since the μ_{max} of the protozoan when consuming
algal culture is 1.1, it is suggested that the algal cells
are able to regrow as fast as they are cropped provided
light is available and the limiting nutrient is supplied
by recycling, (these μ_{max} and K_m values should be accepted
only as approximate and preliminary values). Predator-
prey oscillations were shown for limited time periods in
a continuous culture of the predator (a myxameba, *Dictyo-
stelium descoideum*) and prey (*E. coli*) with a limiting
substrate of glucose (Drake *et al.* 1966).

During the recovery of the protozoan populations in
the first flask, the calculated growth rates were 1.56

per day and 1.17 per day by means of the equation $\mu = D +$ $(1/t) \ln (x/x_o)$ from Jannasch (1969) and on the basis of days 20-24 for the green system and days 20-27 for the red system, i.e., the first day when a measurable population occurred, until population increase ceased. These growth rates were the fastest seen during the entire experiment and are faster than the μ_{max} calculated from the other alga-protozoan experiment. (Higher μ_{max} are found when rich organic mixtures, such as yeast extract proteose-peptone solutions, are used.) The growth rates would have to have been greater than the dilution rates, since the population density rapidly increased in spite of losses from dilution. The substrate for this increased growth rate must be related to the increased algal population since no other major changes occurred (the increase in *E. coli* population cannot explain the increased protozoan growth since the bacterium did not appear in the green system until after the protozoan increase, and a high *E. coli* population did not prevent extinction of the protozoan population during low light intensity in the red system). If such high growth rates of the protozoa can be explained by an increase of algal concentration from ca. 3×10^5 to 2×10^6 cells/ml, why did the protozoan population cease to grow at greater than the dilution rate after day 27, when no significant decrease in algal cells occurred, and why did protozoan growth rates slow down in the second flask and yield reservoirs, where the algal densities were even higher?

Downstream Community

Under conditions of high dilution, algal growth rates were relatively high in the second flask. In the system 1 replicate, the unusual phenomenon of higher growth rates in the second flask than in the first can be explained by its higher dilution rate (since its volume was smaller).

Presumably a relatively high amount of substrate carried
over into the second flask because of the limited production in the upstream community (this is supported by the
nitrate concentration remaining even in the yield; Table
4.6). Growth and production rates for the other organisms
were not consistent between replicates.

Under reduced flow, the algal growth and the new production in the second flask were relatively less. Presumably, the higher production in the first flask used up
more nutrient and more sharply limited nutrient availability. The protozoan populations behaved differently in the
two replicates; in the first, almost no further growth
occurred; in the second replicate, growth continued at
the same rate as in the first flask. In both, bacteria
had significant growth rates.

Under reduced lights, the algal populations continued
to grow almost as rapidly in the second flask as they had
in the first. Presumably, the low production in the first
flask also permitted the carryover of nutrient. In both
replicates, protozoan growth rates and production rates
were high. Growth rates for the *Pseudomonas* and the *E.
coli* in the first replicate were significant but showed
a sharp decrease for *E. coli* in the second replicate.

Correlation of the growth rates between the organisms
within a flask must be attempted with great caution since
the dilution factor, D, is the same for all organisms in
the flask, and D occurs within the calculation for growth
rate

$$\mu = D \frac{(x_2 - x_1)}{x_2}$$

Yield Community

Where algal growth rates were relatively high in the
second flask (high dilution, Table 4.2, or low light,
Table 4.4), slower, but significant growth continued in

the yield reservoir. In the former case, nitrate was still available in the yield flask. Where growth was reduced in the second flask (high light, low dilution), slight losses occurred in the yield reservoir (Table 4.3). No appreciable nutrient was available (Table 4.6).

It is interesting to postulate whether the algal density would have shown significant losses if the condition had prevailed for a longer period of time, i.e., if there had been another growth flask in the series. The decline, if it was not within experimental error, might be caused by lack of growth since there was no available nutrient and there was predation. However, since the protozoan population was not increasing its biomass, its consumption would have resulted in the release of their waste products equal to their consumption. Theoretically, the system could have reached a steady state provided that no significant quantity of products were refractory to recycling.

In general, the agreement between replicates was best in the yield reservoirs. This supports the concept that the communities are tending toward a particular stable point and there is more variation on the rates and exact sequence of events leading to that stability than variation at the stable stage.

The concentrations of total biomass* were highest under conditions of low dilution and high light (Table 4.6), but this condition may not have resulted in the highest total production since a smaller volume of culture was produced than during the high dilution period (Table 4.7). The total biomass production was a function of light, as seen by a comparison of biomass under the two light conditions at low dilution: at half light intensity

_____

*See footnote on Table 4.6.

Table 4.6

Average Yield Reservoir Concentrations in Two Systems Under Three
Conditions of Dilution and Light Intensity

	Dry wt (mg/ml)	Ash wt (mg/ml)	Biomass[a] (ash-free dry wt, mg/ml)	Cox (mg/ml)	Nkj (mg/ml)	C/N
High dilution, high light						
System 1	0.357	n.d.[b]	–	0.028	0.006	4.7
System 2	0.363	n.d.[b]	–	0.031	0.006	5.2
Low dilution, high light						
System 1	0.500	0.079	0.441	0.060	0.008	7.5
System 2	0.505	0.089	0.416	0.056	0.007	8.0
Low dilution, low light						
System 1	0.421	0.173	0.248	0.045	0.006	7.5
System 2	0.395	0.204	0.191	0.043	0.006	7.2

Table 4.6 (Continued)

	NO$_3$ (mM)	NO$_2$ (mM)	Chl a (mg/liter)	Chl b (mg/liter)	Car. (mg/liter)	Car./ Chl a
High dilution, high light						
System 1	0.87	0.000	0.191	0.099	0.064	0.34
System 2	0.75	0.000	0.258	0.113	0.067	0.26
Low dilution, high light						
System 1	0.001	0.000	0.247	0.133	0.052	0.21
System 2	0.001	0.000	0.337	0.182	0.091	0.27
Low dilution, low light						
System 1	0.001	0.001	–	–	–	–
System 2	0.002	0.001	–	–	–	–

[a]These values are probably too high since subsequent work has shown that the inorganic medium has an ash-free dry weight greater than its organic (MDTA) content. The ash loss must be caused by decomposition of mineral components but has been interpreted as biomass. Although the relative values can be compared, they should not be considered absolute. This would account for the unusually low C values ranging from 14 to 22% of the biomass.

[b]n.d. = not done.

Table 4.7

Total Daily Production (Yield Reservoir Concentration x Volume) of Two Systems

Under Three Conditions of Dilution and Light Intensity

	Biomass[a] (mg)	Protein[b] (mg)	% Protein[d] (ash free- dry wt)	Proportion N converted to protein	N recovered[e] (mM)
High dilution, high light					
System 1 (green)	267.2[c]	27.5	10.3	0.84	0.506
System 2 (red)	275.7[c]	30.0	10.9	0.90	0.527
Low dilution, high light					
System 1	172.2	21.0	12.2	1.2	0.589
System 2	167.2	16.6	9.9	0.94	0.475
Low dilution, low light					
System 1	99.5	16.0	16.1	0.91	0.457
System 2	78.4	15.6	20.0	0.87	0.439

[a]See footnote on Table 4.6.

[b]Weight N x 6.25 = weight protein.

[c]No ash weight subtracted.

[d]May be underestimated since biomass may be overestimated.

[e]Greater than 0.5 mM nitrate recovery indicates concentration by evaporation during autoclaving of the medium carboys and during the growth period.

133

approximately half the biomass was produced. This result agrees with that obtained by Pipes and Koutsoyannis (1962) on light-limited *Chlorella* cultures. In all cases, at least 84% of the nitrate was converted to organic (Kjeldahl nitrogen, Table 4.7). These findings also agree with our findings on light- and NO_3-limited *Chlorella*: maximum conversion of NO_3 occurs more rapidly than maximum biomass production so that the protein is proportionally greater when total biomass production has been limited.

When the densities of organisms are multiplied by nominal individual weights*, the biomass can be approximately proportioned. In all cases the algal biomass is at least 85% and sometimes 99% of the total calculated biomass.

CONCLUSIONS

The method of gnotobiotic, continuous cultures has been used here with some degree of success for the purpose of gaining insight into community responses to environmental changes. The use of multiple flasks permitted a greater number of conditions to be examined in the same amount of time and with somewhat less than twice the effort that a single stage would have entailed.

The experiment provided documentation that densities, growth rates, and production rates can vary in different directions. This result has long been known on a theoretical basis, but it is still common to assume that lakes with high standing crops are concurrently highly productive.

-----

*Alga, 4×10^{-8}; protozoan, 1.5×10^{-5}; *Pseudomonas*, 2×10^{-10}; *E. coli*, 2×10^{-10} mg/cell.

A CONTINUOUS GNOTOBIOTIC ECOSYSTEM

The algal populations behaved in a highly predictable manner. Had the NO_3 and NO_2 concentrations been measured in the growth flasks it would have been possible to directly determine kinetic substrate-growth relationships. In the absence of these measurements, it must be assumed that progressive nutrient limitation occurred as the cells traveled downstream. Light limitation was also obviously occurring; in the first flask less mutual shading occurred, but the amount of light received obviously limited production. The further production in the second flask was a function of production in the first.

The inability to determine the factor(s) that limit protozoan density remains a problem. Certainly availability of algal cells does not provide an adequate explanation. It may be possible to explain some of the discrepancy as a result of the variable size, biomass, and nutritional state of the algal cells. Work has been continuing along these lines, and several apparently obvious hypotheses appear to be disproved, but an adequate explanation has not been found.

The greatest weakness of the method has been its failure to show gross production rates—all of the growth and production rates measured here were determined after consumption of algal and bacterial cells had presumably occurred. The use of ^{14}C uptake as a measure of gross algal production, perhaps in combination with radioautography, may show the amount of carbon fixed by the algae and consumed by the protozoa. Suitable controls could provide an estimate of direct ^{14}C uptake by the protozoa and uptake of labeled extracellular algal products.

Work is still being pursued for calculation of the other transient growth rates, and sample variance is being analyzed for determination of confidence limits and more efficient sampling procedures.

Others are invited to use these data, especially for other modeling purposes.

The general method is suggested for use in further experiments on species diversity, including competition, and in pollution experiments with either potentially usable nutritional substrates, such as pulp mill wastes, sugar processing wastes, or sewage effluent, or with sublethal concentration of various toxicants.

ACKNOWLEDGMENTS

This investigation was supported by the Federal Water Pollution Control Agency, Grant WP 00982. Contribution No. 337, College of Fisheries, University of Washington, Seattle. I am grateful to Ruth Hung for chemical analyses and culture maintenance; Sheri Lusk Hamel for enumerations and charts; Fred Palmer for NO_3, NO_2, and pigment analyses; David Drevdahl for computer programming; and Jonathan Heller for general help and stimulating discussions.

A CONTINUOUS GNOTOBIOTIC ECOSYSTEM

REFERENCES

Bendschneider, K., and Robinson, R. I. 1952. A new spectrophotometric determination of nitrite in sea water. *J. Mar. Res. 11*, 87-96.

Difco Manual of dehydrated culture media and reagents for microbiological and clinical laboratory procedures, 1953 (Ninth ed.) Difco Laboratories, Inc., Detroit, Michigan.

Dougherty, E. C. 1959. Axenic culture of invertebrate metazoa: a goal. *Ann. N.Y. Acad. Sci. 77(2)*, 27-54.

Drake, J. F., Jost, J. L., Frederickson, A. G., and Tsuchiya, H. M. 1966. The food chain. In *Bioregenerative systems*. NASA SP-165, pp. 87-95.

Herbert, D. 1964. Multi-stage continuous culture. In *Continuous Cultivation of Micro-organisms*, I. Malek, K. Beran and J. Hospodka (Eds.). Proc. 2nd Symposium held in Prague, June 18-23, 1962. Publ. House of the Czech. Acad. Sci., Prague, 1964, pp. 23-44.

Jannasch, H. W. 1967. Growth of marine bacteria at limiting concentrations of organic carbon in seawater. *Limnol. Oceanog. 12(2)*, 264-271.

Jannasch, H. W. 1969. Estimations of bacterial growth rates in natural waters. *J. Bacteriol. 99(1)*, 156-160.

King, E. O., Ward, M. K., and Raney, D. E. 1954. Two simple media for the demonstration of pyocyanin and fluorescin. *J. Lab Clin. Med. 44*, 301-307.

Monod, J. 1942. Recherches sur la Croissance des Cultures Bacteriennes. Hermann & Cie, Paris, 210 pp.

Novick, A. 1955. Growth of bacteria. *Ann. Rev. Microbiol. 9*, 97-110.

Patten, B. C. 1968. *Ecological Modeling with Analog and Digital Computers*. Tutorial lectures and workshops, BioInstrumentation Advisory Council, AIBS, 3900 Wisconsin Ave., N.W., Wash. D.C. 20016.

Pearson, E. A., Middlebrooks, E. J., Tunzi, M., Adinarayana, A., McGauhey, P. H., and Rohlich, G. A. 1969. Kinetic assessment of algal growth. Preprint of paper presented at 64th nat'l meeting, American Institute of Chemical Engineers, New Orleans, La. 37 mimeo pages.

(Summarized in Joint Industry Government Task Force
on Eutrophication, Provisional Algal Assay Procedure,
1969, P. O. Box 3011, Grand Central Station, New York,
N.Y. 10017.)

Pipes, W. O., and Koutsoyannis, S. 1962. Light limited
growth of *Chlorella* in continuous culture. *Appl. Microbiol. 10*, 1-5.

Sonneborn, T. M. 1950. Methods in the general biology
and genetics of *Paramecium aurelia*. *J. Exp. Zool.
113*, 87-147.

Strickland, J. D. H., and Parsons, T. R. 1968. A practical handbook of seawater analysis. *Bull. Fish. Res.
Bd. Can. 167*, 311 pp.

Taub, F. B. 1969a. A biological model of a freshwater
community: A gnotobiotic ecosystem. *Limnol. Oceanog.
14*(1), 136-142.

Taub, F. B. 1969b. Gnotobiotic models of freshwater communities. *Vehr. Int. Verein. Limnol. 17*, 485-496.

Taub, F. B., and Dollar, A. M. 1965. Control of protein
level of algae, *Chlorella*. *J. Food Sci. 30*(2), 259-364.

Taub, F. B., and Dollar, A. M. 1968a. The nutritional
inadequacy of *Chlorella* and *Chlamydomonas* as food for
Daphnia pulex. *Limnol. Oceanog. 13*(4), 607-617.

Taub, F. B., and Dollar, A. M. 1968b. Improvement of a
continuous culture apparatus for long-term use. *Appl.
Microbiol. 16*(2), 232-235.

Wood, E. D., Armstrong, F. A. J., and Richards, F. A.
1967. Determination of nitrate in sea water by cadmium-copper reduction to nitrite. *J. Mar. Biol. Assoc.
U.K. 47*, 23-31.

Wright, R. T., and Hobbie, J. E. 1965. The uptake of
organic solutes in lake water. *Limnol. Oceanog. 10*,
22-28.

Chapter 5

DIATOM COMMUNITIES

Ruth Patrick

The Academy of Natural Sciences
Philadelphia, Pennsylvania

CONTENTS

ABSTRACT

Results of studies of many diatom communities in
streams have shown that these communities are typically
composed of many species, most of which have relatively
small populations, and the data fit the model of a truncated normal curve. The structure of the curve is very
similar from season to season and from year to year. When
one examines in detail the species composition, one finds
that there are always some very rare species as well as
those more common than the majority of species in a community. Some of the very rare may become more common
under various seasonal conditions. However, some species
never seem to be common. The role of these rare species
in the community is not well understood, but it may be
that their function is similar to that of rate genes in
a gene pool; that is, to maintain the community over time
during periods of great and unpredictable variation in
environmental conditions.

A series of experiments was conducted to try to determine the effects of size of species pool capable of
invading an area, frequency of reinvasion into an area,
and the size of the area on the diversity of the community.
The results of these studies indicate that the numbers of
species capable of invading an area and the frequency of
reinvasion are probably more important than actual size
of area. Analyses of the data by various types of diversity indices have shown that these diatom communities, as
other stream communities, have very low equability. One
probable cause is the great variation in environmental
factors that largely prevents species saturating the environment in which they live. Another contributing factor

is the fact that nutrients rarely become limiting or waste
materials autotoxic, for the nutrients are continually being replenished and the waste products continually removed.
The effects of perturbation on these natural communities
are discussed.

INTRODUCTION

Many studies have been made of the systematics and
ecology of diatoms. Associations of diatoms have been
used to determine water conditions since the latter part
of the nineteenth century. Some of these studies were
concerned with the patterns and influence of glaciation
(Cleve 1899, 1900; Cleve-Euler 1940; Hyyppa 1936; Hustedt
1954; Hanna 1933; Patrick 1946). Other studies used diatoms to differentiate ocean currents (Cleve 1892, 1899;
Gran 1897, 1900, 1902; Aikawa 1936; Hendey 1937). Diatoms
have also been used to show changes in mineral content in
lakes, rivers, and swamps (Hustedt 1936; Kolbe 1932; Cocke
et al. 1934; Patrick 1936, 1968; Pennington 1947; Ross
1950). More recently diatom associations have been used
to trace the nutrient levels and other changes in lakes
in prehistoric times (Patrick 1943, 1954; Pennington 1947;
Hutchinson *et al.* 1956; Stockner and Benson 1967; LaSalle
1966).

Some workers have tried to develop systems using indicator species or shifts in relative abundance of certain
species to denote the degree of pollution of a body of
water (Kolkwitz and Marsson 1908; Nygaard 1949; Foged
1954; Cholnoky 1958; Fjerdingstad 1960).

DISCUSSION

During the last 20 years I have devoted a considerable amount of time to studying the structure of communities of aquatic organisms, particularly diatoms, in streams.

In determining the structure of the community we are concerned with kinds of taxa, numbers of taxa, and relative
sizes of their populations. The results of studies started
in 1948 (Patrick 1949, 1961) showed that diatoms as well
as communities of species of most major groups of aquatic
life were represented by a fairly high number of species,
many of which had moderate to small populations, although
a few might be of common occurrence. Furthermore, the
species dominating the diatom communities changed considerably from season to season and from year to year, and in
some cases a large proportion of the species changed even
though the structure of the communities did not change
very much. When we say that the species change we are, of
course, referring to the collectable species. Further research has shown that species often are present in very
small populations that are not usually collectable at all
times and that with the seasons they become more common.
Others seem to disappear from the community. This result
has been substantiated by our recent studies on White Clay
Creek.

Attempts were made in 1954 (Patrick *et al.* 1954) to
represent these relationships mathematically. It was
found that a fairly large sample of a diatom community
fitted the truncated normal curve as found by Preston
(1948) for other groups of organisms. If one counted a
sufficient number of specimens to always place the height
of the mode in the same interval (i.e., the third interval
to the right of the veil line), the shape of the curve as
measured by height of the mode, σ^2, and intervals covered
were remarkably stable from season to season and from year
to year, if perturbation such as man-made pollution did
not occur (Fig. 5.1, Table 5.1). It was found that the
negative binomial also produced a fair fit, but the truncated normal curve was better for communities in natural
continental streams.

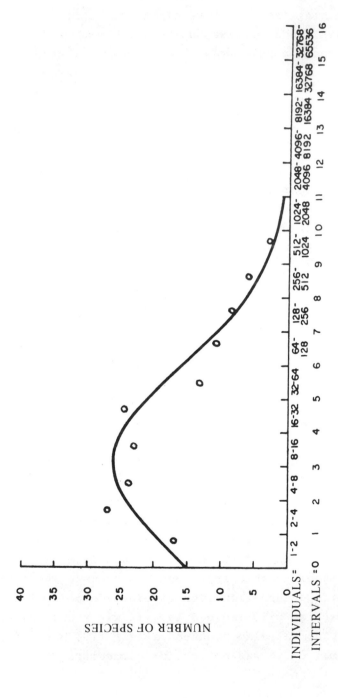

FIG. 5.1. Graph of the diatom population for November, 1951, from Ridley Creek, Chester County, Pennsylvania, a stream not adversely affected by pollution.

Table 5.1

Structure of Savannah River Diatom Communities,
October 1953 to January 1958

Date	Specimen number in model interval	Species in mode	Species observed	Species in theoretical universe
Oct. 1953	4-8	22	150	178
Jan. 1954	4-8	19	151	181
Apr. 1954	2-4	24	169	200
Jul. 1954	2-4	23	153	193
Oct. 1954	4-8	21	142	168
Jan. 1955	4-8	19	132	166
Apr. 1955	2-4	25	165	221
Jul. 1955	2-4	20	132	180
Oct. 1955	2-4	27	171	253
Jan. 1956	2-4	30	185	229
Apr. 1956	4-8	35	215	252
Jul. 1956	2-4	24	147	185
Oct. 1956	2-4	23	149	206
Jan. 1957	2-4	29	177	233
Apr. 1957	2-4	21	132	185
Jul. 1957	4-8	29	181	203
Oct. 1957	2-4	25	157	232
Jan. 1958	2-4	27	152	212
Apr. 1954-1958 averages		24	151	194

The close similarity of the structure of natural diatom communities in streams enabled Patrick and Strawbridge
(1963) to develop confidence intervals for σ^2 and the

145

number of species in the mode for natural fresh soft water
and brackish water streams. The normal bivariate distribution function of the form described by Bennett and
Franklin (1954) was used.

Attempts were made to fit the diatom communities to
the MacArthur type I distribution, but they did not conform. Furthermore, it has been shown that the communities
have very low equability as measured by Lloyd and Ghelardi
(1964). This low equability of distribution of specimens
among species was found to be characteristic of stream
communities of protozoa, invertebrates (including insects),
algae, and usually for fish. This is probably because the
environmental characteristics of streams are unpredictably
variable. Floods occur that greatly reduce species composition. As a result, opportunistic species are often
able to enter and form various sizes of populations, some
of which are fairly large. They are, in turn, cut down
by predator pressure or density-independent variables but
since most of the diatoms that make up stream communities
are opportunistic it often happens that one or two species
are much more common than the rest. Also a factor that
would delay equability is that nutrients are continually
being supplied and waste products removed from an area.
Lloyd in 1968 (Lloyd *et al.* 1968) commented on the fact
that reptile and amphibia populations that spend part of
their life in streams had communities with low equability.

These studies led us to question what were the more
important factors causing high species diversity in diatom communities. The factor first considered was the
effect of size of area on the number of species in a community. To carry out these experiments a series of islands, or small glass squares supported on pedestals, was
placed in a rack in a plastic box. The boxes were changed
each day and the pedestals cleaned to assure that the only
diatoms reaching the slides were from the water flowing

over them. The structure of the diatom communities after various days of exposure indicates that the area size significantly affects the number of species composing the community. The numbers of species on the islands often increased to the first week, increased or decreased a little to the second week; and then subsequently remained about the same or slightly decreased, indicating that saturation as to numbers of species had been reached and some extinction was taking place that was not being replaced by invasion (Table 5.2). An examination of the sizes of populations of species composing the communities shows that populations represented by six or fewer specimens were those most subject to extinction.

A second series of experiments was performed in the stream from Roxborough spring, and a similar series was carried out in Ridley Creek. Previous studies have shown that, in the area of the Roxborough spring stream being studied, about 100 species of diatoms form the community at a given time, whereas in Ridley Creek the area studied supports about 250 and 300 species at a given time. The results of these studies clearly show that the same size area (36 mm^2) supported a very different number of species, 14-29 in Roxborough spring and 160 in Ridley Creek. It therefore appears that the potential species pool is more important than size of area in determining the number of species in a diatom community (Patrick 1967).

The third series of experiments was to determine, if the area and species pool were the same, what would be the effect of varying the reinvasion rate. For these experiments a series of eight boxes was developed so that the boxes would support the same type of diatom communities (Patrick 1968). It has been established that under similar flow rates (about 650 liter/hr) 95% of the specimens were of the same species in all of the boxes. The rate

147

Table 5.2

Experiments in September–October, 1964

| | Roxborough Spring, number of species | | | | Ridley Creek, number of species | |
| | 625 mm² slide | | 36 mm² slide | | 9 mm² slide | 36 mm² slide |
	Box 1	Box 2	Box 3	Box 4	Box 6	Box 7
4 days	46	37	23	23	3	–
1 week	40	32	28	24	–	–
2 weeks	54	35	–	22	10	–
8 weeks	–	–	29	14	14	160

Experiments in Roxborough Spring During Summer, 1964

| | 1 week, 144 mm² slide | | 2 weeks, 144 mm² slide | | 1 week, 625 mm² slide | | 2 weeks, 625 mm² slide | |
	Box 1	Box 2	Box 3	Box 4	Box 5	Box 6	Box 7	Box 8
Number of species	32	28	23	22	47	44	29	28

of reinvasion of stream species was then varied (Fig. 5.2, Table 5.3), although the flow rate was maintained by re-circulation. The results of experiments with an invasion rate of 1.5 liter/hr show that the number of species com-posing the communities is less if repetition or frequency or reinvasion of the various species is reduced; that the communities consisted of fewer rare species; and that a greater proportion of the species had larger populations than under conditions of higher repetition of invasion.

Studies were then carried out in island streams of types similar to those studied in the continental United States in the temperate zone. The main difference between the island streams and the streams in the United States was that the islands were tropical. If the theory that there are larger species pools for diatoms in the tropics is correct, then one expects these islands to support more species than if they were temperate islands and the com-parison with the temperate mainland would show less re-duction caused by isolation than if the islands were temperate islands.

However, the results of these studies do show the effect of isolation. The two streams in Dominica and the one in St. Lucia produced similar results. In all cases the numbers of species in the mode were small (Dominica, five; St. Lucia, eight), and a greater percentage of the species composing the diatom communities had larger popu-lations than one would expect in similar continental streams, usually resulting in σ^2 being larger (Table 5.3). The total number of observed species was also much lower, 46 to 61 as compared to 79 to 129 in continental streams.

Thus we see that size of area is important in main-taining a higher number of species in a diatom community but the potential species pool capable of invading the area is more important. If reinvasion does not take place at a fairly high rate, diversity is reduced because of the extinction of the rarer species.

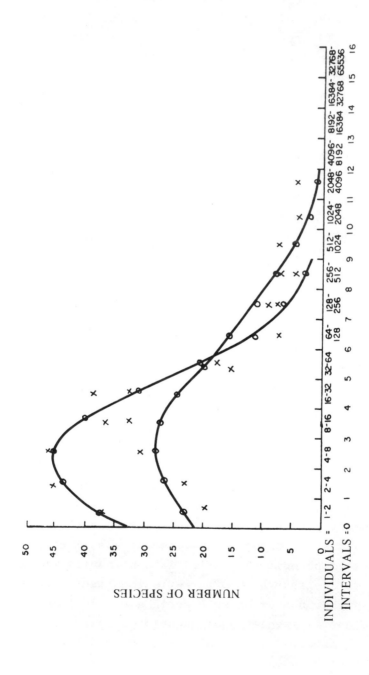

FIG. 5.2. Varied reinvasion rate of stream species of diatom communities under similar flow rates. Upper curve, 650 liters/hr which is natural stream flow; lower curve, 1.5 liters/hr new water, recycled flow 650 liters/hr.

Table 5.3

Structure of Diatom Communities

	Height of mode	σ^2	Theoretical number of species	Observed number of species	Intervals covered by curve
Invasion rate 500-600 liter/hr (Oct.-Nov., 1964) (Darby Creek)	22.4	6.2	140	123	9
Invasion rate 1.5 liter/hr (Oct.-Nov., 1964) (Darby Creek)	15.3	12.0	133	97	15
Invasion rate 550-600 liter/hr (Sept.-Oct., 1964) (Darby Creek)	22.5	6.9	148	129	9
Invasion rate 1.5 liter/hr (Sept.-Oct., 1964) (Darby Creek)	13.9	12.6	124	100	12
St. Lucia, Canaries River	8.4	9.3	64	61	10
Island of Dominica stream (Layou River)	5.3	26.0	67	49	14
Island of Dominica stream (Check Hall River)	5.17	21.6	60	46	14
Maryland stream (Hunting Creek)	12.0	9.1	92	79	14

The important factors in maintaining high species diversity in stream diatom communities seem to be the same
as those for other groups of organisms (Hutchinson 1953;
MacArthur and Wilson 1963). In natural continental streams
a large species pool of diatoms is supplied from upstream
or nearby stream areas. Since diatoms are usually dispersed by current, the turbulence and current in a river
insures a high invasion rate as to number of species and
reinvasion of the same species. Asexual reproduction and
the rapidity of this reproduction (often once a day) insure
a high birth rate and a short maturity time for reproduction. This method of reproduction also eliminates the
necessity of finding a mate and reduces in part the necessity for cohesiveness in distribution. However, as pointed
out by Nalewajko *et al.* (1963) division or reproduction in
algae is more likely to take place if a steady diffusion
gradient of glycolate, which is liberated from algal cells,
has been established around the cells. The presence of
several close together would promote the establishment of
a sufficient diffusion gradient.

Although the death rate is high because of predation
and the precariousness of the environment, the factors
mentioned above insure that normal continental communities
can maintain high numbers of species and many of them with
very small populations. If for any reason the invasion
rate is significantly reduced, the rapid reproductive rate
may compensate, and similarly a high invasion rate may
compensate for a low reproductive rate.

The importance of maintenance of fugitive species or
species with very small populations in a community should
not be overlooked. They may be of great value in maintaining the community over time in an environment that is so
variable. Indeed, their function in a community may be
similar to that of rare genes in the gene pool of a species.

The consequence of perturbation often caused by man-made pollution on diatom communities is to affect the colonization and population sizes of many species in the area and often to bring about a great change in the kinds of species that may live in an area.

Pollution first affects rates of reproduction. As a result certain species, even if they are able to invade the area, are not able to reproduce and soon become extinct. Other species have their reproductive rate reduced, while the more tolerant species, because of less competition for nutrients, become more common. The shape of the truncated normal curve is altered as σ^2 becomes greater and Preston's a becomes less. The curve covers more intervals as some species become excessively dominant. This type of curve usually develops as eutrophication increases. A greater amount of organic pollution, high temperature, and some of the less toxic pollutants produce, in addition to the above changes, a reduction in numbers of species so that the height of the mode is less. There are relatively fewer species with small populations and some of the rate species with narrower tolerances are eliminated. Often such species as *Gomphonema parvulum* and *Nitzschia palea*, which formerly had small populations, greatly increase. Sometimes fairly large changes occur in the kinds of species that are common. Usually there are one or two species that are excessively common.

More severe toxic pollution may have one of the following effects. The invading species are not killed but cannot reproduce. In such cases the number of species may be high, σ^2 may be very small, and the number of intervals covered relatively few. The total biomass is very small. An example of this effect is caused when the pH is greatly reduced in a typical circumneutral stream (Patrick *et al.* 1968).

153

Other types of toxic pollution greatly lower the height of the mode and the curve extends over only a few intervals. Sigma squared is large. This is the result of only a few relatively tolerant species surviving and having variable sizes of population.

Thus it is apparent that the effect of perturbation may or may not destroy the invading species, but it usually alters the reproduction rate of the species in the area. As a result the more tolerant species, often the least desirable as a food source, increase. Perturbation may reduce or eliminate predator pressure by eliminating the predator or by reducing the desirability of the food source to the predator. Thus the population sizes of certain species increase.

Since the evenness of distribution of specimens among the species is greatly reduced, the Shannon-Weiner diversity index is lower. However, this index does not discriminate between certain types of changes. For example, the difference in the indices is relatively small between a community dominated by a few very common species, with only a few species with small to very small populations, and one with a few very common species, and many species with small to very small populations (Patrick 1967; Sager and Hasler 1969). The first type of community is characteristic of much more severe pollution than the latter.

In general, high diversity at the various trophic levels, as evidenced by numerous species with relatively small populations that are able to maintain themselves in associations over time, has been recognized by many as giving stability to the community, for it increases the flexibility of the community to respond to changing environmental conditions. High diversity is probably also important in predator-prey relationships, as laboratory experiments have shown that most organisms have better growth rates if fed on a mixed diet instead of a single food source. If a single species is used as a food source it is usually fortified with nutrients from other species. The importance of the maintenance of a diversified food source for the stability of a higher trophic level is shown by the work of Hairston *et al.* (1968). It should also be noted that less entrophy is introduced into a stream if the standing crop or population of a species is just large enough to insure continuance of the species over time, which means satisfying predator pressure, competition between species, and death caused by density-independent factors. Thus natural communities typically have many species that are primary producers, herbivores, and omnivores, but most of them typically have moderate to small populations. Usually there are fewer carnivore species than species with other types of food preferences. If a large population of a species occurs it is soon cut down by predator pressure or severe changes in density-independent factors.

Perturbations caused by man have upset these relation-ships so that large standing crops or nuisance growths re-sult, which reduce the diversity of the species and change the energetics of the system. It is the meaning of these long-range changes, such as the decrease in the diversity of our ecosystems and the perturbations in the energy re-lationships of communities, that should be researched.

REFERENCES

Aikawa, M. 1936. On the diatom communities in the waters surrounding Japan. *Rec. Oceanog. Wks. Japan 8*(1), 1-160.

Bennett, C., and Franklin, N. 1954. *Statistical Analyses in Chemistry and Chemical Industry*. Wiley, New York.

Cholnoky, B. J. 1958. Beitrag zu den Diatomeenassoziationen des Sumpfes Olifantsvlei südwestlich Johannesburg. *Ber. Deutsch. Bot. Ges. 71*, 177-187.

Cleve, P. T. 1892. Not sur les diatomees trouvées dans la poussiere glaciale de la coté orientale de Gröenland. *Le Diatomiste 1*, 78.

Cleve, P. T. 1899. Postglaciala bildningarnas klassifikation pa grund af deras fossila diatomaceer. *Sv. Geol. Und.*, Ser. C. No. 180, 59-61.

Cleve, P. T. 1900. The plankton of the Nort Sea, the English Channel, and the Skagerak, in 1898. *Kongl. Svenska Vet.-Akad. Handl. 32*, 1-53.

Cleve-Euler, A. 1940. Das Letzinterglaziale Baltikum und de Diatomeenanalyse. *Beih. Bot. Zentralbl.*, Abt. B *60*, 287-334.

Cocke, E. C., Lewis, I. F., and Patrick, R. 1934. A further study of Dismal Swamp peat. *Am. J. Bot. 21*, 374-395.

Fjerdingstad, E. 1960. Forurening af vandløb biologisk bedømt. *Nord. Hyg. Tidskr. 41*, 149-196.

Foged, N. 1954. On the diatom flora of some Funen lakes. *Folia Limn. Scand. 6*, 1-75.

Gran, H. H. 1897. Bemerkungen über das Plankton des Arktischen meeres. *Ber. Deutsch. Bot. Ges. 15*, 132-136.

Gran, H. H. 1900. Diatomacea from the ice-floes and the plankton of the Polar Sea. Norweg. North Pol. Exp. 1893-1896. *Sci. Res. 3*, 1-74.

Gran, H. H. 1902. Das Plankton des Norwegischen Nordmeeres, von biologischen und hydrografischen Gesichtspunkten behandel. *Rep. Norweg. Fish. Mar. Invest. 2*, 1-222.

Hairston, N. G., Allen, J. D., Colwell, R. K., Futuyma, J. D., Howell, J., Lubin, M. D., Mathias, J., and Vandermeer, J. H. 1968. The relationship between species diversity and stability: an experimental appraisal with protozoa and bacteria. *Ecology 49*, 1091-1101.

Hanna, G. D. 1933. Diatoms of the Florida peat deposits. *Fla. State Geol. Surv.*, 23/24 Ann. Rep. (1930-1932), 65-120.

Hendey, W. I. 1937. The plankton diatoms of the southern seas. *Discovery Rep. 16*, 151-364.

Hustedt, F. 1936. Diatoms. *Arch. Hydrobiol. 30*, 1-84.

Hustedt, F. 1954. Die Diatomeenflora des Interglaziale von Oberohe in der Lunenburger Heide. *Abh. Naturw. Ver. Bremen 33*, 431-455.

Hutchinson, G. E. 1953. The concept of pattern in ecology. *Proc. Acad. Nat. Sci. Phila. 105*, 1-12.

Hutchinson, G. E., Patrick, R., and Deevey, E. S. 1956. Sediments of Lake Patzcuaro, Michoacan, Mexico. *Bull. Geol. Soc. Am. 67*, 1491-1504.

Hyyppa, E. 1936. Über die spätquartäre Entwicklung Nordfinnlands mit Ergänzungen zur Kenntnis des spätglazialen Klimas. *C.R. Soc. Geol. Finl. 9*, 401-465.

Kolbe, R. W. 1932. Grundlinien einer allgemeinen Okologie der Diatomeen. *Ergebn. d. Biol. 8*, 221-348.

Kolkwitz, R., and Marsson, M. 1908. Okologie der pflanzlichen Saprobien. *Ber. Deutsch. Bot. Ges. 26*, 505-519.

LaSalle, P. 1966. Lake quaternary vegetation and glacial history in the St. Lawrence lowlands, Canada. *Leidse Geologische Mededelingen 38*, 91-128.

Lloyd, M., and Ghelardi, R. J. 1964. A table for calculating the equitability component of species diversity. *J. Anim. Ecol. 33*, 217-225.

Lloyd, M., Inger, R. F., and King, F. W. 1968. On the diversity of reptile and amphibian species in a Bornean rain forest. *Am. Nat. 102*, 497-517.

MacArthur, R. H., and Wilson, E. O. 1963. An equilibrium theory of insular zoogeography. *Evolution 17*, 373-387.

Nalewajko, C., Chowdhuri, N., and Fogg, G. E. 1963. Excretion of glycollic acid and the growth of a planktonic *Chlorella*. In *Microalgae and Photosynthetic Bacteria*, pp. 171-183.

Nygaard, G. 1949. Hydrobiologische Studien uber danische Teiche und Seen. II. The quotient hypothesis and some new or little known phytoplankton organisms. *Kongl. Dansk. Vidensk. Selsk. Skr. 7*, 1-293.

Patrick, R. 1936. Some diatoms of Great Salt Lake. *Bull. Torrey Bot. Club 63*, 157-166.

Patrick, R. 1938. A flora de quatro acudes de Parahyba. IV. Bacillariophyta. *Ann. Acad. Brasil Sci. 10*, 89-103.

Patrick, R. 1943. The diatoms of Linsley Pond, Connecticut. *Proc. Acad. Nat. Sci. Phila. 95*, 53-110.

Patrick, R. 1946. Diatoms from Patzschke Bog, Texas. *Not. Nat. Acad. Nat. Sci. Phila.* No. *170*, 1-7.

Patrick, R. 1949. A proposed biological measure of stream conditions based on a survey of Conestoga basin, Lancaster County, Pennsylvania. *Proc. Acad. Nat. Sci. Phila. 101*, 277-341.

Patrick, R. 1954. The diatom flora of Bethany Bog. *J. Protozool. 1*, 34-37.

Patrick, R. 1961. A study of the numbers and kinds of species found in rivers in eastern United States. *Proc. Acad. Nat. Sci. Phila. 113*, 215-258.

Patrick, R. 1967. The effect of invasion rate, species pool, and size of area on the structure of the diatom community. *Proc. Nat. Acad. Sci. Phila. 58*, 1335-1342.

Patrick, R. 1968. The structure of diatom communities in similar ecological conditions. *Am. Nat. 103*(924), 173-183.

Patrick, R., Hohn, M. H., and Wallace, J. H. 1954. A new method for determining the pattern of the diatom flora. *Not. Nat. Acad. Nat. Sci. Phila.* No. 259, 1-12.

Patrick, R., Roberts, N. A., and Davis, B. 1968. The effect of changes in pH on the structure of diatom communities. *Not. Nat. Acad. Nat. Sci. Phila.* No. 416, 1-16.

Patrick, R., and Strawbridge, D. 1963. Methods of studying diatom populations. *J. Water Pollut. Control Fed.*
25, 151-161.

Pennington, W. 1947. Studies on the postglacial history
of British vegetation. *Phil. Trans. Roy. Soc. London*
233, 137-175.

Preston, F. W. 1948. The commonness, and rarity of species. *Ecology 29*, 254-283.

Ross, R. 1950. Report on diatom flora from Hawks Tor,
Cornwall. *Phil. Trans. Roy. Soc. London*, Ser. B *234*,
461-464.

Sager, P. E., and Hasler, A. D. 1969. Species diversity
in lacustrine phytoplankton. I. The components of
the index of diversity from Shannon's formula. *Am.*
Nat. 103, 51-61.

Stockner, J. G., and Benson, W. W. 1967. The succession
of diatom assemblages in the recent sediments of Lake
Washington. *Limnol. Oceanog. 12*, 513-532.

Chapter 6

THE INTERRELATIONSHIP BETWEEN
FRESHWATER BACTERIA, ALGAE, AND ACTINOMYCETES
IN SOUTHWESTERN RESERVOIRS

J. K. G. Silvey

Department of Basic Health Sciences
North Texas State University
Denton, Texas

and

J. T. Wyatt

U.S. Army Environmental Hygiene Agency
Edgewood Arsenal
Aberdeen Proving Ground, Maryland

CONTENTS

ABSTRACT

Some specifics and generalities regarding interrelationships of freshwater microflora are discussed. Particular emphasis is given the aquatic actinomycetes and the planktonic blue-green algae of southwestern reservoirs. The occurrence, distribution, enumeration, and some aspects of the overall physiology of aquatic bacteria, actinomycetes, and planktonic algae are reviewed. Also, suggestions are made relative to the role of each organism in such facets of reservoir metabolism as uptake, turnover, nutrient cycling, and organic production. Some speculative observations regarding competition and/or enhancement between these organisms are tentatively offered.

INTRODUCTION

Information on the interrelationships between different varieties of organisms composing the microflora of reservoirs is based largely on data secured from Lake Hefner in Oklahoma City and Garza Little Elm Reservoir near Denton, Texas, although additional reservoirs, streams, and ponds are considered. Extensive studies have been made on Lake Hefner since 1952. During the summer period studies were carried out on a weekly basis and during the winter period on a monthly basis. Perhaps Lake Hefner has been studied in more detail than any reservoir in the Southwest. As early as 1952 investigations were being made on the heat budget of that particular reservoir because of its unusual construction and location. Lake Hefner is an offset reservoir located 4.5 miles from the North Canadian River; it receives its water by way of a canal, which is controlled by gates during flows in the river. Since it is offset and therefore not subjected to floods, the quantity of water added to the lake can be metered. Measurement of the amount that is lost either by way of evaporation or through a venturi as it enters the water plant makes this lake an unusual reservoir for study. Thus it has offered excellent opportunities for investigations on heat budget for evaporation control techniques and for microbiotic cycles of a southwestern reservoir.

Garza Little Elm Reservoir, located approximately 9 miles from Denton, Texas, obtains its water from three streams and exhibits a variety of ecological areas quite different from those found in Lake Hefner. These two reservoirs afforded an opportunity for a comparative study of the microbiotic cycles in an on-stream reservoir contrasted with an offset type over a period of almost 20 years. Moreover, there is diversity in depth between the two reservoirs. Most of Garza Little Elm is relatively shallow, roughly 5-6 m, the deepest area being 15 meters. Lake Hefner, which was excavated into an almost circular basin, has an average depth of 9 m and a surface area of 2500 acres. A comparison of the microorganisms and their cyclic phenomena in these two reservoirs should fairly well establish the ecological parameters obtained from intensive qualitative and quantitative limnological investigations heretofore neglected in this area of the nation.

As pointed out by Cole (1963), limnological information from the Texas area has not been very profound. Some qualitative information, but not very much quantitative data, has been accumulated from a study of its streams and reservoirs. It is the purpose of this chapter to describe the observations and attempt to account for the microbiotic relationships in the warm water reservoirs as demonstrated by detailed studies of planktonic algae, freshwater bacteria, and aquatic actinomycetes. While some attention has been

given to zooplankters, fungi, and benthic organisms, the
data are not sufficiently detailed to be included. It is
perhaps noteworthy that the concentration of protozoa,
microcrustacea, and rotifers in southwestern reservoirs
is very low when compared with data from lakes in Michigan,
Minnesota, and Wisconsin. According to our data, however,
benthic organisms are less sparse. In the shallow regions
of our reservoirs, they comprise a population greater than
is found in similar areas around Douglas Lake in Michigan
or Lake Mendota in Wisconsin.

One of the first comprehensive studies done on reservoirs in the Southwest was completed by Harris and Silvey
(1940) concerning four reservoirs in the northeastern section of Texas. This was followed by a paper published by
Cheatum *et al.* (1942) on a single impoundment in the eastern
part of the state. Deevey (1957) briefly examined some
ponds in the Texas coastal plains and the arid Trans-Pecos
Texas region. In the Oklahoma region, contributions have
been made by Dorris (1956, 1958) concerning areas of the
middle Mississippi and adjacent waters. Since that time,
he has completed a number of studies in the state of Oklahoma. Irwin (1945) contributed information concerning the
precipitation of colloidal particles from turbid waters
in various areas of Oklahoma. There are numerous papers
on fisheries from both states and in other areas of the
Southwest but very little information of a quantitative
nature having to do with the microbiotic cycles and the
interrelationship of the organisms in reservoirs. Those
publications that may be noted are largely concerned with
the identification of species but with no indices of environmental conditions from which they came.

DISCUSSION

The predominant types of algae occurring in the reservoirs of the Southwest are greens, blue-greens, and to

a certain extent flagellates and diatoms. To be certain, there are other varieties but they normally do not comprise a large proportion of the algal population. It is interesting to note that most of the green algae are unicellular or small colonial types. Larger planktonic forms do not normally occur in southwestern reservoirs. It is unfortunate that our data do not include the rather large quantities of sessile algae that are found growing along the perimeter of the reservoirs throughout all seasons of the year. We have not developed a technique for quantitation of these organisms, which perhaps comprise a greater biomass than the planktonic forms and perhaps contribute materially to the microbiotic cycles. We realize this deficiency but have not devised a method of surmounting the difficulty of determining the quantities of this material available in reservoirs. The cyclic phenomena of the various types of algae are described in a later section. Details are presented concerning the genera available, their respective concentrations, and the effects of chemical and environmental conditions upon the growth of these organisms.

The bacteria comprise a group of approximately 10 common genera. Most of these bacteria are also found in soil and frequently in greater concentrations than they are found in the waters of reservoirs. However, the density of the population in reservoirs is much greater than in natural lakes in the Midwest and Northeast. From time to time, various types of pathogens are encountered. This is anticipated in reservoirs in the Southwest since effluent from sewage disposal plants, run off from feed lots, and watershed contamination frequently contribute these organisms, although they occur only on a sporadic basis. This situation is dependent upon the location of the reservoir and the source of its water. More details are given below concerning the cyclic phenomena of the bacteria and their relationship to the actinomycetes and algae.

BACTERIA, ALGAE, AND ACTINOMYCETES

The other group of organisms to be considered is the
aquatic actinomycetes that have been isolated from reservoirs and studied both in laboratory and field investigations. Normally two genera are encountered, namely *Strep-*
tomyces and *Micromonospora*. On rare occasions one may
isolate a member of the genus *Nocardia*, especially from
bottom deposits. The actinomycetes are an interesting
group of organisms and have been studied in our laboratory
since 1948. The above organisms comprise the population
of microflora that is considered in their interrelationships in the two types of reservoirs and other aquatic
environments that have been studied in detail.

Planktonic Algae .

Between population peaks, the planktonic algae of
southwestern reservoirs are usually not particularly conspicuous. Discounting the usual slight variations, about
50% of the total population will be made up of green algae, and the remainder composed of sparse numbers of diatoms, occasional pyrrophytes, with the balance consisting
of a small group of blue-greens. For the most part, the
green algal population consists of unicells or nonfilamen-
tous forms belonging to the genera *Ankistrodesmus, Chloro-*
coccum, Coelastrum, Oöcystis, Pediastrum, Scenedesmus, and
Selenastrum. Upon observation of most of a population in
a particular counting field, notation of more than 10 different green algal genera is unusual. Concerning the other
major phytoplankton component, records indicate the average
blue-green contingent to be fewer in number than the green
algal population and generally composed of relatively in-
conspicuous forms. These are usually the small and somewhat difficult to identify organisms such as *Anacystis,*
Borzia, Gomphosphaeria, Phormidium, and *Pseudanabaena*. We
have not studied the other planktonic algal groups in sufficient detail to merit further elaboration.

Because thorough laboratory investigation of all planktonic algal groups seemed unfeasible, our studies have been almost exclusively limited to the blue-green algae. This restriction is perhaps unfortunate, but it has permitted detailed observation of most of thos "bloom formers," which at various times almost completely dominate the planktonic community. Also, it is quite well known that much reservation must be maintained when projecting data compiled in the laboratory back into the aquatic system. However, some evidence is beginning to accumulate that basic metabolic behavior patterns of phytoplankton may be much the same under laboratory conditions as in situ. For example, Watt (1966, 1969) reports little difference in release of extracellular products from forms grown naturally or laboratory cultivated.

When compared to their terrestrial or amphibious counterparts, the planktonic blue-green tend to be highly active. Measurement shows such abilities as nitrogen fixation (when applicable), photosynthesis, and respiration to be much higher in euplanktonic blue-greens. This may not be caused as much by genetically determined enzymatic differences as by actual growth form. Nonplanktonic blue-green algae seem to invariably grow either attached, or in mats, clumps, or balls. Since only a small portion of the algal mass is usually exposed to light and/or active substrate, physiological and biochemical measurements are generally low per amount of algal material. Planktonic blue-greens, in contrast, generally remain as evenly suspended unicells, hormogones, or microscopic groups. The apparently much greater abundance of gas vacuoles may be a factor in maintaining this mode of life. Nonetheless, this greater activity of planktonic blue-greens is real under our laboratory conditions. From an ecological point of view, the reasons for this phenomenon are probably not too important.

BACTERIA, ALGAE, AND ACTINOMYCETES

We have noted at least eight species of blue-green algae as major components of phytoplankton population peaks in area reservoirs. These include *Anabaenopsis circularis, Oscillatoria agardhii, Nodularia harveyana, Anabaena flos-aquae, Anabaena bornetiana, Anabaena circinalis, Aphanizomenon flos-aquae,* and *Microcystis aeruginosa.* No strain of *Microcystis aeruginosa* yet isolated has ever been reported to fix atmospheric nitrogen. Also, the only culture of *Aphanizomenon flos-aquae* that we have been able to isolate and culture in the laboratory is unable to grow in nitrate-free medium or reduce acetylene (a usually reliable test of nitrogen-fixing ability). Obviously, some strains of *Aphanizomenon flos-aquae* are true nitrogen fixers (Stewart *et al.* 1968, Gentile and Maloney 1969), while others are not (Williams and Burris 1952). Since our particular *Aphanizomenon* is highly susceptible to even the slightest environmental stress, our data may be inconclusive. It may also be interesting to note that the 20 or so species or strains of planktonic blue-green "nonbloom formers" that we have been able to isolate into unialgal culture are all apparently unable to fix atmospheric nitrogen. From this, one might suggest that the ability of an alga to fix atmospheric nitrogen is a prime factor in permitting it to become a dominate part of the phytoplankton. However, our strains of *Microcystis* and *Aphanizomenon* were both isolated from small "blooms" and neither can fix nitrogen.

Any of the aforementioned "bloom forming" blue-green algal species might be almost the only, or most conspicuous, "waterbloom" component. However, our peak populations of blue-green (which would probably never be afforded "bloom" status if compared to northern lakes) are usually mostly composed of three species. *Anabaena circinalis* generally seems to be most often the major component (about 50% of the total), with *Microcystis aeruginosa* and *Aphanizomenon flos-aquae* equally forming the balance of the population.

Although we seem to have made almost no progress in establishing absolute criteria for occurrence of "water-blooms" on southwestern reservoirs, it might be worthwhile to note the following. Observations compiled over the last several years indicate that most "waterblooms" seem to have developed after a severe or abrupt change in reservoir conditions. These changes have often been recorded in the form of rapid additions or losses of large volumes of water after major weather changes.

In laboratory culture, blue-green algal uptake of organic substrate (Allison *et al.* 1953, Carr and Pearce 1966) does not seem to differ significantly from other types of algae (Sloan and Strickland 1966). However, algal uptake of organic sustrate may be relatively unimportant at the low concentrations found in the natural environment (Wright and Hobbie 1966, Hobbie 1966, Munro and Brock 1968). This is in contrast to the substantial amounts of extracellular materials produced by algae (Fogg and Westlake 1955, Fogg *et al.* 1965, Hellebust 1965, 1967, Watt 1966, 1969).

We have not been able to show a significant increase in planktonic blue-green algal growth, either heterotrophically or autotrophically, with a wide variety of organic substrates (acetate, citrate, inositol, glucose, sucrose, and casamino acids). These have been tested at concentrations up to 0.01 M on both axenic and nonaxenic unialgal cultures. These negative results seem to confirm the fact that algal uptake of organic substrate, particularly at the normally low concentrations found in our reservoirs, is not significant. For example, preliminary tests during the spring of 1969 of dissolved glucose concentrations in Garza Little Elm Reservoir only indicated a concentration on the order of 10^{-10} M.

Bacterial enumeration on area reservoirs usually reveals a rather abrupt population increase following algal

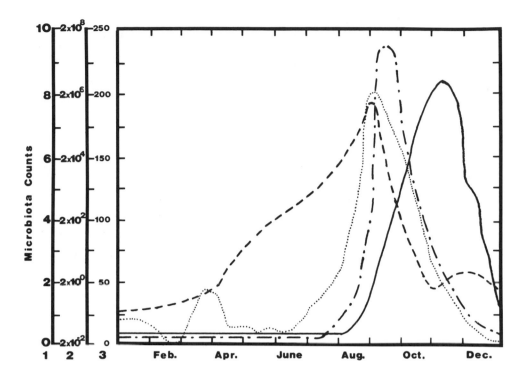

FIG. 6.1. Typical microbiotic cycle of a southwestern reservoir. (· · ·), Blue-green algae (scale 1, ASUs x 10^2); (- - -), gram negative bacteria, (scale 2, ml^{-1}); (—), gram positive bacteria, (scale 2, ml^{-1}); (- · -), actinomycetes, (scale 3, ml^{-1} x 10^3).

population peaks (Fig. 6.1). Whether or not this is presumptive evidence that extracellular organic production is a significant factor in the heterotrophic plankton population has not been established. Laboratory studies (which are discussed in a following section) do not indicate pronounced support of increased bacterial populations by extracellular algal production. Since no accumulation of products is permitted in the culture vessel, this observation may not be pertinent to actual reservoir conditions.

The relatively sudden disappearance of a major component of an algal population from the reservoir has not always been adequately explained. It would seem that

perhaps many interrelated factors must be involved. Observations on laboratory cultures do not seem to offer an
easy solution either. Sometimes seemingly healthy nonaxenic cultures of a generally hardy blue-green alga will
die almost overnight. Instead of the characteristic bright
orange-yellow color of nutrient starved blue-green algae
(Kingsbury 1955, Venkataraman 1960), the cultures will
rapidly become almost colorless with a marked whitish turbidity, which is probably caused by rapid bacterial growth
and indicates disintegration of the algal cells.

In contrast, selective removal of dietary components
from a blue-green algal laboratory culture will cause quick
deactivation of the culture, but not death. Within a few
days, the blue-green culture will become chlorotic, turn
orange-yellow, and finally become a dirty brown. These
minerally deprived organisms along with their associated
flora have been maintained under normal laboratory culture
conditions for periods up to 1 year. When the missing
nutritive element was added complete recovery of the entire cell mass occurred within 3 days.

Would these organisms have been decomposed in the
reservoir during this time? Since actinomycetes were not
present and the surviving bacterial flora might not have
been active decomposers, this observation may not carry
over into the reservoir. However, it serves to indicate
that algal decomposition products may not be as easily
and as quickly available to the heterotrophic population
as is usually implied.

Bacteria

Data obtained over several years of observation indicate that there seems to have occurred some definite and
repeatable patterns in the general microbiotic cycles
(Figs. 6.1 and 6.3) of southwestern reservoirs. Unfortunately, the value of this acquired information is tempered

by some degree of inconsistency. For example, many seasons of sampling the bacterial populations of various habitats in these reservoirs have revealed no "hard" rules concerning either the types of organisms present or the exact magnitude of the total population. Also, when the reservoir as a whole is considered, we have noted few real differences in either vertical or horizontal distribution (Fig. 6.2). This is not meant to imply that such differences never occur (e.g., stratification may have a profound effect on population distribution) but that these apparent variations in population number seem much less pronounced when monitored over extended periods of time. Fluctuations in bacterial populations are common but have been very difficult to correlate with subtle ecosystem changes.

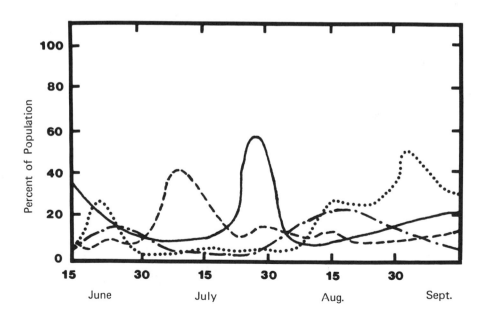

FIG. 6.2. Dominant heterotrophs in Lake Hefner (summer, 1967). (- - -), Fungi; (· · ·), *Flavobacterium*; (- · -), *Brevibacterium*; (——), *Alcaligenes*.

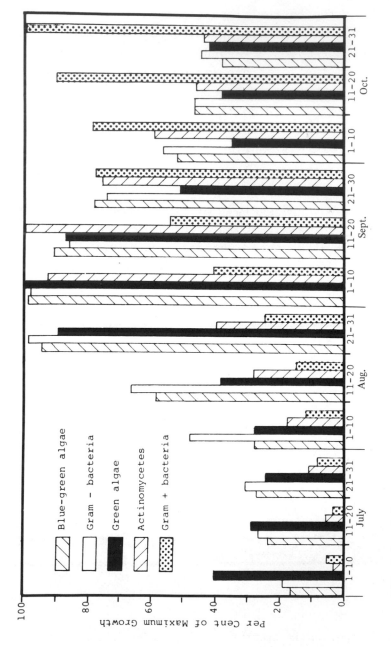

FIG. 6.3. Relative summer population of a typical southwestern reservoir.

BACTERIA, ALGAE, AND ACTINOMYCETES

Although a wide variety of bacterial organisms is usually present, the following types are more common and typical of southwestern reservoirs. The gram-positive variety include *Bacillus cereus, B. cereus* var. *mycoides, Brevibacteria*, and less frequently, the streptococci (group D or enterococci) that are usually recovered most frequently at "influx" stations. The gram-negative organisms usually make up the greater portion of the bacterial flora of reservoirs for most of the year (Figs. 6.1 and 6.2). The dominant forms in this group are normally the genera *Flavobacterium, Pseudomonas, Alcaligenes,* and some members of the *Enterobacter-Klebsiella* group, which, again in their occurrence, usually reflect "influx" stations.

Many articles have discussed the advantages and disadvantages of different types of bacterial enumeration. In this country plate counts have generally been favored, whereas in Europe the direct count method has been used most frequently. Since about 1930 (Potaenko 1968), Russian workers have employed direct count methods in most investigations. As Fig. 6.4 indicates, these methods demonstrate a much higher population of organisms than standard plate count methods. Rodina (1967) stated that only 10^{-4} of the entire population is shown by plate counts in unpolluted waters but 10^{-2} is revealed in polluted waters. However, Straskrabova and Legner (1969) had about equal success using both direct and plate counts in estimating bacterial numbers in running and stagnant waters and correlating these to 5-day biological oxygen demand measurements. Bere (1933) used direct counts to survey the bacterial population of Wisconsin lakes. Later, Henrici (1939) considered Bere's high counts (up to 2×10^6/ml), obtained by the direct method, unreliable. Currently, considerable attention is being devoted to membrane filter-fluorescent-antibody techniques (Guthrie and Reeder 1969) and luminescent microscopy (Rodina 1967).

175

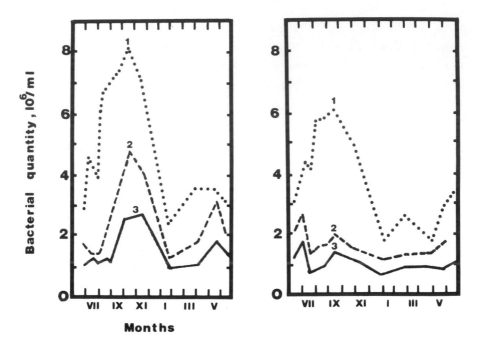

FIG. 6.4. Comparative bacterial population via direct count after Potaenko 1968. (1) Eutrophic lake, (2) intermediate lake, (3) mesotrophic lake.

Since our reservoirs contain substantial quantities of particulate detritus that severely hinders accuracy in direct counting, we generally use the spread-plate technique. Its convenience and avoidance of thermal shock, since many indigenous aquatic microbes may be quite stressed by the 45-50 C temperatures (Wierings 1968) of molten agar used with the pour plate method, has made it more attractive than the standard pour plate method.

When microbial population counts of southwestern reservoirs (Fig. 6.5) made with spread plates are compared to direct counts (Fig. 6.4), the former are usually low and exhibit much the same magnitude as the values obtained

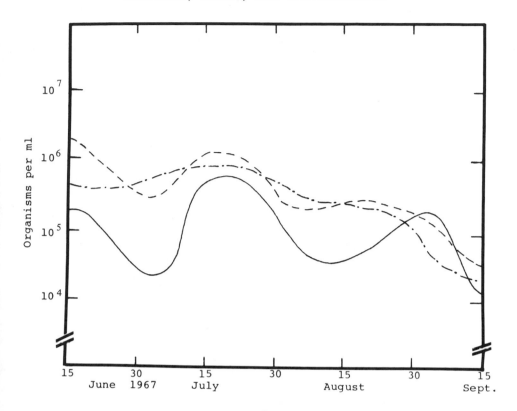

FIG. 6.5. Vertical distribution of bacteria (Lake Hefner). Total
bacteria: (—), top; (- · -), middle; (- - -), bottom.

from plate counts by Henrici (1939) and Potter (1964).
It is apparent that a clearer picture of the nature of
the bacterial population is obtained by incubating at
both 37 and 24 C and maintaining differential counts to
5 days. As Figure 6.6 clearly demonstrates, if plates
were read at the end of a standard 48-hr period, neither
of the incubation temperatures would yield a valid picture
of the actual bacterial population. Even after 5 days,
our ratio of chromogens (which are usually considered as

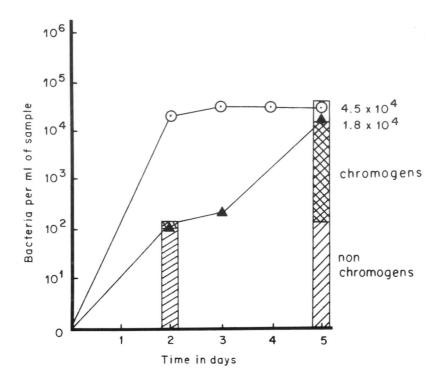

FIG. 6.6. Bacterial enumeration by the spread plate technique with incubation temperature as a counting variable. (○), 35 C; (▲), 24 C.

indigenous flora) does not exceed 50% and is somewhat lower than values obtained by Potter (1964).

The exact extent of bacterial action in the reservoir environment is not easily quantitated. Generally, the process of organic nutrient substance cycling has been indirectly estimated by correlating specific bacterial and actinomycetic counts with reservoir conditions with which they seemed to be associated. It has been only in the last few years that we have attempted measurements of specific nutritive components and/or processes in situ. Some of these deserve specific description. For example, by use of ^{14}C carbonate, we have initiated comparative studies of autotrophic and heterotrophic uptake and extracellular release in several southwestern impoundments. Attempts are being made to correlate many of these relatively recent measurements with the large mass of traditional limnological data gathered over the last two decades. Final computation of data is incomplete at this writing. The recent development (Stewart *et al.* 1967, 1968) of the acetylene-reduction technique has made indirect measurement of nitrogen fixation quite practical. So far, we have not been able to detect any free-living aquatic-heterotrophic bacterial reduction of acetylene. Thus, this preliminary evidence suggests that, at least in southwestern reservoirs, only blue-green algae are important in this regard.

Organic productivity measurements on area reservoirs are made by the carbon dioxide (Beyers *et al.* 1963) and oxygen (Odum and Hoskin 1958) rate of change methods. Generally our southwestern reservoirs cannot be considered high in productivity, which is perhaps limited by excessive turbidity. An extensive study of Garza Little Elm Reservoir produced an average net production of 60 g CO_2/m^2-day fixed (Trotter 1969).

It is now well documented that a considerable portion of the total algal production is released as extracellular products (Fogg 1966), both in culture and natural environments. We have only been able to indirectly follow the effect or utilization of these extracellular products in laboratory cultures of blue-green algae and their naturally associated bacterial flora. To accomplish this, three nitrogen-fixing planktonic species (*Anabaena bornetiana, A. circinalis,* and *A. flos-aquae*) were isolated into unialgal culture by serial dilution in combined nitrogen-free inorganic medium from natural population peaks in area reservoirs. Applying no other purification procedures, we have maintained these cultures in log phase by transfer

179

of 1 ml inocula each time the O.D. 650 reached about 0.006
(generally after about 5 days' growth). Although original
bacterial counts were high (about the same as the natural
water from which the original isolate was made), after
three or four transfers the bacterial count was negligible.
Subsequent tests for antibacterial exudates in the culture
medium proved negative. The bacterial populations increased
only when the algal culture was permitted to age or mature.
We tentatively suggest that, at least for our laboratory
conditions, normal extracellular algal production is not
of sufficient types of concentrations to promote extensive
bacterial development. Also, heterotrophic bacteria may
be more dependent on allochthonous organics and decomposition products of plankton (Fig. 6.7) than extracellular
products. Whether or not this is true in reservoirs remains to be shown.

Aquatic Actinomycetes

One of the early references to aquatic forms of the
actinomycetes was made by Adams (1929). Burger and Thomas
(1934) studied the tastes and odors in the Delaware River
and, while not denying the involvement of actinomycetes,
indicated that the odors resulted from the activity of a
mixture of microorganisms. Thaysen (1936) reported salmon
of the Thames River acquired an earthy or muddy taint as
a result of actinomycete-produced compounds from aquatic
types. Isachenko and Egorova (1944) published the results
of earlier work on several rivers in Russia where earthy
odors were attributed to the actinomycetes predominantly
growing, they thought, on the bottom and sides of the
rivers. Silvey *et al.* (1950) showed that certain actinomycetes were responsible for a variety of tastes and odors
by comparing the odor produced from pure cultures of organisms with the odors found in water supplies. Later
laboratory investigation by Dill (1951), McCormick (1954),

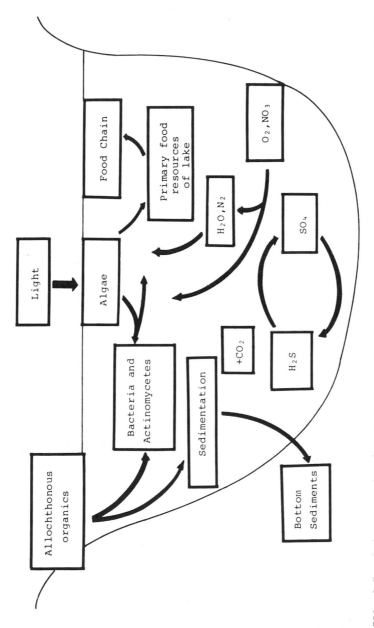

FIG. 6.7. Primary production and microbiological processes in a deep lake after Sorokin (1968).

and Pipes (1955) revealed that certain types of compounds
produced by these aquatic actinomycetes might have some
relation to the microbiotic cycle as it occurred in the
reservoirs, streams, and ponds that they investigated. In
the last 6 years several papers have appeared that describe
methods of growing various species of the genus *Streptomyces,* extracting odor components, and studying them for
chemical identity.

Life History

Detailed studies on the life history of the members
of certain species of the genus *Streptomyces* have been investigated in considerable detail by Klieneberger-Nobel
(1947), Davis (1960), Hopwood (1960), Hopwood and Sermonti
(1962), and Higgins and Silvey (1966).

It is proper to consider the spore as the stage of
transfer and perhaps the overwintering stage of the aquatic
forms in areas where the water temperature reaches 4 C.
The spore is normally spherical, measuring approximately
0.9-1.2 μm in diameter, and germinates by one or two germ
tubes (1.0-0.8 μm). The germination of the spore is dependent upon such factors as (1) temperature, (2) pH, (3)
nutrients, (4) amount of available oxygen, and (5) hydration. Germination will occur between 7 and 36 C. The
optimum appears to be 22-35 C. Temperatures as high as
65 C inhibit the germination process. The pH range at
which the organism may remain viable varies from 2.0 to
12.0; however, the optimum for germination appears to
occur between 6.5 and 8.5. When the richness of the medium
was increased, there was usually an increase in the time
and level of total primary mycelia production. Germination
times appeared to be very little affected by the varying
nutrient strength. It should be noted that germination
and development were inhibited when the total solids exceeded 60 g/liter. Observations made by Higgins and Silvey

(1966) showed that the organism favored an aerobic pathway, since anaerobic conditions greatly reduced the rate of growth. It was noted in laboratory as well as in field work that the germination and development of the primary mycelium required high levels of hydration. Desiccation was found to be the most successful method of controlling the amount of primary growth in all laboratory investigations (Higgins and Silvey 1966).

After spore germination and before branching began in the primary mycelium, there was a lag period. Then the branches started as swellings at the sides of a hyphae. These swellings extended laterally while some maintained a sort of a bulboid appearance. Septation was not common in the primary mycelium, especially in the early growth stages. As the growth of the primary mycelium continued, the diameter of the newly formed hyphae ordinarily diminished to about 0.4-0.6 μm. It is interesting to note that the branching was not limited to the plane of the medium, but that many of the filaments penetrated deeply into the substrate while others became flattened onto the medium so as to maintain sufficient hydration.

The primary mycelium frequently propagated itself vegetatively by fragmentation. Reproduction of this nature may not be commonly observed because the primary mycelium is unusually small. If, however, in the laboratory experimentation, primary mycelia undergo fragmentation by extreme agitation many colonies appear in the liquid medium, indicating an origin of new primary mycelia by this system.

The secondary mycelium differs from the primary in that it is larger and usually carries a black, insoluble pigment in the outermost cellular layer that is less branched and does not attempt to penetrate a solid medium. In addition, one may note that if the secondary mycelium undergoes fragmentation small fragments of the hyphae revert to a primary stage. If several hyphae are involved, new secondary mycelia are produced.

Spores are produced at the apices of the hyphae. All notations on the secondary mycelium indicate that in its final development prior to and during sporulation, it becomes somewhat hydrophobic in liquid culture and floats to the surface. It is interesting to note that as the secondary mycelium develops the primary mycelium, which seems to serve as part of the nutrient for the growth of the secondary ones, undergoes slow autolysis.

In reservoirs the primary stages may occur near the perimeter of the lake and may be associated with emergent, submerged, or floating vegetation. Also, since they are facultative aerobes, they may grow on the bottom mud during periods of summer stratification. As long as temperatures are 20 C or below, the primary stage apparently persists in that condition, although in instances where water is drawn down in a reservoir so as to expose the primary stages near the perimeter of the lake, as the hydration decreases, secondary mycelia are formed. So long as the primary mycelia remain submerged and the temperature remains at about 20 C, they do not appear to develop, and thus secondary mycelial formation seems unlikely. As the temperatures increase in the water, the primary stages that are in the aerobic zone of the reservoir and are associated with such nutrient sources as higher aquatic plants, blue-green algal mats, or organic ooze, produce secondary mycelia around 22 C. As the water becomes warmer, more secondary stages form. During early fall turnover, as the hypolimnion of the lake becomes aerobic, primary mycelia are altered to secondary mycelia. These come to the surface of the water along with mats of floating blue-green algae and are slowly carried shoreward. In reservoirs with dense areas of emergent vegetation, or the remains of submerged trees (quite abundant in the Southwest), the formation of secondary mycelia may be

anticipated during warm periods beginning about the middle
of July and may well continue for some time depending upon
meteorologic conditions (Silvey 1964).

Cultivation and Enumeration

The isolation and enumeration of aquatic actinomycetes
may pose numerous problems because spores and hyphal elements of both the primary and the secondary mycelia all
develop colonies on isolation plates. It should therefore
be understood that the counts of the actinomycete population made from reservoirs or streams reflect a relative
density but not a true number as might be found perhaps
in bacterial isolation, cultivation, and enumeration techniques.

Standard bacteriological techniques are involved in
the collection of samples from reservoirs, ponds, or
streams, and for isolation spread plates are used rather
than poured plates. This is because more rapid appearance
will be noted in the secondary mycelial colonies. The
period of incubation is from 6 to 14 days. Since there
are numerous bacteria mixed in with the actinomycetes it
is desirable to incubate part of the plates at room temperature and others of the same dilution at 32 C. Thirty-
seven degrees centigrade frequently causes the bacteria
to overgrow the plates and dehydration of the medium will
occur prior to the time the actinomycetes appear.

A number of types of media that may be employed for
the isolation of actinomycetes from water samples are
available. One on the market as a commercial product is
Actinomycete Isolation Agar, which appears to be quite
successful. Another variety that we have employed from
time to time is Emerson's modified agar, which contains
20 g of agar, 80 g brown sugar, 8 g peptone, 4 g beef extract, 2 g meat extract, 2 g sodium chloride made to 1000
ml in water. Regular commercial plate count agar, when

enriched with 20 g brown sugar per 1000 ml makes an excellent isolation agar for the aquatic streptomyces. If it is desirable to have a minimal medium that may be used in instances where slow growth is desirable and that will not be highly contaminated by bacteria, a formulation comprised of the following ingredients should be employed: sodium citrate, 10 g; sodium nitrate, 2 g; potassium nitrate, 2 g; calcium chloride, 0.1 g; magnesium sulfate, 0.05 g; bipotassium phosphate, 2 g; agar, 20 g; distilled water to make 1000 ml.

Nutrient Sources for Actinomycete Development

It is well recognized that the actinomycetes have a rate of growth normally slower than most of the fungi and that they do not reproduce as rapidly as most of the bacteria. However, this group of organisms does have the attribute of producing antibiotics and these substances may enable them to compete successfully with more vigorously growing types of aquatic microorganisms. Members of the genus *Streptomyces*, after once colonizing on some type of substrate, are very persistent in their growth for a considerable period of time. In addition, these organisms share with the fungi in the ability to grow into substrates. Although it is recognized their hyphal elements are highly delicate as compared to the fungi, they can obtain sufficient penetration for continuous nutrition.

The role of the actinomycetes in biodeterioration has been well summarized by Williams (1966). A perusal of his paper leads one to conclude that members of the genus *Streptomyces* may make use of simple sugars as well as polysaccharides. The ability of the members of this genus to produce cellulose makes them important organisms in the microbiotic cycle in the reservoirs where cellulose deposits lead to rapid eutrophication. Moreover, the ability to deteriorate lignin is also important in the reduction

of the rate of eutrophication in reservoirs. It is further
noted by Williams (1966) that certain lipids and steroids
may serve as carbon sources and they may be either partially or completely degraded. In addition, tannins of
low molecular weight, including tannic acid and gallotannin,
may be decomposed by soil-inhabiting actinomycetes.

Inorganic sources of nutrition include nitrogen-containing compounds, such as ammonia and nitrates. In the
previously listed media it should be noted that either
ammonia or nitrate is added in order to encourage actinomycetic growth. That actinomycetes apparently need ample
amounts of combined nitrogen is demonstrated by their repeated occurrence as contaminants of only blue-green algal
isolates which are grown in complete medium. We have not
yet detected actinomycetic growth in cultures of blue-green
nitrogen fixers that are routinely grown in nitrate-free
medium. Other nitrogen sources may include organic types,
since proteolytic activity of the actinomycetes appears
to have been well established, as pointed out by Williams
(1966). Organisms that contain such a vast array of enzyme
systems and metabolic pathways are obviously very important
in numerous phases of a microbiotic cycle that occurs in
freshwater reservoirs.

Composition of *Streptomyces* Isolated
from Reservoirs

The different species of the genus *Streptomyces* isolated from the area reservoirs have been identified by
serological and electron microscopic examination (Tables
6.1 and 6.2). No attempt has been made to do detailed
morphological and nutritive studies in order to accomplish
identification. In all instances where serological techniques were involved, known identified species were used
against the unknown in order to obtain information concerning the serological tests. The species involved that

187

Table 6.1

Results of Serological and Electron Microscopic Examination
of Twenty Unknown Species of *Streptomyces*

Culture number	Spore surface	Species identification	Serological similarity[a]	ISP reported spore surface[b]
H1	Smooth	*Antibioticus*	4	Smooth
H9, H20, H21	Smooth	*Odorifer*	4	Smooth
H24, H34	Spiny	*Viridochromogenes*	4	Spiny
H10, H13, H32	Smooth	*Aureofaciens*	4	Smooth
H8	Warty	*Aureofaciens*	3	Smooth
H17	Spiny	*Aureofaciens*	1	Smooth
H2, H3, H21	Warty	*Griseolus*	2	Smooth
H11	Smooth	*Coelicolor*	2	Smooth
H28	Spiny	*Coelicolor*	1	Smooth
H7	Smooth	*Coelicolor*	1	Smooth
H27, H29	Spiny	*Odorifer*	3	Smooth
H30	Warty	*Odorifer*	1	Smooth

[a]Serological relationships, 1 = related, 2 = close, 3 = very close, 4 = identical.

[b]Shirling and Gottlieb 1968, from Taylor and Guthrie 1968.

Table 6.2

Recent Isolates of Odor-producing *Streptomyces* at NTSU

Culture number	Species identification	Serological similarity[a]
SH9	*Streptomyces cinnamoneus*	1
Rat	*Streptomyces coelicolor*	4
62	*Streptomyces antibioticus*	4
NT16	*Streptomyces antibioticus*	4

[a]Serological relationships: 1 = related; 2 = close; 3 = very close; 4 = identical.

were identical in all serological tests are *Streptomyces antibiotics, S. odorifer, S. virivochromogenes, S. aureofaciens, S. cinnamoneus,* and *S. coelicolor.* References made in this chapter to *Streptomyces* may be to any of the above mentioned species since the cultural differences were not great except in the spore form, shape, pigmentation, and the color of the outer cellular layer. No detailed data on any one species can be quoted at this time.

Cyclic Phenomena in Southwestern Reservoirs

The actinomycete population density has been measured for a number of years in Lake Hefner, Garza Little Elm, water supply reservoirs for Wichita Falls, Graham, Abilene, and other west Texas towns. On occasion, in order to obtain information concerning the population density of actinomycetes, samples have also been analyzed from streams that supply the reservoirs. These samples have been collected during times of both high and low water. The population density of these samples has been recorded as a mean and noted on a monthly basis. It may be observed in Table 6.3 that the actinomycete population usually begins to increase in the top and the bottom of reservoirs during the month of July, reaching a peak in August. Obviously the high counts obtained from bottom deposits result mostly from spores. Also, many of these may be of recent origin because of the diversity in the population density as demonstrated by the colony count. These cycles demonstrate rather clearly a period of increased growth and a period of diminution. During September most reservoirs in the Southwest "turn over." This gives more equalization in the distribution of the actinomycetes in the reservoirs. In November the population begins to decrease, so from December through spring the concentration of organisms is relatively low.

Table 6.3

Actinomycete Population Density Average Southwestern Streams and Reservoirs

	Jan.	Feb.	Mar.	Apr.	May	June
Avg. temp. (C)	7.2	5.6	14	16	19	22
Top (organisms/ml)	100	80	90	120	340	760
Middle (organisms/ml)	40	20	40	38	46	60
Bottom (organisms/ml)	180	160	200	260	460	840

	July	Aug.	Sept.	Oct.	Nov.	Dec.
Avg. temp. (C)	26	29	27	24	18	10
Top (organisms/ml)	1160	120,000	30,000	12,000	800	190
Middle (organisms/ml)	65	400	35,000	20,000	1200	200
Bottom (organisms/ml)	1250	680,000	90,000	23,000	3400	900

There is an obvious periodicity in the algal and actinomycetic blooms that occur in southwestern reservoirs.
This may be modified from time to time by unusual flows
during the month of June. Over a 15-year average, the
June bloom appears to be relatively unimportant. If rainfall occurs in February, there is likely to be a sporadic
increase in the blue-green algae, followed by a growth of
diatoms. This is again a rather seasonal phenomenon, but
on an annual basis shown in Figure 6.8 the tendencies are
duplicated year after year. It is observed that blue-green
algal peaks generally precede the "bloom" of actinomycetes,
which is sometimes followed by an increased diatom population. The interrelationships between the blue-green algae
and the actinomycetes are well demonstrated during this
period and are duplicated year after year. Even in laboratory culture of blue-greens from these reservoirs it
is quite common for secondary mycelia of the actinomycetes
to grow and float to the surface of the culture chamber.
This further indicates the close association that exists
between these two groups of organisms, for in many cases
it has proved almost impossible to rid the contaminated
blue-green of its actinomycete "associate."

The bacteria also associated with the blue-green algae
and the actinomycete blooms in southwestern reservoirs
(Figs. 6.1 and 6.3) illustrate the average annual cycle of
gram-negative heterotrophic bacteria that reach their maximum density at approximately the same time as the blue-
green algae. Their association, of course, particularly
in artificial culture, is very well recognized (Gorham
1964, Vance 1965). Following the blue-green algal bloom,
the actinomycetes reach their maximum population density
and in due time the gram-positive heterotrophic organisms
undergo a rapid bloom that is attenuated normally by the
last of December or early in January. These associations
and cyclic phenomena have been observed over sufficient

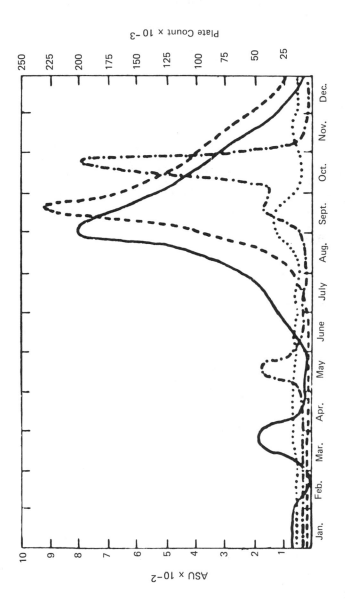

FIG. 6.8. Cyclic phenomena in southwestern reservoirs. (· · ·), greens; (- · -), diatoms; (—), blue-greens; (- - -), actinos.

periods of time to be recognized as a possible interrelationship that bears further scrutiny. For example, in Figure 6.9 the relationship between the gram-positive bacteria and the actinomycetes is very definite. As the actinomycete bloom begins to decrease, the gram-positive

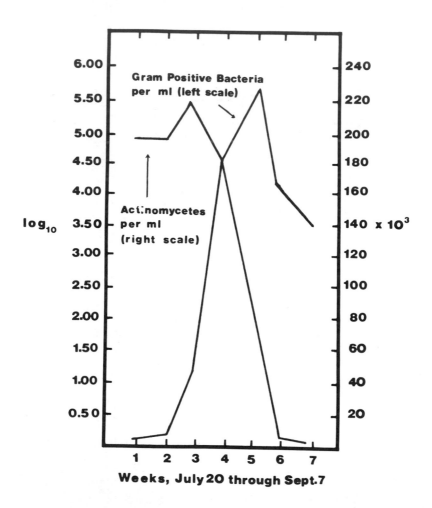

FIG. 6.9. Inverse relationship between actinomycetes and gram positive bacteria.

population increases in a logarithmic curve from the early part of July to the middle of September. Work in the laboratory has demonstrated that the gram-positive bacteria, *Bacillus cereus*, is capable of metabolizing the odor components produced by the actinomycetes. We have not been able to show definitely whether they actually attack the hyphal elements of the streptomyces or metabolize the organic components produced by these organisms. Detailed microscopic examinations reveal that the mycelial mass appears to be undergoing disintegration in proportion to the increase in population of *B. cereus*. After the actinomycetes have reached their peak of logarithmic growth, if there is an absence of these bacteria in laboratory culture, autolysis of the mycelia is very slow indeed. Since the organic components produced by the actinomycetes are recognized by their odors and the fact that they disappear from the water in the presence of high population concentrations of *B. cereus*, it is obvious from our work that there is a metabolic and enzymatic relationship between these two groups of organisms. We are hesitant to state precisely where or how it occurs.

SUMMARY

The reservoirs of the Southwest offer a variety of ecosystems because of the diversity in construction and maintenance. Since many areas are confined largely to agricultural practice while others are more or less industrialized, many varieties of nutrient materials enter the feed streams to the reservoirs. In general, one must consider these bodies of water as having relatively high quantities of nutrients and consequently should be unusually fertile. Added to these characteristics, the growing seasons are long and warm, encouraging certain species of the

microflora to increase sporadically. There is not normally
a period of stagnation during the relatively mild winters.
Therefore, the reservoir is in a "turnover" state from the
middle of September until the middle of June. Perhaps the
greatest factor limiting productivity in southwestern reservoirs is turbidity, which is normally attributable to
various types of suspended inorganic matter, although from
time to time organic detritus from the remains of submerged
vegetation or inundated trees will produce both color and
turbidity. Notwithstanding all of these various characteristics, one may consider southwestern reservoirs to contain a rather rich and varied planktonic microflora.

The phytoplankton are comprised of greens, blue-greens,
diatoms, and occasionally, certain types of flagellates.
The green algae appear to have an almost constant level of
population density, with some variations in the spring and
early fall. Diatoms may express a sudden increase or
"waterbloom" on occasions following high flows in the early
part of the spring. Otherwise, their population is somewhat
constant, is rather diverse, and does not compose a large
portion of the total population. In general, the phytoplankton picture in the southwestern reservoirs reveals
that blue-green algae exert the major influence upon the
entire aquatic ecosystem, particularly during peak populations. These blue-green algae that exert such an important
effect are dominated by nitrogen-fixing species. The euplanktonic blue-green algae have accelerated metabolic
patterns when compared to other blue-green forms. Although
these organisms do not seem to be highly involved in uptake
of the organic substrate, they may produce substantial
amounts of extracellular products. However, studies on
young cultures of selected blue-green algae and associated
bacteria indicate that the extracellular production by the
algal population may be of limited use to the associated

gram-negative heterotrophic flora. In laboratory culture, minerally starved blue-green algal organisms deactivate readily but fail to die and decompose within a period up to a year. The gram-positive heterotrophic bacteria do not exhibit an increase in population density until the termination of an actinomycete bloom. Data secured from southwestern reservoirs show that the aquatic actinomycetes follow the blue-green algae bloom and laboratory observations indicate a more rapid deterioration of blue-green algae that have reached the logarithmic peak of growth in the presence of actinomycetes than in the absence of these forms; it may therefore be postulated that the bacteria in our aquatic environments are not contributing as much to the decomposition of the blue-green algae as the streptomycetes which, in some manner, influence the reduction and perhaps degradation of the blue-green algae. In turn, from our findings, it appears that the gram-positive heterotrophic bacteria, such as *Bacillus cereus*, which invariably increases after an actinomycetic bloom, may contribute to either the biogradation of the mycelial mat or to the byproducts produced by the actinomycetes.

In addition, it must be recognized that the allochthonous organics and particulate detritus of an oxidizable nature that is brought into the reservoirs by way of the feed streams must be important in furnishing nutrients for the gram-negative and gram-positive heterotrophic bacteria. When the interrelationships of the microflora of the reservoir are better annotated, the problem of eutrophication may be more rapidly solved.

ACKNOWLEDGMENTS

The investigations were supported in part by Federal Water Pollution Control Administration Grants WP-00785, WP-00805, and 5T1-WP-107; Bureau of Reclamation Contract

14-06-D-5298; the City of Oklahoma City and North Texas State University water research laboratories.

BACTERIA, ALGAE, AND ACTINOMYCETES

REFERENCES

Adams, B. A. 1929. Odors in the waters of the Nile River. *Water and Water Eng. 31*, 309-314.

Allison, R. K., Skipper, H. E., Reid, M. R., Short, W. A., and Hogan, G. L. 1953. Studies on the photosynthetic reaction. I. The assimilation of acetate by *Nostoc muscorum*. *J. Biol. Chem. 204*, 197-205.

Bere, R. 1933. Numbers of bacteria in inland lakes of Wisconsin as shown by the direct microscopic method. *Int. Rev. Ges. Hydrobiol. Hydrograph. 29*, 248-263.

Beyers, R. J., Larimer, J. L., Odum, H. T., Parker, R. B., and Armstrong, N. E. 1963. Directions for the determination of changes in carbon dioxide from changes in pH. *Pub. Inst. Mar. Sci. Univ. Texas 9*, 454-489.

Burger, J. W., and Thomas, S. 1934. Tastes and odors in the Delaware River. *J. Am. Water Works Assoc. 26*, 120-127.

Carr, N. G., and Pearce, J. 1966. Photoheterotrophism is blue-green algae. *Biochem. J. 99*, 28.

Cheatum, E. P., Longnecker, M., and Metler, A. 1942. Limnological observations on an East Texas lake. *Trans. Am. Microscop. Soc. 61*, 336-348.

Cole, G. A. 1963. The American southwest and middle America. In *Limnology in North America*, D. G. Frey (Ed.). University of Wisconsin Press, Madison, pp. 393-434.

Davis, G. H. G. 1960. The interpretation of certain morphological appearances in a *Streptomyces* sp. *J. Gen. Microbiol. 22*, 740-743.

Deevey, E. S. Jr. 1957. Limnological studies in middle America with a chapter on Aztec limnology. *Trans. Connecticut Acad. Arts Sci. 39*, 213-328.

Dill, W. S. 1951. The chemical compounds produced by actinomycetes and their relation to taste and odors in a water supply. M.S. thesis, North Texas State University, Denton, Texas.

Dorris, T. C. 1956. Limnology of the middle Mississippi River and adjacent waters. II. Observations of the life histories of some aquatic Diptera. *Trans. Illinois Acad. Sci. 48*, 27-33.

Dorris, T. C. 1958. Limnology of the middle Mississippi River and adjacent waters. Lakes on the leveed floodplain. *Am. Midl. Nat. 59*, 82-110.

Fogg, G. E. 1966. The extracellular products of algae. *Oceanog. Mar. Biol. Ann. Rev. 4*, 195-212.

Fogg, G. E., and Westlake, D. F. 1955. The importance of extracellular products of algae in freshwater. *Verh. Int. Verin. Limnol. 12*, 219-232.

Fogg, G. E., Nalewajko, C., and Watt, W. D. 1965. Extracellular products of phytoplankton photosynthesis. *Proc. Roy. Soc.,* Ser. B *162*, 517-534.

Gentile, J. H., and Maloney, T. E. 1969. Toxicity and environmental requirements of a strain of *Aphanizomenon flos-aquae* (L.) Ralfs. *Can. J. Microbiol. 15*, 165-173.

Gorham, P. R. 1964. Toxic algae. In *Algae and Man*, D. F. Jackson (Ed.). Plenum, New York, pp. 307-336.

Guthrie, R. K., and Reeder, D. J. 1969. Membrane filter-fluorescent-antibody method for detection and enumeration of bacteria in water. *Appl. Microbiol. 17*, 399-401.

Harris, B. B., and Silvey, J. K. G. 1940. Limnological investigation on Texas reservoir lakes. *Ecol. Monogr. 10*, 111-143.

Hellebust, J. A. 1965. Excretion of some organic compounds by marine phytoplankton. *Limnol. Oceanog. 10*, 192-206.

Hellebust, J. A. 1967. Excretion of organic compounds by cultured and natural populations of marine phytoplankton. *Am. Assoc. Adv. Sci. Publ. 83*, 361-366.

Henrici, A. T. 1939. The distribution of bacteria in lakes. In *Problems of Lake Biology*, F. R. Moulton (Ed.). Am. Assoc. Adv. Sci. Publ. 10, Science Press, Lancaster, Penn., pp. 39-64.

Higgins, M. L., and Silvey, J. K. G. 1966. Slide culture observations of two freshwater actinomycetes. *Trans. Am. Micros. Soc. 85*, 390-398.

Hobbie, J. E. 1966. Glucose and acetate in freshwater: Concentrations and turnover rates. In *Chemical Environment in the Aquatic Habitat*, H. L. Golterman and R. S. Clymo (Eds.). Proceedings of an I.P.B. Symposium in Amsterdam and Nieuwersluis 10-16, Oct. 1966. N.V. Noord-Hollandsche Uitgevers Maatschappij-Amsterdam, pp. 245-251.

Hopwood, D. A. 1960. Phase-contrast observations of *Streptomyces coelicolor*. *J. Gen. Microbiol. 22*, 295-302.

Hopwood, D. A., and Sermonti, G. 1962. The genetics of *Streptomyces coelicolor*. *Adv. Genet. 11*, 273-342.

Irwin, W. H. 1945. Methods of precipitating colloidal soil particles from impounded waters of central Oklahoma. *Bull. Oklahoma Agr. Mech. Coll. 42*, 16 p.

Isachenko, B. L., and Egorova, A. 1944. Actinomycetes in reservoirs as one of the causes responsible for the earthy smell of their waters. *Microbiologiya 13*, 224-230.

Kingsbury, J. M. 1955. On pigment changes and growth in the blue-green alga, *Plectonema nostocorum* Bornet ex Gomont. *Biol. Bull. 110*, 310-319.

Klieneberger-Nobel, E. 1947. The life cycle of sporing actinomycetes as revealed by a study of their structure and septation. *J. Gen. Microbiol. 1*, 22-32.

McCormick, W. C. 1954. The cultural, physiological, morphological and chemical characteristics of an actinomycete from Lake Waco, Texas. M.S. thesis, North Texas State University, Denton, Texas.

Munro, A. L. S., and Brock, T. D. 1968. Distinction between bacterial and algal utilization of soluble substances in the sea. *J. Gen. Microbiol. 51*, 35-42.

Odum, H. T., and Hoskin, C. M. 1958. Comparative studies on the metabolism of marine water. *Publ. Inst. Mar. Sci. Univ. Texas 5*, 16-46.

Pipes, W. O. 1955. An investigation of naturally occurring tastes and odors from fresh waters. M.S. thesis, North Texas State University, Denton, Texas.

Potaenko, Yu.S. 1968. Seasonal dynamics of total bacterial number and biomass in water of Narochan lakes. *Microbiology 37*, 441-446.

Potter, L. F. 1964. Planktonic and benthic bacteria of lakes and ponds. In *Principles and Applications in*

Aquatic Microbiology, H. Heukelekian and N. C. Dondero (Eds.). Wiley, New York, pp. 148-166.

Rodina, A. G. 1967. On the forms of existence of bacteria in water bodies. *Arch. Hydrobiol. 63*, 238-242.

Shirling, E. B., and Gottlieb, D. 1968. Cooperative description of type cultures of *Streptomyces*. II. Species descriptions from study. *Int. J. Syst. Bacteriol. 18*, 69-189.

Silvey, J. K. G. 1964. The role of aquatic actinomycetes in self-purification of fresh water streams. In *Advances in Water Pollution Research*, Vol. 1 B. A. Southgate (Ed.). Pergamon, London, pp. 227-243.

Silvey, J. K. G., Russell, J. C., Redden, D. R., and McCormick, W. C. 1950. Actinomycetes and common tastes and odors. *J. Am. Water Works Assoc. 42*, 1018-1026.

Sloan, P. R., and Strickland, J. D. H. 1966. Heterotrophy of four marine phytoplankters at low substrate concentrations. *J. Phycol. 2*, 29-32.

Stewart, W. D. P., Fitzgerald, G. P., and Burris, R. H. 1967. In situ studies on N_2 fixation using the acetylene reduction technique. *Proc. Nat. Acad. Sci. U.S. 58*, 2071-2078.

Stewart, W. D. P., Fitzgerald, G. P., and Burris, R. H. 1968. Acetylene reduction by nitrogen-fixing blue-green algae. *Arch. Mikrobiol. 62*, 336-348.

Straskrabova, V., and Legner, M. 1969. Bacterial and ciliate quantity related to water pollution. Fourth International Conference on Water Pollution Research. Prague, Pergamon Press (Reprint).

Taylor, G. R., and Guthrie, R. K. 1968. Characterization of cytoplasmic antigens for serological grouping of *Streptomyces* species. *Bacteriol. Proc.* p. 20.

Thaysen, A. C. 1936. The origin of an earthy or muddy taint in fish. I. The nature and isolation of the taint. *Ann. Appl. Biol. 23*, 99-105.

Trotter, D. M. 1969. A comparison of the carbon dioxide and oxygen rate of change methods for measuring primary productivity. M.S. thesis, North Texas State University, Denton, Texas.

Vance, B. D. 1965. Composition and succession of cyano-phycean water blooms. *J. Phycol. 1*, 81-86.

Venkataraman, G. S. 1960. Growth and pigment changes in *Scytonema tenue* Gardner. *Indian J. Plant Physiol. 3*, 203-211.

Watt, W. D. 1966. Release of dissolved organic material from the cells of phytoplankton populations. *Proc. Roy. Soc.,* Series B. *164*, 521-551.

Watt, W. D. 1969. Extracellular release of organic matter from two freshwater diatoms. *Ann. Bot. 33*, 427-437.

Wierings, K. T. 1968. A new method for obtaining bacteria-free cultures of blue-green algae. *Antonie van Leevwenhoek J. Microbiol. Serol. 34*, 54-56.

Williams, A. E., and Burris, R. H. 1952. Nitrogen fixation by blue-green algae and their nitrogenous composition. *Am. J. Bot. 39*, 340-342.

Williams, S. T. 1966. The role of actinomycetes in biodeterioration. *Int. Biodetn. Bull. 2*, 125-133.

Wright, R. T., and Hobbie, J. E. 1966. Use of glucose and acetate by bacteria and algae in aquatic ecosystems. *Ecology 47*, 447-464.

Chapter 7

BACTERIAL TYPES AND INTERACTIONS IN A THERMAL
BLUE-GREEN ALGAE-EPHYDRID FLY ECOSYSTEM

Robert W. Gorden

Department of Microbiology
South Colorado State College
Pueblo, Colorado

Richard G. Wiegert

Department of Zoology
University of Georgia
Athens, Georgia

CONTENTS

ABSTRACT

Algal-fly mat ecosystems in thermal streams located in Yellowstone National Park support bacterial populations that increase at a rate comparable to the growth rate of the algae during early stages of succession. Dominant bacteria, mainly chromogenic, gram-negative rods differ in type and number down the thermal, CO_2, O_2, and temperature graidents of the 1 m x 24 m experimental boards. Bacteria contribute to the decomposition of the cooled algal mat only when actively feeding fly larvae (ephydrid) are present. Experimentally cooled areas of the mat show small increases in bacterial numbers but little loss of biomass in the absences of fly larvae. Experimental data demonstrate that bacterial growth rates are potentially greater than the demonstrated washout rate. Uptake rates of radiocarbon substrates suggest that biologically active bacterial populations are present in oxygenated and anaerobic portions of the mat during all stages of succession and at maturity.

INTRODUCTION

Description

Aquatic thermal ecosystems consisting of a mat of thermophilic blue-green algae and dependent consumer organisms are distributed worldwide (Brock 1970). Thus, despite the limited total area occupied by such ecosystems, they are accessible to many ecologists and provide information on the differences resulting when a few groups of organisms adapt to a wide variety of chemical and physical environments, similar only with respect to temperature.

The living components of aquatic thermal ecosystems range from simple (perhaps monospecific) filamentous bacterial mats at temperatures above 90 C (Brock 1967; Bott and Brock 1969) to the complex systems found where the effect of the thermal input is so minor that only slight differences exist between the thermal system and the "normal" aquatic community. From the systems ecology point of view, the former are taxonomically too simple to reward study. The latter are too complex for such studies to result in useful generalizations (at least they offer no advantage over study of the normal aquatic community).

There is a class of natural thermal communities, however, that provides sufficient trophic diversity for ecological modeling studies together with the advantages of few species, small size, ease of sampling, and rapid (as few as 2 months) progression from a bare substrate to steady state (see Wiegert and Mitchell 1973).

Systems of this type in Yellowstone National Park are characterized by: (1) temperature (40-55 C) and pH (6-10) values permitting the growth of a mat of thermophilic filamentous blue-green algae, (2) the presence of biophagous brine fly species (Ephydridae) feeding on the algal mat wherever local cooling brings the temperature below 40 C, and (3) predaceous and parasitic arthropods dependent, at least in part, on the fly population. These include lycosid spiders, dolichopodid flies, tiger beetles, and red mites (*Partuniella*). Together these groups constitute a viable association that exhibits a constant primary successional pattern.

Primary Succession

The development of a steady-state algal-fly community plus associated predators and parasites proceeds rapidly in its initial stages. Fraleigh (1971) and Wiegert and

Fraleigh (1972) described the process for communities developing on artificial board substrates (1 m wide and 8-24 m long) in thermal effluents characteristic of the Lower Geyser Basin of Yellowstone National Park. In addition to the temperature and pH characteristics mentioned above, these waters have high levels of dissolved CO_2 (15-90 ppm) and total phosphorus (0.1-0.3 ppm). These high nutrient levels permit very rapid growth of the algae during the first few days of the primary succession. Growth begins to be slowed as soon as the mat of algal filaments becomes thick enough (1-2 mm) to impede seriously the turbulent flow of water that constantly replenishes the CO_2 and phosphorus levels in the interstitial water. The phosphorus level decreases little as the water flows downstream because the amount going into new production is only a small fraction of the total available. The dissolved CO_2, however, degasses rapidly from the flowing water since the saturation values for these waters at temperatures of 40-55 C are on the order of 1 ppm or less. Therefore, the level of dissolved CO_2 determines the amount of carbon diffusing into the thickening mat and thus permits prediction of the level of production occurring during the successional development (Fraleigh and Wiegert 1975).

There are some species changes in the blue-green flora during succession. However, the species *Mastigocladus laminosus* becomes dominant during the first week and remains the dominant alga (>70% of the total biomass) in those mature mats discussed in this chapter.

As the algal mat increases in thickness, it reaches the point where parts of it begin to project from the water. At this point the mat for the first time provides a resting, feeding, and oviposition site for the grazing flies.

Algal-Fly Interactions

In the experimental systems discussed in this chapter only one species of grazing fly, *Paracoenia turbida* Cresson, is numerically important. As the algal mat grows it changes the water flow into a series of channeled flows, leaving occasional islands of cooler algae. These islands attract large numbers of adult flies that feed and oviposit on the algae. At first these islands may be cooled below 40 C only in the surface millimeter or two, rendering the habitat unsuitable for complete development of the larvae, which are killed by temperatures in excess of 40 C. Later the cool stagnant patches of algae become larger, thicker, and less transient. The feeding of large numbers of larvae then quickly destroys the structure, i.e., solubilizes the mat. Some of the oldest larvae pupate and emerge as adults; the remainder are killed when, as inevitably happens, the destruction of the algal mat allows the hot water to again flow through, starting the microsuccessional sequence anew with fresh algal growth and no flies (for a more detailed discussion of this sequence, see Wiegert and Mitchell 1973).

The Role of Bacteria

Unicellular bacteria are present in the interstitital spaces between the algal filaments during all stages of the successional sequence.* Our objectives in undertaking

---

*Filamentous bacteria (flexibacteria) are also present but are quantitatively important only at the higher temperatures (upper 50s C) where the filamentous algae give way to a unicellular form (*Synechococcus*). The role of the flexibacteria in the *Mastigocladus*-dominated mat community is not considered in this chapter.

the research reported in this chapter were:
1. To describe quantitatively and qualitatively the status and functional role of these unicellular heterotrophic bacterial groups as the algal-fly succession and later destruction process occurred, and
2. To evaluate, on a preliminary basis, certain parameters of the important bacterial groups that would permit their inclusion in a detailed predictive model of the algal-fly system (Wiegart 1975). These parameters include such values as maximum specific growth rate and generation time under field conditions.

METHODS

The Study Site

All of the studies reported in this chapter were carried out on the two board substrates described by Fraleigh (1971) and Wiegert and Mitchell (1973). These are two troughs, each 1.2 by 24 m (28 m^2 per trough) located in a small meadow (Serendepity) just south of Firehole Lake Drive in the Lower Geyser Basin of Yellowstone National Park. Water at 55-57 C and a pH of approximately 6 flowed onto the upper end of each board from a plastic pipe at 30 liters/min. Prior to the initiation of the described studies, the mat on each trough had been allowed to develop undisturbed for 1.5-2 years.

Succession

To study bacterial changes during succession, the algal-fly community was first removed from one of the two plywood troughs by scraping and scrubbing with a stiff push broom. After allowing the board to dry for several days, the flow of water was restarted. The second board was retained as a stable "control" community.

Stratified random samples of the mat community were taken; the location of each sample point was measured to the nearest centimeter. Six strata were delineated, each one being a section of board 1 x 4 meters. Usually five samples were taken randomly within each stratum using a brass cork borer 0.237 cm^2 in area. The borer containing mat plus water was emptied into a sterile Whirlpak bag and placed on ice for transport to the laboratory. No more than a 3-hr delay between sampling and processing was experienced. Instruments were flooded with alcohol and flamed between samples.

Aliquots of ground core or duplicate cores placed in tared aluminum weighing pans were dried, weighed, ashed at 500 C for 2 hr, and reweighed to determine total biomass and ash-free dry weight.

Both direct microscopic counts and dilution plate counts were made from the cores taken in the field. Cores were ground with a Teflon tissue homogenizer (in 1971-1972) or a Polytron homogenizer (in 1973), with grinding times and speeds predetermined to give the best dispersion and highest counts per sample. Dilutions for counting and plating were made with sterile spring water. Direct microscopic counts of unicellular bacteria wère made with Petroff-Hauser counting chambers and a Wild UV microscope, using acridine orange stain (Strügger 1948). Plate counts were made on 0.1% tryptone-0.1% yeast extract agar, nutrient agar, and thioglycollate agar. The highest colony counts and most consistent colony pigmentation were obtained from the dilute tryptone-yeast extract agar. This agar was then used for laboratory cultures of the bacterial isolates. Algae and filamentous bacteria did not grow on any agar used. A temperature range of 38-43 C was used for incubation of plates. Different bacterial colonies were quite easily distinguished on the dilution agar plates.

Pigmentation and colony morphology of 7-10 day colonies
were viewed under stereoscopic magnification. Duplicate
plates were counted and averaged to get the numbers of
each colony type. These were then designated as the dominant, abundant, present, and rare bacteria. Only aerobic
and facultative, heterotrophic bacteria were counted.

Periodic measurements were made of uptake of three
radiocarbon compounds, ^{14}C-u-glucose, ^{14}C-u-acetate, and
$^{14}CO_2$ ($Na^{14}CO_3$). Small cores of mat were placed in vials
and 1.0 ml of sterile spring water containing 1.0 µC of
the isotope was added. The vials were capped and replaced
for incubation in the stream at the site where the core
was taken. One-hour incubations with glucose and acetate
and one-half hour incubations with $Na^{14}CO_3$ were used. To
inactivate and preserve the sample for counting, 1.0 ml
of 3% formalin was added at the conclusion of incubation.

In the laboratory the cores and liquid were ground
as described above for dilution and direct counts. Then
0.1 ml of the ground, diluted material was pipeted onto
a 0.45 µm Millipore filter and washed with sterile spring
water. Filters were removed, dried, and placed in liquid
scintillation vials containing water, PPO, POPOP, and
toluene. Counts were made on both Beta-Mate and Packard
Tricarb Beckman scintillation counters.

Grazing and Decomposition Changes

Upon reaching a thickness where some flow diversion
and mat cooling occurs, grazing by flies and decomposition
of the alga-bacterial system begins. In 1972 we began a
series of field experiments designed to clarify the interacting roles of flies and bacteria in this process.

Light weight stainless steel strips 6-7 cm wide were
formed into squares approximately 20 cm on a side. Several
holes to accept No. 00 stoppers were drilled in the "upstream" and "downstream" sides. These diverters were

placed at selected sites in the mature mat on the south
(or control) board and driven down to form a tight seal
with the plywood substrate. The algal mat inside the diverter was removed. Subsequent regulation of the water
flow (and temperature) in the diverters was possible by
removing the desired number of upstream and downstream
stoppers.

To start an experiment a 20 x 20 cm square of algal
mat was cut out of a suitable part of the mat and placed
inside the diverter. These sections were carefully in-
spected to make sure they were free of fly eggs or larvae.
Initially only a small flow of water, just enough to keep
the mat warm and wet, was admitted to the diverters. Thus
the diverter-enclosed mat simulated the cool, ungrazed
mat. By placing a screen over the top of the diverter,
we were able to study the decomposition process both in
the presence and in the absence of flies. As a control
we used the naturally fly-free, hot, flowing areas of the
board. Thus, we had available:

1.   Hot, fly-free mat (control)
2.   Cooled, fly-free mat
3.   Cooled, grazed mat

Sampling was conducted on a daily basis from each of
these types of mat for 15 days. Measurements of algal
and bacterial numbers, biomass, and uptake of radiocarbon
were made using the methods described earlier. Following
day 9 in the grazed-mat diverters, the mat was (1) allowed
to continue being eaten until days 15-20, or (2) more
stoppers were removed, allowing hot water to flow through
the diverter, killing the larvae, initiating larval decomposition, and eventually resulting in regrowth of a fly-
free mat.

BACTERIAL TYPES AND INTERACTIONS

Bacterial Population Parameters

If the unicellular bacteria in the climax or steady-state algal mat are themselves in steady state (an assumption borne out by our preliminary counts), then the gain by inflow plus the growth rate in situ must balance losses by washout and mortality. An indirect measure of the necessary rate of growth is derived simply by measuring (or estimating) these gains and losses and summing them.

The inflow and washout rate of bacteria was measured simply from direct counts made on inflow and outflow water. No corrections were made for bacterial losses associated with large pieces of mat washing out of the system. Although this source of loss can be important under unusual conditions such as heavy rain or hailstorms (see Brock and Brock 1969), such conditions were not encountered during our sampling of the steady-state algal mat. Sources of mortality of unicellular bacteria by ingestion are minimal because fly larvae densities were low in the steady-state mat studied. These losses were estimated from data on fly densities and ingestion rates.

Direct measures of bacterial growth rate were obtained by first dividing a core of algal mat into its three most distinct layers; a thin orange to green surface layer including filamentous blue-green algal filaments, a thick (up to 2 cm) middle layer composed of blue-green filaments in a gelatinous matrix of colloidal silica, and a bottom layer of detritus, some blue-green filaments, or, at the higher temperatures nearer the source, flesh-colored filamentous bacteria.

After separation, the three layers were placed in sterile petri plates and washed repeatedly in sterile spring water. Small cores cut from these washed layers of mat were placed in 1-dram vials containing sterile spring water. Sterile cellulose strips were placed over the open ends of the vials and held tightly in place with rubber bands. These cores were taken to the field and suspended inverted in the position from which the original sample had been taken.

The idea behind this procedure was to reduce the bacterial density in each core below the steady-state density in the field and to replace the core in the field, exposing it to temperature, pH, and nutrient levels as close as possible to the natural condition during the measurement of bacterial growth response.

RESULTS AND DISCUSSION

Succession

Numbers and Biomass

During the initial stages of the algal-bacterial succession the water proceeds down the board as a thin sheet, cooling and rising in pH as a result of excess CO_2 degassing to the atmosphere. From an initial temperature of 55-56 C, the water cools to 30-40 C at the outflow, the precise temperature depending on the weather; cool and windy days cause a cooler outflow temperature as opposed to warm and calm days. The pH gradient is also affected by the weather and to some extent differs between day and night once an algal mat has become established. Generally the outflow pH range is from 8 to 9, an increase of 2-3 pH units from the value at the inflow. As the biomass increases, all or part of the mat usually floats and the water flows underneath. This causes a decrease in the rates of cooling and degassing and thus results in somewhat higher temperatures and pH values at the outflow. For a more detailed discussion of the interaction of physical factors with successional change, the reader should

216

consult Fraleigh (1971) or Fraleigh and Wiegert (1975).

Changes in algal biomass and the associated changes in bacterial numbers and density with succession are shown in Figure 7.1. These successional changes were measured on the north board in 1971.

Initially the algal biomass increased at an exponential rate that itself decreased as the maximum asymptotic limit was approached (Wiegert and Fraleigh 1972; Fraleigh

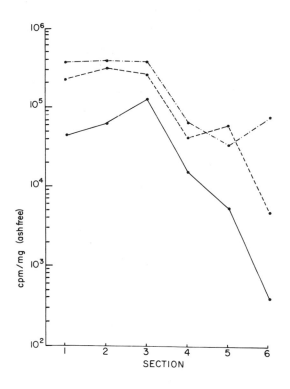

FIG. 7.1. Graph of successional change in algal bacterial biomass and bacterial numbers on the north board (1 x 24 m) during 1971 and 1973. Time is given in weeks from start of succession. The "mature" were taken from the undisturbed south board. This represented a state that could be assumed to exist from week 10 onward. Successional data from 1971 ended at week 5; 1973 data on numbers ended with week 7. (- · -), Glucose; (- - -), acetate; (——), NaHCO₃.

and Wiegert 1975). Standing crops on the order of 100 g
or more per m^2 are reached within a few weeks. The "mature"
samples came from the south or control board where the mat
had developed undisturbed for more than a year. As sug-
tested by Figure 7.1, however, this level of biomass could
probably be expected after 10 weeks or more of successional
growth.

The number of bacteria (direct counts) per square
meter also increased at a constant specific rate during
the early stages of succession. (The bacterial censuses
were repeated in 1973 with essentially similar results,
as shown by Figure 7.1.) The rate itself was similar to
the growth rate of the algae. Thus the bacterial density
per gram of mat rose only slightly and the increase in
bacteria during this period was directly related to the
increased biomass of mat. The bacterial numbers per square
meter soon leveled off, however, causing a steady decline
in the density of bacteria per unit biomass of mat as succession proceeded. Plate counts of bacteria showed the
same successional trends as did direct counts, but numbers
were on the average two to three orders of magnitude lower
for plate counts.

The unicellular bacteria measured by the plate and
direct count techniques are found singly and in clumps in
the interstitial water between algal filaments, on the
filaments themselves, and on other particules in the mat.
During the early stages of succession the productivity of
the algae in the first millimeter or two of the mature
mat surface that has direct exchange with the flowing water
is at a maximum for a given temperature, flow, and free
CO_2 concentration (Fraleigh and Wiegert 1975). Thus bacteria dependent on a source of organic matter produced
by the algae would be expected to increase directly with
an increase in the algae as long as the latter were photo-
synthesizing at or close to their maximum rate, and less

slowly than the algae as more and more of the algal mat
becomes suboptimum with respect to available resources,
and perhaps also in terms in pH and available oxygen.
According to Fraleigh and Wiegert (1975) the threshold
density at which algal growth is first limited is when the
mat is approximately 1-2 mm thick (roughly corresponding
to 10-20 g/m^2 ash free dry wt). This in turn suggests
that the downturn in rate of successional change of algae
(and presumably also bacteria) had already begun during
the first week.

Heterotrophic Metabolic Activity

After the successional sequence had continued for
approximately 1 month we measured the metabolic activity
of the mat in each section by using radiocarbon compounds
(Fig. 7.2). Uptake of all the nuclides followed similar
patterns, a high rate in the upper warmer sections of the
mat, decreasing with distance from the source after the
8 meter point. (Uptake of all materials except glucose
was depressed in the sparse, cool, almost moribund, mat
growing at that point.) We have no explanation other
than possible error for the apparent "order of magnitude"
increase in the uptake of glucose in this section. Only
a repeat experiment will be able to confirm or refute the
reality of this curve. Later in succession, when the
floating mat had extended downstream and insulated the
water, a viable mat developed on this lower section.

Uptake of $NaHCO_3$ in the dark vials was always at
least one (and sometimes almost three) orders of magnitude
lower than uptake in the light. Dark uptake was probably
caused in part by bacterial incorporation and in part by
absorption and/or physical entrapment. It was subtracted
from the light uptake values to give net uptake in the
light.

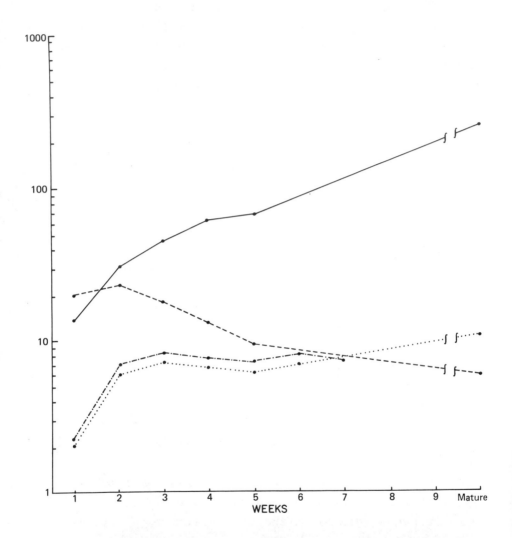

FIG. 7.2. Uptake of radiocarbon substrates by a thermal algal-bacterial mat down a temperature gradient. The mat developed in an artificial board trough 1 x 24 m in size. (——), Green algae x M^{-2}; (-–- -), number of bacteria/g x 10^{10}; (- · -), number of bacteria x M^{-2} x 10^{12} (1973); (· · ·), number of bacteria x M^{-2} x 10^{12} (1971).

The relative rate of primary production, measured by uptake of $NaHCO_3$, was maximal at about 8 meters, in a temperature of 50 C. This was approximately the temperature regime found to yield optimum biomass and photosynthesis in the thermal algae mats studied by Brock and Brock (1969). Without detailed quantitative knowledge of the free CO_2 and carbonate buffering system, uptake of $NaHCO_3$ provides only a relative measure of productivity. The values reported in Figure 7.2 were obtained under uniformly low conditions of free CO_2 because of the limited amount available in the incubation water used in the vial. Fraleigh and Wiegert (1975), using larger cores in ground glass weighing bottles, calculated that the available free CO_2 was on the order of 10 mg/liter. We considered their calculations to be valid for our experiments because both the size of the cores and the amount of incubation water in our experiments were reduced by relatively the same amount.

Dominant and Minor Colony Types

The sorting and classification of bacteria is difficult. Unicellular bacterial populations in the algal mat include a variety of gram-positive and gram-negative bacteria. More than 15 distinctly different and identifiable colony types were observed on dilution plates. Of the bacteria isolated and grown in axenic culture, eight were chromogenic with yellow, orange, red, and pink the dominant shades. Other types were cream, tan, brown, and cream sheen when grown on tryptone-yeast agar. There is a strong tendency for the dominant types in the algal mat to be either chromogens or spore formers. Gram-positive and gram-variable rods were the most common morphological types. Two spirochaete types were observed in samples from cooled mat. These organisms are only rarely observed using direct counting methods and have not been observed

Table 7.1

Some Morphological and Biochemical Characteristics of Bacterial Isolates from the Thermal Springs System[a]

Organism	Symbol	Gram reaction	Morphology	Motility	Endospores
Orange	O	+	Small rods	−	−
Pink	P	+	Paired short rods	+	−
Bacillus	Bac	+	Rods	−	+
White	W	+	Rods	−	+
Yellow	Y	−	Rods	+	−
Tan	T	+	Short rods	+	−
Cream	C	−	Small rods	+	−
Brown cotton	BC	+	Rods	−	+
Grey	G	+	Rods	−	+
Fried egg		−	Small rods	−	−
Cream swirl	Swirl	+	Pleomorphic	−	−
Pink	P	65% G+C	Long, thin rods	Opt. temp., 60 C	Gen. time, 50 min pH, 7.5−10

Table 7.1 (Continued)

Organism	Biochemical Tests								
	F	L	M	Gal	D	LM	N	H$_2$S	C
Orange	A-G-	A-G-	A-G-	A-G-	A-G-	-	-	+	-
Pink	A-G-	A-G-	A-G-	A-G-	A-G-	-	-	-	-
Bacillus	A+G-	A-G-	A+G-	A-G-	A+G-	red coag pep	+	+	-
White	A+G	A-G-	A+G-	A-G-	A+G-	red coag pep	+	+	-
Yellow	A-G-	A-G-	A-G-	A-G-	A-G-	-	-	+	-
Tan	A-G-	A-G-	A-G-	A-G-	A-G-	-	-	+	-
Cream	A-G-	A-G-	A-G-	A-G-	A-G-	red turbid	+	gas	+
Brown cotton	A-G-	A-G-	A-G-	A-G-	A-G-	alk-ali	-	+	-
Grey	A+G-	A+G-	A+G-	A+G-	A+G-	-	-	-	-
Fried egg	A-G-	A-G-	A-G-	A-G-	A-G-	red pep	-	+	+
Cream swirl	A-G								
Pink	-								

Growth on Different Media				Pigment formed at
Nut	Yeast	Ext.	Glut.	15% CO$_2$
-	7.8+	8.5+	8.5+	

223

Table 7.1 (Continued)

[a]F = Fructose; L = lactose; M = maltose; Gal = galactose; D = dextrose; LM = litmus milk; N = nitrate reduction; H_2S = hydrogen sulfide production; C = citrate utilization; Nut = nutrient agar; Yeast = yeast extract tryptone agar, pH 7.8; Ext = yeast extract tryptone agar, pH 8.5; Glut = glutamine, pH 8.5; + = positive test; - = negative test; A+ = acid production; G = gas production; red = milk reduced; pop = peptonization; coag = coagulation; turbid = turbidity increase.

in the plate culture. It is probable that the medium was
unsuitable for their growth and/or grinding caused lysis.
Table 7.1 provides limited data regarding the bacterial
isolates.

To document changes, each colony type was counted on
duplicate plates for each trough section. Dominance was
defined as the most plentiful colony type followed by
abundant, present, and minor types. Table 7.2 lists the
colony types (in order of relative abundance) found at each
successional stage and at maturity, classified by gradient
within the board trough as well as by time. Orange was
most abundant in 24 out of the 35 sets of samples, followed
by white (six), pink (three), and yellow and tan (one
each). Clearly the type of heterotroph represented by the
orange colony type was an important component of these
communities, although its dominant position based on the
plate counts must be accepted with caution when weighed
against the possible biases involved in this sort of assay.

In terms of the major dominant, orange, Table 7.2
provides no indication of any major shift with successional
time nor with distance from the source. The lesser domi-
nants, however, are not so consistent. Most of the pink
colonies were found in material from the upper, hotter
sections and appeared during the middle period of the successional sequence. No doubt other patterns could be detected by a more thorough restudy of the successional
process in these thermal mats, emphasizing more samples,
different media, and more frequent censusing.

Functional Associations

In microbial populations, as among the higher organisms, the organism present in the highest numbers or biomass may not actually be the dominant member of the system
in a functional sense. This is especially true in microbial

Table 7.2

Summary of the Dominant Bacterial Types on the Thermal Board Systems, Arranged by Section of Board (Thermal-Chemical Gradient) and by Successional Time

Board Section	Dominant bacterial types					
	Week 1	Week 2	Week 3	Week 4	Week 5	Mature
I	Orange White Yellow	Orange Red	Orange Pink	Orange Pink	White Yellow	White *Bacillus* Orange
II	Orange Cream-sheen	Pink Orange	Pink Orange	Orange "Red Pink	Yellow Orange Tan	Orange White Tan
III	Orange Cream-sheen White	Pink Yellow	Orange Pink	Orange Pink	White Tan	Orange White Tan
IV	Orange	Orange Red	Orange Pink	Orange Pink	White Tan	Orange White Tan
V	No data	Orange Red	Orange *Bacillus* Tan White Pink	Orange	Tan White	White- tan Orange
VI	Orange White	Orange Red	Orange	Orange Pink	Orange Yellow	White- tan Orange

populations where a particular species may be maintained at a low standing stock but in an extremely active metabolic stage by a predator. We have conducted some studies designed to show the type of interaction between bacteria and the major functional dominant organisms of these ecosystems; filamentous blue-green algae and ephydrid flies.

A *Bacillus* sp. was closely associated with living larvae and pupae retrieved from the algal mat. When incubated in sterile inflow water both larvae and pupae supported far more *Bacillus* than any other type of bacteria. However, orange and yellow colony types were quite capable of growing on the dead larvae and pupae in abundance. The P-type (pink) colonies were often isolated from among the filaments of the under portions of the mat, were present among the flesh-colored, filamentous bacterial mats of the upper temperature areas, and were always found in the inflow water. The Y-type (yellow) was abundant in fresh grazed mat, less abundant in fresh intact mat, and present in regrowth of a previously cooled-grazed mat.

Assignment of definite specific roles for these bacteria in the algal mat is not now possible. However, some forms, such as O and P, seemed to be closely associated with intact mat, possibly utilizing algal excretions as food sources. *Bacillus* sp. and Y probably assimilate animal protein when possible and W (white) may aid in the algal decomposition process. Normally, 90-95% of the bacterial populations consisted of five or six colony types; many of the other isolates appeared only occasionally.

Occasional colony forms appeared only under restricted conditions and therefore may eventually be assigned a functional role. For example, brown cotton (BC) appeared only during a larvae decomposition experiment; cream sheen (CS) grew in abundance only during the first week of succession; and spirochaetes were found only in silicated, cooled surface mat.

Grazing and Decomposition

Cooling and Solubilization of Algal Mat

An outstanding feature of the thermal algal-fly-bac-
terial ecosystem is the rapid reduction in structure,
stratification, and presumably metabolic activity following cooling and invasion by ephydrid flies. The sequence
of cooling of the algal mat, egg laying by adult flies,
and larval grazing takes less than 4 days. The pupal stages
begin at about the sixth day of larval life (Wiegert and
Mitchell 1973) and adults emerge 14 days after eggs are
first deposited in and on the cooled mat of filamentous
algae.

Grazing involves chemical and mechanical fragmentation
of the algal mat. Decomposition is the biological-chemical
transformation of materials into other forms; ultimately
organic nutrients are degraded to CO_2 and H_2O plus excreted
wastes under aerobic conditions. Decomposition can be
separated into three types or phases. Phase 1 constitutes
the rapid metabolic assimilation of dissolved organics,
such as sugars. Phases 2 and 3 involve the mechanical and
chemical breakdown of the more resistant insoluble materials, such as cellulose, hemicellulose, and lignin (Jewell
and McCarthy 1971).

The earlier studies with radiocarbon compounds showed
a rapid uptake of glucose and acetate by bacteria in the
intact mat. We subsequently tried to measure the rate of
cellulose breakdown by suspending strips of pure cellulose
(dialysis tubing) at various points on the board systems.
In all cases we found no weight loss or visual evidence of
decomposition of the cellulose strips. One major reason
for this may have been the complete absence of free-living
fungi in these warm to hot systems (Mike Tansey, pers.
comm.). Cooney and Emerson (1964) stated that no fungi
have been isolated from thermal springs. Sparrow (1968),

228

citing their results, was of the opinion that fungi are
present but undetected. If present, cellulose decomposers
must be scarce and relatively unimportant.

When subjected to grazing by flies, however, the mat
is "solubilized" quickly and that portion which does not
immediately wash downstream is further used until the dry
weight biomass is only a fraction of the original weight
of mat. In our diverter experiments we first let the fly
population multiply until the mat structure was completely
destroyed (solubilized) and the larvae were living in a
"soup" of algal cells and water. In one series of preliminary experiments the total biomass (dry weight) in
g/m^2 declined from 159 on day zero to 90 g on day 4 and
then 45 g on day 10; the last decrease was measured after
hot water had reentered the system and washed out the
majority of the solubilized material. These results were
similar to those Wiegert and Mitchell (1973) reported for
entire ecosystems in which virtually all the algae were
made available by shutting down the flow of hot water to
the mat.

One 15-day diverter study is summarized in Table 7.3.
During days 0-4 the covered and grazed diverters were
cooled and the latter was exposed to grazing and egg laying by adult flies. Extensive solubilization of the mat
in the grazed diverters occurred during days 5-9 as a result of larval grazing. Following the reestablishment of
some hot water flow to the grazed diverters on day 9, the
larvae were killed and mat reconsolidation and regrowth
began to occur rapidly. Unfortunately, the volume of hot
water flow was not substantial enough to remove the solubilized mat, possibly accounting for the failure to statistically demonstrate an effect of grazing on biomass.

The data in Table 7.3 are averages of the daily samples. Using these daily data as replicates within each
time interval, we performed two-way ANOVAs with replicates

Table 7.3

Means of Radionuclide Uptake, Cell Counts, and
Biomass from the Diverter Experiments on a Mature
Thermal Algal-Bacterial Mat on an Artificial Board
Trough (1 x 24 m)[a]

	Grams biomass (g/m^2)			NaHCO$_3$ (cpm)		
Day	Fly free	Covered	Grazed	Fly free	Covered	Grazed
0-4	179	226	203	2.00	1.72	2.26
5-9	186	192	235	3.24	0.89	1.59
10-15	224	242	160	0.43	2.08	1.81
	Acetate (cpm/g)			Glucose (cpm/g)		
0-4	3.86	3.64	3.65	17.87	10.39	13.09
5-9	6.39	11.13	3.75	20.28	13.51	21.52
10-15	4.74	5.57	4.78	16.77	22.24	11.62
	Plate counts (No./g x 10^8)			Direct counts (No./g x 10^{10})		
0-4	0.58	0.82	0.61	6.34	5.78	8.08
5-9	1.75	0.90	4.28	8.20	7.75	9.56
10-15	0.63	1.89	4.73	5.42	6.72	10.20

[a]Data are means of daily measurement taken during
the periods indicated. Biomass is measured as ash-free
dry weight.

after first ascertaining that the variances did not differ significantly at the $\alpha = 0.05$ level.

In only one instance (plate counts) were we able to show a highly significant effect (F = 7.1; df = 2; $\alpha < 0.01$) of time. Grazing by fly larvae was accompanied by a marked increase in viable bacterial cells or spores. This increase was apparently not diminished by the subsequent infusions of hot water. The fly-free and covered diverters had a highly significant lower mean value averaged over all times (F = 10.5; df = 2, 36; $\alpha < 0.01$). In addition, the lower mean for the initial time period in the grazed diverter (before larval grazing action became large) caused a highly significant interaction between time and treatment (F = 5.6; df = 4, 36; $\alpha < 0.01$).

The interaction of time and treatment as it affected biomass just barely missed significance at the $\alpha = 0.05$ level, undoubtedly because we adjusted the hot water flow in a manner that precluded the washout on day 9 of much of the already solubilized mat.

The effect of treatment on direct counts, showing generally higher values in the grazed diverters, also almost reached significance at the $\alpha = 0.05$ level (F = 2.4; df = 2, 36).

In general the experiments with the diverters have provided us with the suggestion of greater numbers of unicellular heterotrophic bacteria under a grazed and possibly under a cooled regime. Biomass decreased greatly during solubilization by grazing, and later when it was again subjected to a flow of hot water. The actual decrease in grams ash-free dry weight consumed by the fly larvae and bacteria may be more than one-half of the standing crop in 10 days.

The experimental design that we employed in this study shows promise. However, several changes that need to be made are (1) to obtain less variability in the control

samples by use of a hot flowing system in a diverter instead of reliance on the constancy of a section of the intact mat; (2) to take more replicate samples on a given
day instead of a single set on each of 15 days; (3) to
allow the experiment to run longer before the grazing is
terminated with the reentry of hot water; and (4) to insure a more uniform grazing perturbation by covering all
treatments with screen and introducing a known number of
flies into the cooled, grazed diverters. Our present knowledge of the bacterial-fly interaction as it is expressed
in the grazing and decomposition of algae is not sufficient
to separate the roles of the bacteria in these two processes.

The role in the decomposition process in the absence
of grazing by the fly larvae (cooled-uncovered mat) appears
a passive one. Without the intervention of the mechanical
and/or chemical perturbations induced through ingestion
by ephydrid larvae, the algal mat, even when cooled, re-
tains its structure and the bacterial populations expand
to utilize the available organics secreted by the algae.
In this respect the cooled portion of the algal mat resembles a "carcass" that has become available to "carrion"
insects by reason of cooling, not death. The analogy extends also to the relative importance of insects in the
breakdown process. Payne (1965) studied the role of insects versus microorganisms in the breakdown of baby pig
carcasses and found the insects vital to speedy decomposition.

The role of other microscopic animals (protozoa and
rotifers) may also be important in this grazing-decomposition sequence. Jewell and McCarthy (1971) found exclusion
of protozoa as well as bacteria to deterrent decomposition
of algal cells. Although protozoa are uncommon in the
intact mat at temperatures in the 40-50 C range, they increase rapidly in the cooled mat even though protected

232

from grazing (Wiegert, unpubl. data). Future refinements
in experiments to determine the role of the bacteria in
the grazing-decomposition process in the mats must consider the protozoa.

Decomposition of Fly Tissue

Fly mortality contributes small amounts of chitin to
the system. We have demonstrated the presence of chitin-
decomposing bacteria in the system by a simple in situ experiment with chitin contained in dialysis tubing. When
inoculated with spring water, chitin decomposed at a faster
rate under light (oxygenated) conditions than under dark
(anaerobic) conditions. However, the overall rate of disappearance is slow, as proved by the rather long (several
days to several weeks) retention time of pupal cases in
the mat community, particularly near the bottom of the mat
where photosynthesis is low or absent, presumably leading
to low or zero levels of oxygen.

The bulk of the material contributed by fly mortality
is protoplasm. Table 7.4 summarizes the results of bacterial growth in infusions made with fly larvae and pupae
in sterile spring water, inflow water, and outflow water.
In contrast to the results from the intact mat, plate counts
did not differ greatly from direct counts, suggesting that
most cells in these experiments are active metabolically.
Although populations of bacteria were generally higher when
nonsterile spring water was the inoculum, the total amounts
of material consumed were similar in all experiments (49%
for larvae, 53% for pupae). However larvae material supported three times more bacteria per gram of substrate remaining than did pupal material (3.6 versus 1.2).

Expressed as a weight-specific rate of decomposition,
fly protoplasm, both larval and pupal tissue, decomposed
at the rate of approximately 0.1 g/g-day. This rate is

Table 7.4

Bacterial Decomposition of Ephydrid Larvae and Pupae
in Both Inflow and Outflow Water[a]

Medium	Sample material	Bacterial cells (per g x 10^12)		g/material consumed	Percent consumed	Rate of Decomposition (g/g-day)
		Plate	Direct			
Sterile H_2O	Larvae	0.157	0.200	0.0093	45.8	0.087
Inflow H_2O	Larvae	0.240	0.552	0.0098	48.3	0.094
Outflow H_2O	Larvae	0.547	0.326	0.0105	51.7	0.104
Means		0.315	0.359	0.0099	48.6	0.095
Sterile	Pupae	0.030	0.182	0.0309	56.2	0.117
Inflow	Pupae	0.101	0.142	0.0297	54.0	0.111
Outflow	Pupae	0.146	0.048	0.0273	49.6	0.098
Means		0.092	0.124	0.0293	53.3	0.109

[a]All weights of animal material are on an ash-free dry weight basis. Suspensions were made in sterile spring water, unfiltered water from the inflow to the boards, and outflow water. Experiments were run for 1 week, starting with 0.055 g of dry pupae or 0.0203 g of dry larvae.

less than half as great as the maximum specific growth
rate of the fly populations (0.25 g/g; see Wiegert 1975).
The measured decomposition rate may have been because of
the lag necessary before the bacterial populations entered
the log or exponential phase of growth.

We regard the population density censused at the termination of the experiment as an asymptotic equilibrium
density for the organisms utilizing this rich infusion of
animal protein under conditions of no washout by flowing
water. The converse hypothesis, that the bacterial population is still growing exponentially after 1 week, can be
discarded on the basis of two independent calculations.

First, the specific growth rate computed on the assumption of continuous exponential growth for 7 days would
be unreasonably low for organisms growing in such a rich
infusion of organic material at 38-43 C. The inoculum
bacterial content ranged from 0 (sterile spring H_2O) through
0.1 ml (inflow water) to 25 x 10^4/10 ml (outflow water).
In separate experiments we found the mean density of bacterial cells per larva or pupa to be 7 x 10^5. The number
of larvae or pupae used in the growth experiment was about
30. Thus the number of cells initially ranged from 21 x
10^6 (sterile inflow) to 21.3 x 10^6. The maximum final
number in Table 7.4 was calculated as 5.8 x 10^9 cells and
the minimum final number was 0.72 x 10^9. Thus the specific
growth rate, assuming constant log phase of growth, ranged
from 0.5 to 0.8 cells/cell-day, i.e., only slightly exceeding one doubling per day at maximum. Clearly, as a
maximum estimate this is unreasonably low for organisms
cultured in such a rich infusion at these temperatures.

A second analysis of the data of Table 7.4 showed
the maximum weight of cells produced per gram of substrate
utilized to be 0.015, or 1.5% (using a measured mean volume
per cell of 0.13 μm^3 and assuming dry weight equal to 20%
of wet weight, sp. g. = 1.0). This is much lower than the

value found by Payne (1970) for exponentially growing
cells in a rich medium.

Two independent calculations suggest that the cells
in these experiments were at some equilibrium density at
the conclusion of the experiment and were operating under
limiting, nonoptimal conditions of crowding. Thus the
decomposition rates computed from the data of Table 7.4
should approximate the values expected under field conditions, where bacterial turnover times would be considerably longer than the minimum turnover time that the populations are capable, physiologically, of achieving.

Field Growth Rates

In the intact steady-state mat bacterial growth rates
may be measured in two different ways: (1) the losses from
washout and predation are compared with inflow to calculate
the number of cells that must be produced in order to maintain the steady state; (2) the steady-state turnover may
be estimated by reducing the bacterial population in cores
of the mat by a small amount and measuring the rate at
which the density returns to the equilibrium level. Both
of these approaches involve technical difficulties and
require a number of assumptions; therefore, our results
here must be considered preliminary. However, the construction of a predictive model of the thermal ecosystem
incorporating the unicellular bacteria required an estimate
of both maximum specific rate of growth and realized rate
of growth in the field. An attempt to obtain such an estimate in two independent ways is described.

The number of cells (plate counts) in the stream inflow water was low and relatively consistent (70-240 cells/
liter), with a mean equal to 100, mainly the white (W) and
pink (P) forms. Effluent waters of mature mat contained
an average of 2.5×10^6 cells/liter (direct counts) with

a high reading of 6.4 x 10^6. As expected, after flowing
over and around the intact mat the colony types (from
plate counts) were more varied, with Y, O, W, P, and *Bacillus* types predominating, not necessarily in that order.
Using the estimate of 11 x 10^6 unicellular bacteria per
square meter on the mature mat (Fig. 7.1), we computed the
total bacterial population on the board (24 m^2) to be 26.4
x 10^{13}. A mean loss of 25 x 10^6 cell/liter was used.
Multiplying by the daily flow (73,000 liters) gave a net
washout loss from the board system of 1.82 x 10^{12} bacterial
cells per day. From data on predation loss of larvae from
the board (Kuenzel and Wiegert 1973, and unpublished data),
we calculate that we must correct for the bacteria in these
larvae by adding to the washout an additional daily loss
of 3 x 10^9 bacteria. Then the total daily loss of 1.823
x 10^{12} represents 0.69% of the standing crop. Using the
maximum value for the outflow counts (64 x 10^6) raises this
estimate to only 1.8%. Obviously, the constant losses from
the system form only a small fraction of the standing stock
and their replacement uses only a minor fraction of the
growth potential. To replace this loss the bacterial population would need a realized specific rate of growth a-
mounting to only $\log_e$ 1.018 or 0.018 cells/cell-day.

We were able to demonstrate with the dialysis tubing
experiments that this realized specific rate of growth was
well within the field-measured maximum. Plate counts were
made of bacteria from cores washed with sterile water and
then incubated in flowing water in the field behind dialysis tubing (Table 7.5). In the surface and middle areas
of the mat the specific rate of growth during the first
24 hr reached $\log_e$ (10.4/1.7) = 1.8 cell/cell-day. In the
bottom layer the cells remaining after the washing procedure at time zero decreased during the first 24 hr but increased at a rate of 0.65 cell/cell-day during the following 48-hr period. The initial decline could have been

Table 7.5

Changes with Time in Plate Counts of Bacteria From
Samples of Mat at Surface, Middle, and Bottom Layers
of Mature Thermal Algal Bacterial Mat[2]

| | Time | | |
	0	24 hr	72 hr
Surface	8.9	26.0	53.5
Middle	1.7	10.4	11.2
Bottom	65.8	6.0	22.2

[a]All counts are means of two replicates and represent
numbers of bacteria per square meter x 10^8. Zero time
begun with washed cores (see text for discussion).

caused by the difference between the normal dark, low-
oxygen environment of the bottom part of thick mat and the
lighted aerobic conditions of incubation. The latter were
designed to simulate conditions in the surface and middle
layers of the mat. In the latter, at least, the data of
Table 7.5 suggest a maximum rate of increase far above the
small increment necessary to maintain the heterotrophic
unicellular bacterial populations of the thermal blue-green
algae-ephydrid fly community.

SUMMARY

The bacterial populations of these thermal algal mat
ecosystems are diverse in both species and functional types
and are likely to be associated with certain types of algal
mat. In the mature system the bacterial growth rate potential is higher than is required for their replacement.
During succession bacterial populations increase rapidly
in numbers/square meter but decrease in numbers/g AFD wt

after early successional stages.

Aerobic bacteria attacked the dead larvae and pupae and chitinous bacteria decomposed their chitin exoskeletons within days under light growth conditions. However, the role of bacteria in the decomposition of the algal mat may be insignificant in the absence of mechanical maceration of the mat by fly larvae. In the presence of feeding larvae, the mat is readily decomposed and the bacterial populations increase in numbers/g AFD wt.

Specific functional roles cannot be assigned to dominant bacterial types, but these data suggest that microbial activities in the thermal algal mat system are not totally passive. The mat is constantly being cooled, invaded by fly larvae, macerated, decomposed by bacteria, and then recycled through the successional stages. Bacteria are thus present and involved in every stage of the thermal system. Different bacterial types are present at each section of the gradient, but some dominant types are capable of growth at every section of the gradient.

ACKNOWLEDGMENTS

Members of the departments of Microbiology and Plant Pathology, Montana State University, generously provided assistance, advice, and the use of equipment and materials. We particularly appreciate the use of liquid scintillation counting equipment in the laboratories of Gary Strobel, MSU, and T. D. Brock, University of Wisconsin. Field and laboratory technical assistance was provided by Linda Young and Ralph Bonham. We thank the staff of Yellowstone National Park for permission to conduct the field work and assistance with administrative problems. Support for this study was provided by NSF Grants GB 7683 and GB 21255.

REFERENCES

Bott, T. L., and Brock, T. D. 1969. Bacterial growth rates above 90 C in Yellowstone Hot Springs. *Science* *164*, 1411-1412.

Brock, T. D. 1967. Life at high temperatures. *Science* *158*, 1012-1019.

Brock, T. D. 1970. High temperature systems. *Ann. Rev. Ecol. Syst. 1*, 191-220.

Brock, T. D., and Brock, M. L. 1969. Recovery of a hot spring community from a catastrophe. *J. Phycol. 5*(1), 75-77.

Cooney, D. G., and Emerson, R. 1964. *Thermophilic Fungi*. Freeman, San Francisco, California.

Fraleigh, P. C. 1971. Ecological succession in an aquatic microcosm and a thermal spring. Ph.D. thesis, University of Georgia, Athens.

Fraleigh, P. C., and Wiegert, R. G. 1975. A model explaning successional changes in standing crop of thermal blue-green algae. *Ecology 56*(3), 656-664.

Jewell, W. J., and McCarthy, P. L. 1971. Aerobic decomposition of algae. *Environ. Sci. Tech. 5*, 1023-1031.

Kuenzel, W. J., and Wiegert, R. G. 1973. Energetics of a Spotted Sandpiper feeding on brine fly larvae (*Paracoenia*; Diptera; Ephydridae) in a thermal spring community. *The Wilson Bull. 85*(4), 473-476.

Payne, J. A. 1965. A summer carrion study of the baby pig *Sus scrofa* Linnaeus. *Ecology 46*(5), 592-602.

Payne, W. J. 1970. Energy yields and growth of heterotrophs. *Ann. Rev. Microbiol. 24*, 17-52.

Sparrow, F. K. Jr. 1968. Ecology of freshwater fungi. In *The Fungi*, Vol. III. G. C. Ainsworth and A. S. Sussman (Eds.) Academic, New York, pp. 41-93.

Strügger, S. 1948. Fluorescence microscope examination of bacteria in soil. *Can. J. Res. Sci.*, C *26*, 188-193.

Wiegert, R. G. 1975. Simulation modeling of the algal-fly components of a thermal ecosystem: effects of spaital heterogeneity, time delays and model condensation.

In *Systems Analysis and Simulation in Ecology,* Vol. 3 B. C. Patten (Ed.). Academic, New York, pp. 157–181.

Wiegert, R. G., and Fraleigh, P. C. 1972. Ecology of Yellowstone thermal effluent systems: Net primary production and species diversity of a successional blue-green algal mat. *Limnol. Oceanog. 17*(2), 215–228.

Wiegert, R. G., and Mitchell, R. 1973. Ecology of Yellowstone thermal effluent systems: Intersects of blue-green algae, grazing flies (*Paracoenia*, Ephydridae) and water mites (*Partnuniella*, Hydrachnellae). *Hydrobiologia 41*(2), 251–271.

Chapter 8

LITTERS AND SOILS AS FRESHWATER ECOSYSTEMS

Stuart S. Bamforth

Newcomb College
Tulane University
New Orleans, Louisiana

CONTENTS

ABSTRACT

Litters and soils constitute a special freshwater ecosystem where the water is held in surface films and pore
spaces. These microhabitats contain large amounts of organic matter, and their volumes depend upon the fluctuating
conditions of precipitation and evaporation.

Although several hundred species of limnetic microorganisms can be found in the litter-soil system, most
species, with the exception of the higher fungi, must be
considered transients rather than residents. The presence
of these "aquatic" species, however, indicates that the
moisture fluctuations are not extreme.

Although large numbers of algae may be found at the
soil surface, their role is poorly understood. Higher
plants dominate the soil ecosystem by furnishing organic
matter, especially celluloses and lignins. The fungi have
diversified in this habitat by exploiting these compounds
and are therefore the primary decomposers. Bacteria compete with and supplement the fungi, and bacterial activities are enhanced by phagotrophic protozoa.

Litter-soil ecosystems furnish colonizing species to
organically enriched fresh waters. Hence, an abundance
of "soil species" servces as an indicator that the initial
stages of decomposition are occurring.

INTRODUCTION

Although the term "freshwater" usually calls to mind
ponds, lakes, and streams, the interstitial ecosystems of
moist vegetation, plant litters, and soils must be included
because these habitats have been colonized by limnetic

microorganisms and meiofauna, and an ecological continuum extends from the aquatic vegetation of lakes through bryophyte vegetation to plant litters and soils.

The microbiota typical of the litter-soil ecosystem appears in other freshwater systems during the initial stages of organic enrichment; hence, consideration of this ecosystem provides perspectives in studies of pollution and sewage disposal. Conversely, the latter studies aid in understanding the litter-soil ecosystem.

The physical structure of the soil-litter ecosystem disperses the biota into discontinuous microhabitats and precludes the usual methods of limnetic study. Quantitative and qualitative data, obtained by examining half a gram or more of litter or soil material, provide important insights into soil life, but these standard approaches blur the many microhabitats that exist, often discontinuously, in soil. New methods, such as the ampulla diver technique used in microrespirometry (Stout 1973) and scanning electron microscopy (Gray 1967), are needed to examine the microhabitats in order to elucidate their structure and function. In spite of these criticisms, the "gross" methods of soil biology, correlated with studies on heterotrophic succession in fresh waters, have enhanced our knowledge of both soil and freshwater ecosystems.

DISCUSSION

The Litter-Soil Environment

Litters and soils consist of irregular surfaces and interstices that hold small volumes of water in surface films and pore spaces. Each of these films constitutes a microhabitat that may contain several microorganisms. Sometimes these tiny ecosystems connect when moisture is abundant, or if the milieu is displaced by animal movement (e.g., burrowing), or by bacterial action, that disintegrates the humus-separating particles.

LITTER-SOIL ECOSYSTEMS

Most litters and soils are subject to moisture fluctuations. Evaporation will reduce the size of the aquatic system, breaking it up into discontinuous microhabitats; and the loss of water increases the concentration of salts, thereby producing osmotic stress. Precipitation causes the opposite effect but may introduce anaerobic conditions by filling up the air spaces between leaf and soil particles (waterlogging). The drainage patterns of interstitial habitats may favor one or the other of these phenomena and thereby can affect the organisms present; e.g., waterlogging excludes most fungi.

The physical structure and chemical nature (organic constituents and clays) of litters and soils protects organisms against desiccation because evaporation is retarded, allowing the organisms time to adjust to drying. An example of this may be seen in the evolution of over a dozen edaphic ciliate species, but only one has been able to successfully colonize the more exposed surfaces of leaves.

Litter-soil ecosystems are dominated by higher plants because they furnish leachates and decaying organic matter (which decomposes further, adding a greater amount of nutrients to soil water than in many limnetic systems). This decomposition, together with root respiration, produces a higher concentration of carbon dioxide in water films. Thus, these films constitute a detritus-based ecosystem, and the microbial population is greatly influenced by the type of flora above the ground.

Microorganisms invading the litter-soil milieu not only have to adapt to moisture fluctuations and high organic matter, but also to the shallowness of the moisture films that presents a large "benthic" surface in relation to the volume of water; hence many organisms grow along the substrate. Not surprisingly, many nonmotile unicells remain attached to each other after cell division, seen

247

in bacterial colonies adhering to soil particles in agar
film slides and in scanning electron microscope photo-
graphs. Growth in a linear direction results in filaments,
which can bridge the distance between soil particles and
exploit new resources; hence it is not unexpected to see
many filamentous species among soil algae, the evolution
of actinomycetes among bacteria, and prevalence of the
filamentous growth habit in fungi. The mycorhizal association of basidiomycetes and the roots of most seed plants
emphasize the extension advantage of the filamentous habit
in procuring nutrients. The secretions of microorganisms
for substrate attachment can also bind soil particles together, hence the microbial community can influence soil
structure (Cameron and Devaney 1970; Baily *et al.* 1973).

Another spatial factor besides the microhabitat of
the water film is gradients of moisture and organic matter.
Deeper layers are more moist than upper layers and support
different species (Schönborn 1964; Bamforth 1970). Decomposing organic particles, such as fecal pellets, can create
chemical gradients that enhance microorganisms. A very
important gradient is the rhizosphere surrounding the roots
of plants. These regions support large numbers of microorganisms, and an exchange of nutrients exists between the
roots and microbiota (Darbyshire and Greaves 1973; Nikolyuk
and Geltzer 1972).

The Litter-Soil Microbiota

Several hundred species of freshwater microorganisms
have been reported from litters and soils, but in view of
the exacting requirements of these ecosystems, most must
be considered transients rather than residents. The presence of these less edaphic species, however, indicates
that a litter or soil contains a large amount of moisture
and a certain degree of stability, i.e., water fluctuations
will be less extreme.

LITTER-SOIL ECOSYSTEMS

From the standpoint of energy flow, the soil is a heterotrophic ecosystem: Only the upper few millimeters of soil can admit light, but here large populations of algae can be found. Cells may be washed down into the lower levels, but show little, if any, growth there. Many algae occur accidentally in the soil but several (e.g., many chlorococcalean algae, such as *Bracteacoccus* spp., *Protosiphon* and *Botrydium* spp., *Hanzschia amphioxys*) occur essentially only as soil organisms, and six genera contain as many terrestrial as freshwater species (Shields and Durrell 1964). Blue-green algae contribute to nitrogen fixation in tropical soils and to binding of surface particles in arid soils. Parker and Bold (1961) showed that in 143 two-membered cultures of soil isolates, positive interactions occurred between the heterotroph organism and the algal member, but the role of algae in soils and their relation to other soil microorganisms remains poorly understood.

The filamentous growth habit of fungi enables them to enter the cell walls of plant litters and better attack complex compounds, such as celluloses and lignins, whereas cellulolytic bacteria can only erode the surfaces of plant tissues. Consequently the fungi may be considered the primary decomposers in most litter-soil ecosystems. Their biochemical abilities parallel phylogeny: Decomposition of chitin and keratin is common among the aquatic phycomycetes, whereas cellulose decomposition is common among the edaphic ascomycetes and basidiomycetes. Lignin degradation has been developed among the more advanced families of the basidiomycetes, some of which have developed mycorhizal relations with roots of most seed plants (Garrett 1963).

The bacterial community encompasses a wide variety of heterotrophs and chemotrophs (e.g., *Nitrosomonas*). From the standpoint of ecosystem function, bacteria include two groups: a stable autochthonous population, containing such genera as *Arthrobacter* and *Nocardia*, which is little influenced by added organic matter; and a numerically fluctuating zymogenous population, made up largely of Pseudomonadaceae, Enterobacteraceae, and many actinomycetes, which respond rapidly to additions of readily available nutrients, such as the amino acids and sugars of fresh plant residues or animal bodies.

Phagotrophic microorganisms, the protozoa, are usually smaller than their limnetic counterparts, and their motile patterns emphasize interface locomotion to exploit their food on the benthic surface.

Among flagellated protozoa, soil species belong mainly to the genera *Bodo, Cercobodo,* and *Oicomonas.* These organisms often tend to glide along substrates, and some species of the latter two genera show amoeboid tendencies, which are further developed in *Mastigamoeba.* Predominantly amoeboid genera are usually small: *Hartmanella, Naegleria, Nuclearia, Thecamoeba, Vahlkampfia. Biomyxa* is a larger, more spreading organism, whereas the enigmatic Mycetozoa (slime molds) have evolved to grow and flow over decaying wood and vegetation.

The shell-bearing amoebae, or testacea, can be arranged into an ecological series showing influence of thickness of water film and resistance to desiccation (Bonnet 1964). Flattened *Arcella*-like and vaulted *Nebela-Euglypha* genera are found in aquatic vegetation and moist forest litters. Spherical (globose) shells, e.g., *Phryganella*, are more typical of soils, while such wedge-shaped genera as *Trinema, Corythion,* and *Centropyxis* are found in soils and drier litters where the water film is thinner and more transitory.

Soil ciliates typically display flattened bodies and crawling habits: *Colpoda* (the overwhelmingly prevalent genus), *Chilodonella, Leptopharynx,* and hypotrichs. With the exception of the tiny *Vorticella microstoma*, which can readily detach itself and swim, attached ciliates (such as peritrichs and *Stentor*) are rare and their ecological equivalent may be found in bdelloid rotifers (e.g., *Philodina*), which creep rather than swim and are very resistant to desiccation. Such planktonic species as *Halteria* and *Blepharisma*, among ciliates, and monogont rotifers are present only under abundant moisture conditions.

Nutritionally, two types of protozoa are found in litter-soil ecosystems: naked, fast-growing microphagous bacteria feeders (flagellates, amoebae, and ciliates), and slow-growing testacea, which are dependent upon humus and similar materials (Stout 1965).

The testacea are a polyphyletic group that can apparently live on highly resistant plant materials, such as lignin (Schönborn 1965). Their generation time may be a week or longer (Lousier 1974) and they are most abundant in regions of low decomposition—coniferous forests and tundras.

Microphagous protozoa feed primarily on zymogenous bacteria (Stout 1973) but must be considered more than predators. Bacteria accumulate nutrients in their bodies and recycle nutrients only through autolysis or after being ingested by small animals (e.g., protozoa, nematodes), which in turn excrete the nutrients in a soluble form that can be used by the plant community. Studies by Nikolyuk and Geltzer (1972) have shown that the activities of soil bacteria are enhanced when protozoa are present. These workers also found that the same antibiotics secreted by fungi and actinomycetes against bacteria also antagonize protozoa.

Function of the Litter-Soil Ecosystem

The functioning of the litter-soil ecosystem may be considered in three stages (Alexander 1961): decomposition

251

of litter, synthesis of microbial protoplasm, and excretion of metabolic end products. Fungi and bacteria colonize the litter, and the latter also colonize dying fungal hyphae as decomposition proceeds. Protozoa and other small animals then feed on the bacteria (Tribe 1961). The importance of the three stages is shown in Jenkinson's (1971) radiocarbon enrichment studies, where 60% of the organic matter of decaying rye roots was decomposed in the first year but only 10% was converted into microbial biomass. The rates of loss in succeeding years were greater in the litter than in the microbial biomass. This resistance to organic loss by the microbial community, along with the 50% difference between litter and microbiota the first year, and the previously cited studies of Nikolyuk and Geltzer (1972) indicate a frequent turnover of microbial biomass.

Decomposition in Soils and Waters

Fresh waters differ from litter-soil ecosystems in the greater permanence of water and the greater proportion of water to organic substrate, thereby favoring a greater role of bacteria (rather than fungi) in decomposition activities (especially cellulose). The bacteria and phagotrophic protozoa are the prominent soil organisms that first appear in organically enriched fresh waters. Their ubiquity in the soil and their facile transport through air (Maguire 1963) almost insure their invasion of fresh waters. Their rapid encystment-excystment abilities, necessary for survival in soil life, enable them to quickly exploit fresh waters when stimulated by the presence of large quantities of organic material.

Enriched fresh waters, e.g., hay infusion (e.g., Woodruff 1912), lake and stream pollution, or a sewage disposal plant, follow a similar pattern, which has been simulated by Bick (1973) in laboratory aquaria. Bacterial populations grow rapidly and are accompanied a day later by a population of small zooflagellates, especially *Oicomonas* and *Bodo*, and the ciliate genus *Colpoda*, the dominant protozoa in soils. As the organic matter continues decaying, the flagellates and *Colpoda* are displaced by other ciliates, mainly hymenostomes, but among these are several more soil species, such as *Chilodonella uncinata, Leptopharynx sphagnetorum,* and *Vorticella microstoma.* Later these are succeeded by hypotrichs and peritrichs. If this succession reverts back to the initial stages, the soil species, especially *V. microstoma*, reappear and dominate.

Most of the hymenostomes in the second stage of the succession described in the previous paragraph are at least as capable of exploiting the rich bacterial milieu as are the soil ciliates, but they lack the encystment-excystment mechanisms to extend their distribution into the organically rich litter-soil ecosystem.

In other experiments, Bick (1973) found that *Colpoda* continued its dominance if a high organic content was maintained in the medium. He also heated some cultures to 50 C to eliminate nonsoil species from the medium and found that *Colpoda* was succeeded by the other soil species when the organic matter began to decompose. Studies on vegetation litters and soils support Bick's conclusion that *Colpoda* can colonize habitats where extreme environmental conditions exclude competitors and predators. Numerically, 50-95% of the ciliates in soils belong to the genus *Colpoda* (Bamforth 1973), the high values correlating with drier and colder habitats. Common soil species are *Colpoda steini* and *C. maupasi*. The latter has been reported from the Namib Desert (Dietz-Elbraechter 1973), where noon temperatures can reach 60 C and the only other microorganism is a blue-green alga. *Colpoda maupasi* survived experimentally produced extremes of temperature, desiccation, anoxia, and shortwave radiation (Losina-Losinsky 1969). A third species, *C. cucullus*, is also found in soils but more abundantly on the more exposed surfaces of living leaves, where it dominates the entire protozoan population (Mueller and Mueller 1970; Bamforth 1971).

Thus, a litter-soil ecosystem exemplifies the first stage of decomposition because of the continued input of organic matter by the litter despite interruptions of drought. This extreme habitat has selected certain micro-organisms; visually, these can be seen among the protozoa—soil flagellates and *Colpoda*. As organic matter migrates from upper to lower layers it becomes more degraded and the protozoan population diversifies to include more complex ciliates, e.g., *Leptopharynx sphagenetorum, Vorticella microstoma,* and hypotrichs. Where moisture conditions are more stable, such aquatic protozoa as *Heliozoa* and *Halteria* may appear. This succession may be seen in a culture where a bit of litter or soil is placed in water over nonnutrient agar in a petri dish and observed for several weeks.

The preceding discussion explains the presence of "soil organisms" in initial stages of organically enriched waters. If abundant, the easily identified soil protozoa can serve as indicators of "polysaprobity" in the saprobic system and hence can be used in evaluating conditions of pollution and sewage disposal. For example, prominence of *Vorticella microstoma* and *Chilodonella uncinata* in a sewage disposal plant is associated with a high biochemical oxygen demand (BOD) (Curds 1969) and indicates that the operation of the plant must be adjusted to achieve better efficiency.

REFERENCES

Alexander, M. 1961. *Introduction to Soil Microbiology.* Wiley, New York, 472 p.

Bailey, D., Mazurak, A. P., and Rosowski, J. R. 1973. Aggregation of soil particles by algae. *J. Phycol. 9*, 99-101.

Bamforth, S. S. 1970. Distribution of ciliates in deciduous litters. *J. Protozool. 17* (Suppl.), 15.

Bamforth, S. S. 1971. Population dynamics of leaf-inhabiting protozoa. *J. Protozool. 18* (Suppl.), 75.

Bamforth, S. S. 1973. Population dynamics of soil and vegetation protozoa. *Am. Zool. 13*, 171-176.

Bick, H. 1973. Population dynamics of protozoa associated with the decay of organic materials in fresh water. *Am. Zool. 13*, 149-160.

Bonnet, L. 1964. Le peuplement thecamoebien des sols. *Rev. Ecol. Biol. du Sol. 2*, 123-408.

Cameron, R. E., and Devaney, J. R. 1970. Antarctic soil algal crusts; scanning electron and optical microscope study. *Trans. Am. Micros. Soc. 89*, 264-273.

Curds, C. R. 1969. An illustrated key to the British freshwater ciliated protozoa commonly found in activated sludge. Water Pollution Research Technical Paper No. 12. H. M. Stationery Office, London, 90 p.

Darbyshire, J. F., and Greaves, M. P. 1973. Bacteria and protozoa in the rhizosphere. *Pest. Sci. 4*, 349-360.

Dietz-Elbraechter, G. 1973. On the occurrence of ciliates under quarcite stones with "window-algae" in vegetationless regions of the Namib Desert. In *Progress in Protozoology*. Universite de Clermont, Clermont-Ferrand, p. 114.

Garrett, S. D. 1963. *Soil Fungi and Soil Fertility*. Macmillan, New York, 165 p.

Gray, T. R. G. 1967. Stereoscan electron microscopy of soil microorganisms. *Science 155*, 1668-1670.

Jenkinson, D. S. 1971. Studies on the decomposition of C^{14} labeled organic matter in soil. *Soil Sci. 111*, 64-70.

Losina-Losinsky, L. K. 1969. Behaviour of ciliates in imitated Martian conditions. In *Progress in Protozoology*, Int. Conf. Protozool. Nauka, Leningrad, p. 181-182.

Lousier, J. D. 1974. Effects of experimental soil moisture fluctuations on turnover rates of testacea. *Soil Biol. Biochem. 6*, 19-26.

Maguire, B. 1963. The passive dispersal of small aquatic organisms and their colonization of isolated bodies of water. *Ecol. Monogr. 33*, 161-185.

Mueller, J. A., and Mueller, W. P. 1970. *Colpoda cucullus*, a terrestrial aquatic. *Am. Midl. Nat. 83*, 1-12.

Nikolyuk, V. F., and Geltzer, J. G. 1972. Pochvennye Prosteyshie CCCP (Soil Protozoa of the USSR) Tashkent, USSR: Academy of Sciences. 310 p. (In Russian)

Parker, B. C., and Bold, H. C. 1961. Biotic relationships between soil algae and other microorganisms. *Am. J. Bot. 48*, 185-197.

Schonborn, W. 1964. Lebensformtypen und Lebensraumwechsel der Testaceen. *Limnologica 2*, 321-335.

Schonborn, W. 1965. Untersuchungen über die Ernahrung bodenbewohnder Testaceen. *Pedobiologia 5*, 205-210.

Shields, L. M., and Durrell, L. W. 1964. Algae in relation to soil fertility. *Bot. Rev. 30*, 92-128.

Stout, J. D. 1965. The relation between protozoan populations and biological activity in soils. In *Progress in Protozoology,* Excerpta Med. Int. Cong. Ser. No. 91, 2nd Int. Conf. Protozool., London, July-August 1965, p. 119.

Stout, J. D. 1973. The relationship between protozoan populations and biological activity in soils. *Am. Zool. 13*, 193-201.

Tribe, H. T. 1961. Microbiology of cellulose decomposition in soil. *Soil Sci. 92*, 61-77.

Woodruff, L. L. 1912. Observations on the origin and sequence of the protozoan fauna of hay infusions. *J. Exp. Zool. 12*, 205-264.

Chapter 9

FACTORS AFFECTING THE NUMBER OF
SPECIES IN FRESHWATER PROTOZOAN COMMUNITIES

John Cairns, Jr.

and

William H. Yongue, Jr.

*Biology Department and Center for Environmental Studies
Virginia Polytechnic Institute and State University
Blacksburg, Virginia*

CONTENTS

ABSTRACT

Although protozoans are affected by changes in the
chemical, physical, and biological environment in the same
general ways as other organisms, their probable cosmopolitan distribution is shared with only a few other groups.
One might reasonably assume that cosmopolitan distribution
has resulted in maximum development of diversity in all
types of habitats since essentially all species are available as potential colonizers. Therefore, all habitats
should be maximally partitioned and the stable number of
species for each represents the maximum diversification
possible. If the distribution of species is as general
as some evidence indicates, this coupled with the ability
of many species to encyst and excyst at appropriate times
provides an opportunity for interspecific relationships
that are not possible for many other groups of organisms.
Thus species with similar environmental requirements might
occur together anywhere in the world as frequently as appropriate ecological conditions materialize. Under these
conditions, complex relationships might develop that would
produce a variety of interactions between species, including interdependence and exclusion. Thus, one might view
an aggregation of protozoan species as either (1) a structured community with a complex series of interlocking
cause-effect pathways, or (2) an assortment of species
present because their arrival or excystment coincided with
acceptable ecological conditions. The composition of the
community would be determined in the former case primarily
by a combination of internal and external factors and,
in the latter case, by random chance or probability (i.e.,
the "right" species being deposited, by wind or animal,

259

etc., or already being present as a cyst in the right place
at the right time). In order to explore these possibilities the following factors were considered: (1) chemical-
physical environment, (2) interspecific relationships,
(3) substrate, (4) colonization-extinction rates, (5)
changes in community composition through space and time,
and (6) perturbations that drastically reduce the number
of species in a community. Available evidence indicates
that both internal and external factors regulate the number of species present in much the same way as for higher
organisms.

INTRODUCTION

The investigations covered in this chapter span 26
years and have been published in a variety of journals
throughout that period. For a few years, we have been
urged to present this information in condensed form in
one publication. This opportunity to do so, therefore,
was irresistable. However, even with condensing, space
limitations have reduced or eliminated our coverage on
several important areas, such as transport. Some of these
are covered in other chapters in this book. As Oppenheimer
(1974) correctly noted in his review of *The Structure and
Function of Freshwater Microbial Communities*, the interactions of bacteria and other microorganisms that might
result in protozoan changes were not considered. This
weakness remains in this chapter and must remain uncorrected until we find a competent bacteriologist willing
to work with us on this problem. Several other weaknesses
are noted at the end of the chapter.

Many biologists view free-living protozoan communities
as haphazard assemblages of species thrown together by the
whims of a capricious nature. Although they recognize
that protozoans often have quite specific environmental

requirements, the ability to predict which species will be present after a thorough environmental analysis is lower than for terrestrial plants. Finally, most teachers of elementary biology have collected water samples containing protozoans and identified the species present the day before the proposed class use only to find the species composition markedly changed the following day. These and other similar experiences have tended to foster the belief that free-living protozoan communities are not structured as are communities of higher organisms and that they are not regulated by the same constraints that form the operational prerequisites of communities of higher organisms. The purpose of this chapter is to provide evidence that freshwater protozoan communities are structured and that this structure is determined and regulated by many parameters similar to those regulating communities of higher organisms.

In 1948, the Academy of Natural Sciences began a series of river surveys under the direction of Dr. Ruth Patrick. Most of us participating in these surveys were impressed by the regular occurrence of only a slight fluctuation in the number of species in many major groups of aquatic organisms, despite considerable variation in the kinds of species that comprised these aggregations (Table 9.1). Perhaps the term "dynamically constant" should be used, and one may ask why the differences are around 20-30% instead of two or three orders of magnitude. These results have been reported in a number of papers; Patrick's (1949, 1961) are representative of this series. This relatively constant number of species was particularly surprising for protozoans because of (1) the continual changes in kinds of species quite common in freshwater protozoan communities; (2) the strong possibility that most protozoan species have a cosmopolitan distribution and may be found

Table 9.1

Total Number of Systematic Entities for Each Study in Several Rivers

	Soft-water rivers						Hard-water rivers						Mean all rivers	Mean soft rivers	Mean hard rivers
	Escambia	Savannah 54	Savannah 55	North Anna	White Clay	Flint	N. Fork Holston	Rock Creek	Ottawa 55	Ottawa 56	Potomac 56	Potomac 57			
Algae	77	105	101	98	73	79	63	65	76	58	105	103	84	89	78
Protozoa	38	61	40	58	56	51	–	86	–	48	85	68	59	51	72
Insects	29	58	51	61	57	–	83	48	59	61	89	99	63	51	73
Fish	39	19	35	21	20	13	21	24	18	28	18	29	24	25	23

wherever appropriate ecological conditions develop; (3) the aerial and ground transport systems described by Brown *et al.* (1964), Schlichting (1961, 1964), Maguire and Belk (1967), Cairns and Ruthven (1972), and others that suggest that many protozoans (and other microorganisms) are being deposited all over the world nearly continually; (4) the large number of species, collectively tolerating a wide range of ecological conditions, which are available to colonize various habitats; and (5) the continual and often abrupt changes in environmental conditions in a particular habitat. Given these conditions, a variation in number of species of several hundred percent on either side of a mean would not be surprising, yet such variation would be exceptional in an unpolluted temperate zone river or stream in North America. A discussion of some of the factors affecting the number of species in freshwater protozoan communities follows, together with an attempt to explain the observed small variation in the number of protozoan species.

The word "community," when used for organisms found in rivers or streams, usually means that group of aquatic organisms found in a particular area at a particular time. The organisms, however, particularly protozoans, usually form a continuous gradient through both time and space, with few or no sharp lines of demarcation that enable one to perceive distinct aggregations of organisms that one may call "communities." Actually, although there may be considerable differences in the types of protozoan species found in different areas of the same stream (Fig. 9.1 and Table 9.2) at approximately the same time, there is no sharp boundary between the areas. The percentage overlap of species found at a single area through time, also may be comparatively small, since abrupt changes in weather can cause swift changes in species composition (Table 9.3).

Table 9.2

Percentage of Species Overlap at Four Stations Sampled
from Nine Different Surveys on the Savannah River (Fig. 9.1)

Survey 1

Station	1	3	5	6
1	100	10	21	27
3	11	100	23	23
5	25	25	100	38
6	22	19	28	100

Average overlap, 22.67

Survey 2

Station	1	3	5	6
1	100	63	45	50
3	50	100	47	42
5	32	41	100	45
6	29	30	35	100

Average overlap, 42.42

Survey 3

Station	1	3	5	6
1	100	17	33	45
3	27	100	33	60
5	29	18	100	53
6	22	19	30	100

Average overlap, 32.17

Survey 4

Station	1	3	5	6
1	100	50	37	60
3	40	100	35	53
5	34	43	100	43
6	30	35	23	100

Average overlap, 40.25

Survey 5

Station	1	3	5	6
1	100	-	-	25
3	-	-	-	-
5	-	-	-	-
6	25	-	-	100

Average overlap, 25.00

Survey 6

Station	1	3	5	6
1	100	42	42	30
3	33	100	38	31
5	37	40	100	51
6	22	30	45	100

Average overlap, 36.75

Table 9.2 (Continued)

Survey 7

Station	1	3	5	6
1	100	40	37	34
3	40	100	40	31
5	38	40	100	57
6	27	24	42	100

Average overlap, 37.50

Survey 8

Station	1	3	5	6
1	100	21	25	25
3	22	100	27	30
5	22	22	100	29
6	24	28	31	100

Average overlap, 25.50

Survey 9

Station	1	3	5	6
1	100	40	43	34
3	38	100	43	40
5	39	40	100	40
6	29	34	38	100

Average overlap, 38.17

Table 9.3

Percentage of Species Overlap Between Surveys on a Given Station (Savannah River)

	Station 1								
Survey	1	2	3	4	5	6	7	8	9
1	100	26	10	10	10	5	15	15	42
2	22	100	22	13	22	22	22	37	28
3	9	20	100	54	20	38	12	25	20
4	10	13	60	100	22	37	13	22	28
5	5	12	12	12	100	25	18	18	20
6	2	13	23	21	27	100	21	21	29
7	5	10	5	5	13	15	100	31	21
8	5	14	10	10	12	14	30	100	14
9	14	11	10	11	14	20	20	14	100

Average overlap, 18.49

	Station 3							
Survey	1	2	3	4	5	6	7	8
1	100	18	–	18	5	18	18	30
2	10	100	10	25	25	29	14	32
3	–	20	100	67	33	27	27	20
4	10	25	35	100	29	21	18	14
5	1	13	10	15	100	13	18	20
6	8	21	10	15	19	100	19	27
7	7	9	9	10	19	14	100	29
8	9	15	5	7	18	18	24	100

Average overlap, 17.86

Table 9.3 (Continued)

	Station 5							
Survey	1	2	3	4	5	6	7	8
1	100	25	19	19	12	38	25	25
2	12	100	17	22	25	25	22	35
3	10	18	100	42	40	18	18	29
4	13	30	52	100	43	34	13	40
5	4	18	23	21	100	20	10	20
6	17	21	13	21	24	100	24	32
7	7	11	9	5	9	15	100	20
8	7	18	12	14	14	20	20	100

Average overlap, 20.89

	Station 6								
Survey	1	2	3	4	5	6	7	8	9
1	100	28	13	22	13	19	31	40	13
2	15	100	15	25	23	23	18	23	12
3	7	12	100	34	20	17	20	22	20
4	11	23	40	100	27	20	27	23	14
5	5	17	19	20	100	30	22	29	27
6	9	19	17	17	32	100	22	17	29
7	12	12	19	20	22	20	100	39	24
8	13	13	17	15	22	12	31	100	17
9	5	10	19	11	25	25	24	20	100

Average overlap, 20.11

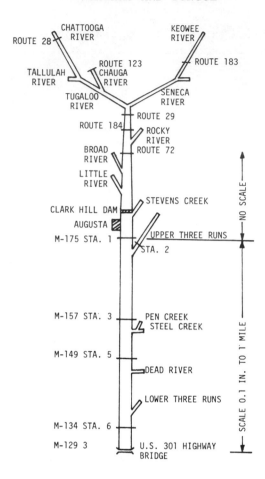

FIG. 9.1. Diagrammatic map of Savannah River basin (from Patrick *et al*. 1967). Shows location of station furnishing the data in Tables 9.2 and 9.3.

Because of these factors, it is often difficult to delimit a protozoan community in a flowing water system in either time or space. Cairns *et al*. (1974), in a paper that probably should have been called "The Failure of the Habitat Gestalt" (Robert McIntosh, personal communication, reported

that the proximity of one rock to another (regardless of
water flow direction) did not result in greater similarity
in kinds of species present. Even more surprisingly, visually quite dissimilar habitats (e.g., mud, rocks) had
communities no more different in numbers and kinds of species than those on two rocks. Communities on polyurethane
foam units (PF units) in two geographically proximate but
chemically dissimilar ponds were not appreciably more different than those on two different PF units in either of
the two ponds (Yongue et al. 1973). We believe environmental conditions influence protozoan distribution and
also that the degree of influence of the gross chemical-
physical conditions (e.g., hardness, pH) has been somewhat
overestimated. In the past we often attributed difference
in kinds of species present to habitat differences. Then
we began to use series of identical PF units in a fairly
homogeneous environment (e.g., the epilimnion of Douglas
Lake, Michigan), and found comparable differences between
these, even though they represented identical habitats
(Cairns et al., in press). Earlier one of us speculated:

> Presumably there are limitations on interac-
> tions, for example, while upstream protozoans
> may influence downstream protozoans, the re-
> verse seems quite unlikely. In addition, al-
> though water flow in a river system is not
> laminar, it seems unlikely that a protozoan
> could have any great influence on other pro-
> tozoans more than a few centimeters on either
> side of it at right angles to the direction of
> the flow. In view of the shifting patterns of
> micro-turbulence, any significant long-lasting
> effect upon downstream protozoans may be un-
> likely, unless the upstream protozoans are
> fairly numerous in all parts of the stream.

269

When one considers the half-life of many organic compounds, long-lasting indirect effects
through time and space seem unlikely except
when massive blooms occur. [Cairns 1971, p.
221]

DISCUSSION

Effects of Chemical-Physical Environment

One of the most critical factors in determining both
diversity (number of species) and density of protozoan is
water velocity. Zimmerman (1961) has shown that current
velocities also affect the kinds of species found.   In
rivers, the most diverse and dense protozoan populations
developed in pondlike situations where the velocity of the
current was not great (Cairns 1965a). This was evident
both from the comparison of stations with varying rates
of flow and from the comparison of different flow rates
within a single station. Laboratory experiments with
natural lake water flowing through plastic troughs (Cairns
and Yongue 1968) suggest that there is an optimal flow
rate that produces the greatest number of species. Although these experiments were not designed to determine
the effects of flow rate, the average number of species
was lowest in the system with the greatest flow, next
highest in that with the least flow, and highest in those
with an intermediate flow. It is possible that this represents a balance between a current too swift to permit
certain species to become established and one too slow to
bring sufficient nutrients past a given point to sustain
other species. However, considerably more experimental
evidence than this is needed. Probably it will be necessary to determine the velocity to which each individual
protozoan is exposed, a fairly difficult task.

FRESHWATER PROTOZOAN COMMUNITIES

Determining the number of protozoan species present in a sample is affected by their density. There is a threshold density below which finding even a single specimen of a species becomes difficult. Large species are easier to locate than small ones so there is a bias in their favor. Perhaps this is not entirely bad since biomass is often important. However, probably only a fraction of the low-density species are ever reported, and this may prove to be a major weakness in the data just discussed. Some of the difficulties in concentrating specimens and other sampling difficulties are discussed at length in Cairns (1965b).

Much of the chemical-physical data on environments in which protozoans have been found is too fragmentary to provide useful information on the range of conditions tolerated by most species. Even when a large amount of data is available, it is not usually from the microhabitat in which the protozoan has been found but from a nearby area which may or may not have been the same. Probably the latter is the rule rather than the exception. An example of this is given in Noland and Gojdics (1967, p. 229). Furthermore, the most critical environmental conditions influencing the kinds of species present may have been those that occurred some time before the sample was taken. However, the greatest obstacle is the low frequency of occurrence typical of most species. Even in a fairly extensive survey, such as that of the Savannah River (Patrick *et al.* 1967), many species were reported only once, or at best, only a few times; hardly sufficient evidence to provide acceptable ranges of tolerance. Those species found frequently in another study had a range of environmental conditions approximating the range of most North American temperate zone streams (Cairns 1964). Although nearly 20 different types of determinations were included in this study, many important parameters (vitamin B_{12}, carbohydrate, etc.) were omitted and were probably major determinants for many species (for example, Chu 1942; Rodhe 1948; Hutner *et al.* 1949; Parker 1968). At present we can only assume that most protozoan species are similar to, and possibly more sensitive than, the comparatively few species that have been carefully studied (since many of these many be "weed" species), and that most species have sufficiently complex requirements to insure optimal conditions to be rare, suboptimal conditions more common, acceptable conditions frequent, and inadequate conditions common. A diagrammatic sketch of this model is given in Figure 9.2.

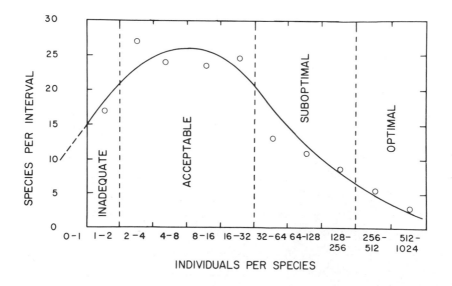

FIG. 9.2. A model for the truncate normal curve distribution. The four zones relate to conditions for survival and colonization.

FRESHWATER PROTOZOAN COMMUNITIES

Species-Species Relationships

Many protozoan species collected today are remarkably similar morphologically to those described in the early days of light microscopy. Even if we discount these descriptions as inadequate, the similarity to descriptions of specimens examined by microscopists in the early part of this century is impressive. Apparently, many species have been morphologically stable for many generations. Coupled with a cosmopolitan distribution, various transport systems, and the common ability to encyst and excyst, this apparent stability appears to provide an opportunity for interspecific relationships to develop. In other words, if species have similar environmental requirements and a cosmopolitan distribution, it may reasonably follow that the probability of interspecific relationships developing is high. This possibility was explored by Cairns (1965b). In this study the frequency of occurrence of nearly 1200 protozoan species for 202 areas of various rivers and streams mostly in the United States was determined. Approximately 75% of the species occurred in three or less of the areas sampled, or less than about 1.6% of the time. The possibility of relationships developing between species with a high frequency, so this possibility was examined with those that occurred four or more times in the 202 areas sampled. Of these, only 20, or roughly 6% of those found four or more times, occurred in at least 25% of the areas studied. The number of areas in which each of these 20 species occurred was as follows: 125, 87, 80, 69, 67, 66, 64, 64, 62, 60, 60, 60, 55, 53, 52, 49, 49, 48, 47, 47. Naturally, these ubiquitous species occurred together quite often. In order to determine whether or not pairs of these species occurred together more frequently than expected from chance alone, an association matrix was made for the 20 sampling areas where these

species were most common. A chi-square test of significance was run on the 190 possible associations of species
pairs and, of these, 44 occurred together more frequently
than expected from chance alone at the 5% level of confidence. An examination of the environmental ranges for
these species (Cairns 1964) indicated: (1) Pairs or larger
groups of associated species always had virtually identical
ranges of environmental conditions; (2) these species always tolerated rather broad ranges of environmental conditions; and (3) having identical ranges of tolerance to
environmental conditions did not insure that species would
be associated more often than would happen by chance a-
lone, since only 44 pairs out of a possible 190 were associated significantly beyond chance despite the fact that
all species had broadly overlapping environmental ranges.

There are indications that some protozoans modify
their environment in a manner beneficial to others (for
example, Robertson 1921, 1924; Picken 1937; Mast and Pace
1938; Kidder 1941; Lily and Stillwell 1965; Stillwell
1967; Cairns 1967). It even seems probable that communities of protozoans may modify microhabitat conditions to
their collective benefit (Yongue and Cairns 1971). Cairns
(1967) speculated that if beneficial interactions existed,
one might expect natural selection to have shaped these
interactions into synergistic associations or patterns of
species. Once such a synergistic pattern developed for a
given type of habitat, it would be advantageous for a pro-
spective colonizer to be able to take advantage of that
pattern, to be able to fit into it. To be associated with
another pattern would be of less use to it, since successful invasion would then depend on the simultaneous invasion of several species. Thus, as a consequence of the
high probability that protozoans have a cosmopolitan distribution, we would expect that once such a synergistic

pattern or "community strategy" developed for a given type of habitat it would tend to become nearly universal for that type of habitat and the pattern would have acquired the property that each role or niche in it could be occupied by any one of a number of species (Fig. 9.3). The species found to occupy that niche in a particular community at a given time would depend on some combination of environmental conditions and historical accident. This model would account for the relative constancy in number

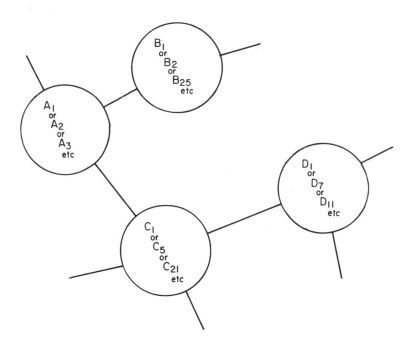

FIG. 9.3. Role A might be temporarily filled by a species designated A_1, or as far as species temporarily filling role B is concerned, A_7. Thus community structure would be relatively stable despite species replacing each other because the "role relationships" remain constant. Division of role A among species might be determined by differential tolerances to pH, temperature, etc. The number of roles would, of course, greatly exceed those depicted in this simplified diagram.

of species, the low frequency of association between particular species, and the apparent successional pattern
that is characteristic of most freshwater protozoan communities. Unfortunately, there is little evidence to
support or reject this model. One of the major obstacles
is that species taxonomy is determined morphologically
instead of functionally. It is possible that morphologically similar species may have dissimilar functions,
while two morphologically dissimilar species may have
nearly identical functions. However, if the model just
described is valid, then the relatively constant number
of morphological species observed suggests a significant
correlation between form and function. The synergistic
pattern hypothesis has several attractive features and
appears to be worth further testing despite the considerable difficulties involved.

Substrate

Based on studies of the Potomac River from 1956 to
1963, Cairns (1966) reported that although the types and
amounts of various types of substrate remained remarkably
constant, variations in kinds of species present and the
total diversity of species were common. This suggests
that these changes are initiated by environmental factors
other than substrate quality. However, there is little
doubt that unstable substrates, such as shifting sand,
are less favorable habitats than algal mats growing on
rocks. River substrates might thus be divided into two
groups: (1) unfavorable—shifting sand, rocky wave-splashed
shorelines, turbulent areas with swift currents, etc.,
and (2) favorable—algal mats, flocculent mud surfaces,
mild currents, etc. In unpolluted situations, the former
usually have few individuals and few species and the latter have many species with a wide range of individuals

per species. The substrate does not appear to directly
determine the nature of succession or the kinds of species
present unless some material is exuded into the environment. Cairns *et al.* (in press) suspended various substrates, such as pine needles, apples, and polyurethane
foam, in the well-mixed epilimnion of Douglas Lake,
Michigan, and found that some of these had markedly different numbers of species than others. Curiously the inert
substrates (e.g., polyurethane foam) had the highest number of species. In the same paper, preliminary evidence
was provided that indicated that for plastic foam units,
pore size and color did not influence the colonization
process or the number of species present.

Cairns and Ruthven (1970) found that smaller substrates generally had fewer species than the larger substrates and a linear relationship was shown between log
volume and the number of species (Figs. 9.4 and 9.5).
Some of the smaller substrates had disproportionately
large numbers of species in the beginning of the study.
It may be that optimal substrate size for a freshwater
protozoan community is partly a function of the types of
species present. If the arrival of appropriate colonizing
species is largely a matter of chance, one might expect
certain aggregations of colonizing species to be better
suited to small substrates than others.

Colonization-Extinction Curves

MacArthur and Wilson (1963, 1967) have proposed that
islands with a continuing colonization of new species and
a continuing extinction of established species may reach
an equilibrium point between these processes that result
in a relatively constant number of species being present
despite succession. Cairns *et al.* (1969) studied two
series of polyurethane substrates (10 substrates per

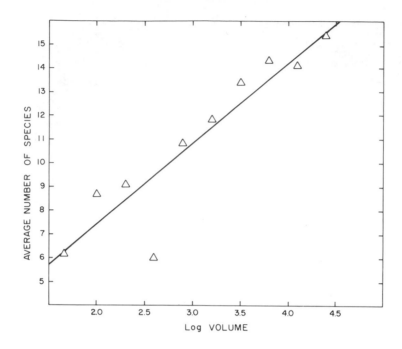

FIG. 9.4. Relation between number of species of freshwater protozoans and log substrate size. $\gamma = 0.64 + 3.39 \log x$. (From Cairns and Ruthven 1970.)

series) of identical size suspended in Douglas Lake, Michigan. Although the protozoan communities colonizing each of the substrates were not identical, the colonization process was remarkably similar for the entire 20 substrates. In the early stages of invasion of a new substrate, the colonization rate greatly exceeded the extinction rate as proposed by MacArthur and Wilson. However, as the number of species on each substrate increased, extinction and colonization rates approached but did not quite reach equilibrium. Equilibrium might have been reached had the summer session at the University of Michigan Biological Station been a few weeks longer, but it

278

is more likely that environmental changes constantly upset
species relationships so that a rigid equilibrium point
is not very probable. A more likely possibility is that
most biological systems have numbers of species oscillating
about an equilibrium point or points. For a protozoan
community, one might conceive of this as a feedback system
in which changes in the rate of one process inevitably
affect the other, causing the number of species to vary
constantly but within a specific range (barring some major
and unusual environmental stress).

One of our graduate students (James L. Plafkin) used
some of the data from Cairns *et al.* (1969) for a class

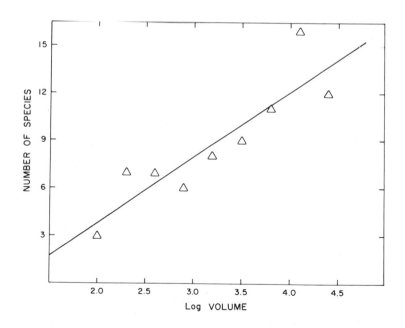

FIG. 9.5. Relation between number of species of freshwater protozoans
and log substrate size. $y = -4.35 + 4.10 \log x$. (From Cairns and
Ruthven 1970.)

279

exercise and discovered an error that should be corrected. Calculations for some of the rates were off because a zero was used for the species number for some of the substrates that were lost. The basic conclusions were sound and have been supported by subsequent investigations.

In both the colonization-extinction studies (Cairns *et al.* 1969) and the substrate volume-number of species studies (Cairns and Ruthven 1970), sudden increases in the number of species occurred (Figs. 9.6 and 9.7). These did not appear to be correlated to external environmental events but rather to events associated intimately with the colonization of the substrate. This was particularly true for the colonization-extinction studies. Following an initial period of colonization, there was an increase in the numbers of individuals per species but no marked increase in the number of species. This was followed by a secondary increase in the number of species, possibly resulting from microenvironmental changes produced by the protozoans already inhabiting the substrate. If this hypothesis is correct, it would mean that the presence of some species may enhance the suitability of the substrate for other species. Since Fogg (1965) and Fogg and Boalch (1958) have shown that microorganisms may produce extracellular products that are useful to other species, this is quite possible.

We have noted with interest a gradual decrease in the time required to reach equilibrium on PF units in Douglas Lake or the rate at which new species colonize the substrates. In 1966, about 40-50 days were required to obtain a community of approximately 35 species. In 1974, approximately 50 species had colonized the substrates in 14 days or less in the same area. Substrates were suspended in this area in all intermediate years, except 1967 and 1971, and a cursory examination of the data indicates

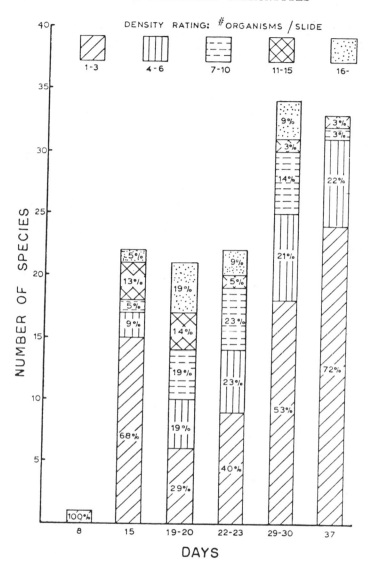

FIG. 9.6. Number and percentage of protozoan species in each of five density-rating groups on artificial substrate at sampling location 5 in Douglas Lake, Michigan. (From Cairns *et al.* 1969.)

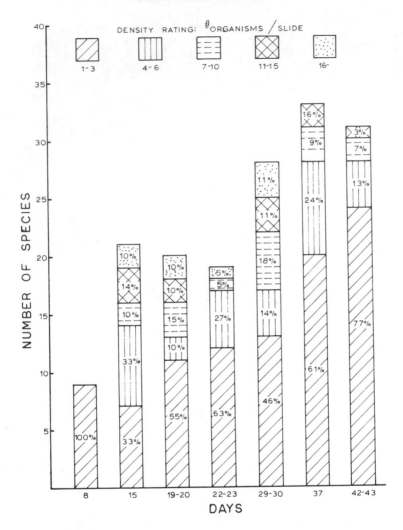

FIG. 9.7. Number and percentage of protozoan species in each of five density-rating groups on artificial substrate at sampling location 8 in Douglas Lake, Michigan. (From Cairns *et al*. 1969.)

a continuing trend of increasing colonization rates. The
possibility of a long-term fundamental change is too
tempting to ignore entirely, but the evidence cannot currently justify a conclusion.

Yongue and Cairns (1971) found that 54.8% of the
species found in the initial samples (harvesting) of PF
units in a small pond in North Carolina were "resident"
species (i.e., a group that persisted throughout the study
period). After the first 2 weeks, there were 3 weeks in
each of which 9 or 10 new species colonized the foam
units. Most of these were lost rather quickly (e.g., of
the 10 new species on week 10, seven were lost in the
weeks immediately thereafter). It was not surprising to
find the fugitive or transient species, but the large number of species that persisted for the entire study period,
December 23, 1969, to June 2, 1970, was startling. Apparently, these were able to resist displacement despite a
substantial seasonal change in environmental conditions.
It was possible that this was not unique for protozoans
but only reflected the fact that a specific microhabitat
(the PF unit) was studied over a period of many months.
Most continuing field studies of protozoan involve the
same general area but not the specific microhabitat.

Perturbations

There are few environments on the face of the earth
free from the effects of man's activities. These range
from direct effects of insecticides, thermal pollution,
etc., to "indirect" effects, such as atmospheric contaminants, which alter the earth's energy budget and change
the climate of habitats occupied by various organisms.
As a result man many be the major determinant in the survival of many species. It would be foolish to assume that
because of their ability to encyst, small size, and

possible cosmopolitan distribution protozoans are exempt from manmade stresses, although they are certainly less vulnerable than species with limited geographic distribution.

Initially organic enrichment of unpolluted streams usually results in an increase in both the density of individuals and the diversity of species (Cairns 1965a). The increase in the number of species may merely mean that some species with densities so low that they were missed have become sufficiently numerous to be detected. However, further increases in organic enrichment will usually produce other environmental changes (low dissolved oxygen concentration, sulfides, increased suspended solids) unfavorable to many of the species present, which will eliminate some of them, thus reducing the total number of species although the numbers of individuals of tolerant species may actually increase (Cairns 1965a). This simplification of a complex community (i.e., many species to few species) is the typical response of other major groups of aquatic organisms to environmental stress or pollution (Patrick 1949). These responses were noted in aquatic communities in rivers and streams of the Conestoga Basin, Pennsylvania.

Similar responses (i.e., reduction in number of species) to high- and low-pH shocks as well as temperature shocks were noted in plastic troughs kept indoors with controlled photoperiod but with natural water from Douglas Lake flowing through them (Figs. 9.8, 9.9, and 9.10). These shocks lasted only a few minutes, and since the original environmental conditions were quickly restored once the stress ceased, the fairly rapid recovery rates were not surprising. Note that the number of species at the end of the recovery period approximated that found before the shock was applied and was approximately that of the control (which often varied a few species plus or minus).

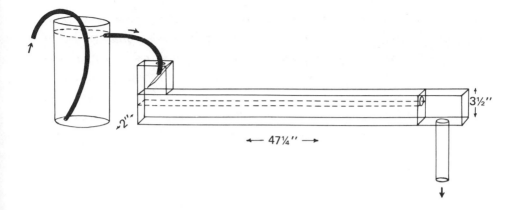

FIG. 9.8. Plastic reservoir and trough used in temperature stress experiments. (From Cairns 1969.)

Nonselective, periodic, removal of protozoans from polyurethane substrates suspended in Douglas Lake resulted in an increased number of species when these substrates were compared to undisturbed controls exposed at the same time to similar environmental conditions (Cairns *et al.* 1971b). This was accomplished by suspending 56 substrates in the epilimnion of Douglas Lake and squeezing certain selected substrates to collect a sample. These substrates were then reexamined after various time intervals and compared to control substrates on which no previous examinations had been conducted (and therefore no squeezing, which reduced the number of protozoans present when a sample was taken). The substrates that had periodic removal of portions of the protozoan community had higher numbers of species than those left undisturbed. Of course, collecting a comparison sample removed them from the undisturbed class. When the latter substrates were reexamined after a sample had been taken, the number of species had generally increased. These results indicate that potential colonizers may be excluded from an established protozoan community, and that periodic removal of portions of the community favors establishment of new species. It is also possible that habitat complexity was increased as a consequence of sucking surface organic material into the sponge interior following squeezing. This would account for the lack of further increase in diversity following subsequent

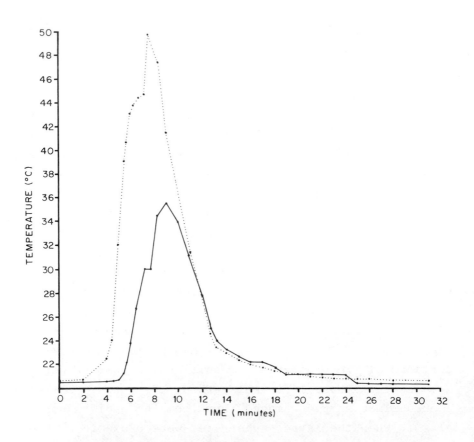

FIG. 9.9. Pattern of thermal shock from about 20 to 50 C. Note that the water flowing over the surface film (· · ·) has a much higher temperature than the bottom film (—) which contains most of the protozoan species. (From Cairns 1969.)

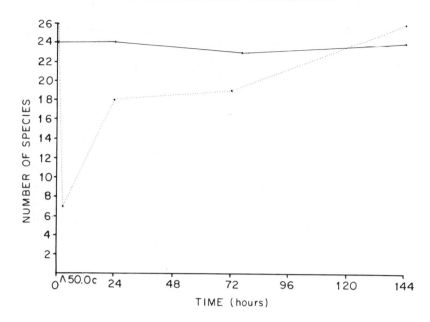

FIG. 9.10. Changes in protozoan species diversity following the
thermal shock in Figure 10.9. (——), Control; (· · ·), experimental.

squeezings. Since these substrates were suspended in
Douglas Lake for a much longer period of time than substrates in previous experiments, it would appear that
these earlier experiments were terminated approximately
when peak species diversity was reached, and that an oscillation in number of species follows even when portions
of the community are removed periodically. When the system reached a certain level of complexity or species diversity, it oscillated near this level for the remaining
period of observation (about 6 weeks for most substrates).
Certain substrates sometimes had rather dramatic changes
in number of species present, at times nearly 30%. During
this period, the usual successional replacement of species
occurred. Results also indicate that undisturbed sub-
trates (those not previously sampled) first increase in

287

number of species present and then the number of species present declines and oscillates about a mean that is significantly lower than the mean for disturbed substrates (Table 9.4). Whether either of these should be called

Table 9.4

Number of Protozoan Species Found on
Each Substrate for Each Sampling Period
(Douglas Lake, Michigan)

Substrate series number	Substrate number in series	Number of times sampled							
		1	2	3	4	5	6	7	8
46 days in lake									
I	1	28	44	39	40	29	31	31	54
	2	27	34	27	27	36	29	29	37
II	1	23	29	19	21	31	31	27	32
	2	20	23	30	18	29	31	24	35
III	1	18	27	18	17	24	35	–	–
	2	24	23	15	19	27	32	31	33
53 days in lake									
I	3	14	15	31	40	42	35		
	4	12	15	22	22	42	33		
II	3	15	25	20	25	31	35		
	4	13	21	19	21	19	22		
III	3	16	23	28	24	26	31		
	4	19	18	37	30	37	36		
60 days in lake									
I	5	34	28	28	–	31			
	6	22	22	26	34	31			
II	5	21	30	31	29	31			
	6	16	20	25	–	–			
III	5	21	26	26	25	32			
	6	14	21	27	27	27			
67 days in lake									
I	7	16	–	–	–				
	8	12	23	37	37				
II	7	14	19	–	–				
	8	16	21	27	29				
III	7	23	29	30	27				
	8	24	26	26	31				

Table 9.4 (Continued)

Substrate series number	Substrate number in series	Number of times sampled							
		1	2	3	4	5	6	7	8
74 days in lake									
I	9	19	24	24					
	10	18	23	-					
II	9	26	30	30					
	10	16	25	24					
III	9	20	32	21					
	10	17	31	30					
81 days in lake									
I	11	21	30						
	12	26	-						
II	11	19	27						
	12	15	29						
III	11	21	28						
	12	24	-						
88 days in lake									
I	13	22	24						
	14	25	37						
II	13	17	28						
	14	25	31						
III	13	23	29						
	14	33	35						
95 days in lake									
I	15	24							
	16	-							
II	15	21							
	16	12							
III	15	13							
	16	14							

"stable communities" is questionable, but each seems to oscillate within a range of ±6 species about 75% of the time.

Cairns *et al.* (1971a), reporting on the response of protozoans to detergent enzymes, noted that the effects of concentrations of Axion causing substantial reductions in the number of species (e.g., 35-78% reduction) disappeared in from 140 to 240 hr after exposure ceased. Since the species in the exposed communities were not the same, some communities were probably more vulnerable than others. Cairns and Lanza (1972) found no readily discernable difference in number of species exposed to simulated passage through a steam electric power plant condensor system, but effects were noted on population density. Unfortunately, normal variability is sufficiently large to obscure subtle changes, a recurring problem in studies of this sort. Cairns and Plafkin (1975) found that free-chlorine concentrations above 0.66 ppm administered every 20 min over a 2-hr period produced significant decreases in number of species. Community diversity, determined by the Wihlm and Dorris (1973) diversity index ($\bar{d}$), was also significantly reduced by three exposures to concentrations above 1.15 ppm. Exposure every 20 min produced significant reductions in diversity at five of the seven exposure concentrations between 0.58 ppm and 2.85 ppm. Some species were more tolerant of chlorine stress than others.

Ruthven and Cairns (1973) reported that the percent survival of colonizing species in an artificial system was affected by copper and zinc, but again "background noise" (i.e., oscillation in number of species) made precise comparisons difficult. However, since variability is a general characteristic of natural systems, we must learn to cope with it instead of avoiding problems with variable systems. The same paper suggests that the sensitivity of protozoans to toxicants may be either more or less than that of macroinvertebrates and that published information is not adequate for predicting the relationship without experimental evidence. A more extended discussion of toxic effects may be found in Alexander (1971);

Cairns *et al.* (1972); Mitchell (1974); and Cairns *et al.*
1975).

Factors Affecting Colonization
of Artificial Substrates

Some of the experiments involving artificial substrates already reported in this chapter inevitably led
to further questions about the colonization process. Was
location an important factor? What was the relationship
between length of substrate immersion and number of species? Would comparable results be obtained from standing
bodies of water other than Douglas Lake? Could the number of species in an artificial system be increased by
increasing the invasion rate? Were the kinds of species
on artificial substrates the same as those on nearby natural substrates? What influence does substrate quality
have on the colonization process? These are questions
for which we and our colleagues have been able to get some
preliminary answers. More difficult questions on natural
invasion rates and spatial distribution for the systems
we are studying remain unanswered, although investigations
are underway as this is being written. However, superb
information on invasion rates is available elsewhere in
this volume.

Cairns *et al.* (1973) reported that (1) pioneer colonizers appear to persist longer than colonizers of substrates already inhabited by established communities; (2)
the same species may colonize artificial substrates placed
in different areas of a lake but may do so at different
times; and (3) natural substrates and artificial substrates
may have many species in common.

Yongue and Cairns (1973) studied artificial substrates
suspended for a year in a pond in Montgomery County,
Virginia, and compared the results with those obtained

from substrates suspended in the same pond for 13 days
at the end of that year. The number of species found was
comparable to those found in Douglas Lake. Of the 47
species common to the natural habitats in the pond only
seven were not seen in the PF units. Thirty species were
found in both mature (1-yr immersion) and new (13-day immersion) substrates. Although this experiment was not
designed to test the MacArthur-Wilson (1963) equilibrium
model, the close number of species in 1-yr and 2-week
substrates indicates that the hypothesis was sound. This
paper also confirmed the persistence of pioneer species.
Further confirmation was provided by Yongue *et al*. (1973),
although the primary and surprising result of this latter
study was that the environmental conditions in the water
surrounding the PF units had less influence on the kinds
of species present than the "habitat gestalt" suggests,
since 13 species were common to both ponds throughout the
study.

   Cairns *et al*. (1973) found that while the number of
species may be dependent upon length of time of immersion
for a certain period after placement, the number of species eventually becomes quite similar on each unit. Cairns
and Yongue (1973), using plastic containers in a laboratory, found that an influx of new species increased the
diversity in a system low in protozoan species but not
in one that was comparatively rich in species. Thus, one
may conclude that invasion pressure does not increase the
number of species beyond a certain point. In this experiment about half the invading species were successful in
the absence of competition. A somewhat similar result
was obtained by Cairns and Yongue (1974) with artificial
substrates suspended at different depths in Douglas Lake.
The substrates in the epilimnion had an average of 35.4
species, while those in the hypolimnion (generall 11-20+

meters) had 11.4 during summer stratification. Following
the fall overturn, the substrates in the 11-20 m depths
had an average of 31.4 species per substrate. Possibly
this was the result of increased invasion rate during the
overturn or the environment changes that accompanied it.
The latter seems less likely, however, because Yongue and
Cairns (1971) found that substantial pH shifts were not
accompanied by comparable changes in the protozoan communities associated with artificial substrates.

In a set of experiments begun in 1966 but only recently prepared for publication, Cairns et al. (in press)
found that in a series of natural and artificial substrates, polyurethane foam was among the highest in number of species present. Hohn and Hellerman (1963) reported that Styrofoam was a better substrate for diatoms
than glass slides so that texture differences may cause
differences in the colonization of these inert substrates
by diatoms. Chemical inertness and an uneven surface seem
to be ideal characteristics for accumulating a large number of species. Of course this inertness is quickly lost
as the surfaces become coated, but externally applied
"character" appears to be a stronger determinant than
preexisting "character" for acquiring large numbers of
species.

Spatial Relationships

One of the most intractable problems in the understanding of protozoan communities is that of spatial relationships of individuals and species. Except for a few
papers, such as Spoon and Burbanck (1967), we have not
progressed much beyond the work of Picken (1937). There
is certainly not enough information to permit the construction of any sort of a spatial relationship model comparable to those for some higher plants and animals. The

factors that seem to support the hypothesis that protozoans have some sort of spatial relationships in a community are (1) that they seem to respond to forces in a manner similar to that of higher organisms (e.g., MacArthur-Wilson equilibrium model) in which spatial relationships are not uncommon; (2) that protozoan communities seem to have the ability to exclude invaders (e.g., Cairns and Yongue 1973); and (3) that the size of the substrate below a particular threshold seems to affect both the number and kinds of species present (e.g., Cairns and Ruthven 1970). We have attempted several times to address this problem but without pronounced success. Some evidence indicates that there are no spatial relationships comparable to those exhibited by some higher organisms. Cairns and Yongue (1968) found that on a presumably homogeneous substrate (a plastic substrate) the individuals of various protozoan species were not uniformly distributed. Furthermore, no patterns were evident to suggest a cause for the distributions observed. Unfortunately even with micro-pipets it is highly probable that the process of collecting the same distorts the distribution appreciably. Furthermore, since we have no idea of "territory," it is quite possible that our sampling pattern is not adequate. We are presently trying to resolve this problem by physically cutting polyurethane substrates into smaller sizes and adding to polyurethane substrates already colonized an uncolonized piece of substrate.

Invasion Rate

Another frustrating problem is the determination of invasion rates. It has been possible to measure colonization rate with reasonable success, although it has not been possible to clearly identify fugitive species that may, when favored by a shift in conditions, be regarded

as colonizers. Species once abundant that disappear and
subsequently become abundant again may have done so the
second time as the result of reinvasion or as a result of
rapid reproduction of a few individuals present below the
level of detectability. Ruth Patrick and other diatomists
studying similar problems have a substantial advantage
because of the rather indestructable nature of diatom
frustules, which permits counting and identification of
large numbers of individuals without the concern for
changes in the same caused by death or reproduction that
haunt protozoologists forced to work with living material.
Perhaps the elegant work of Patrick and others (covered
elsewhere in this volume) will have to serve as a model
for similar problems with protozoan communities, but one
hopes to obtain direct evidence. We are presently attempting to get at the invasion rate problem by trying
to estimate the number of potential invaders of a substrate
in the Douglas Lake water surrounding it. Since most of
the species in the surrounding water are true planktonic
organisms and most of the species on PF units are substrate
associated species, we may assume any species commonly
associated with substrates to be a potential invader if
found in the water near a PF unit. At the present time
the sheer magnitude of this effort is overwhelming and
the results are too meager to justify the effort. It is
scant comfort to know that others working with similar
problems using insects have found such information equally
difficult to get and have only half jokingly considered
such desperate measures as erecting a dome of fly paper
over a Carribean Island (Daniel Simberloff, personal communication). Cairns and Yongue (1973) have shown that
ersatz invasions may frequently result in unsuccessful
colonization. Successful colonizers might be generated
in four ways: (1) from cysts already present on the substrate; (2) from arrival via waterborne currents that were

seeded from other submerged substrates or by airborne
organisms or other dissemules; (3) from being carried
there by other organisms, such as snails, birds, or insects; and (4) from development from low-density species
suddenly favored by changed conditions. Superb invasion
rate studies with diatoms have been carried out by Patrick
(1967), and these are a probable indication of the nature
of the invasion process for protozoans.

Behavior

Although there are some very elegant papers on protozoan behavior (e.g., Salt 1967), there seem to be none
dealing with protozoan behavior in such very complex aggregations of species as communities. Most papers involve
studies of protozoans either in single species culture or
with only a relatively few species involved. Although
excellent, these studies only give us limited insights
into the behavior of protozoans when 30, 40, or more species are present in varying numbers of individuals per
species. One might expect on the basis of structural
differences alone the behavioral patterns of *Chlamydomonas*
(a phytomonad), *Vahlkampfia* (an amoeba), *Stylonychia* (a
hypotrich); and *Paramecium* to be quite different. Despite
individual behavioral differences, it is quite likely that
such differing species when found together in a community
might have a variety of interactions including predator-
prey relationships, territoriality, and the like, and
these might result in displacement of some species, enhancement of the prospects of others, and perhaps have no
effect on some. One of our students, James L. Plafkin,
carried out some preliminary investigations with *Spiro-
stomuum intermedium* that indicate that the amount of activity of this species is influenced by the number of individuals of that particular species already present.

Another aspect of the same investigation suggested that direct physical contact was not necessary to reduce activity as density increased, but that some material found in the culture fluid would accomplish this. It was never determined whether this unknown material would influence the activity of other species, particularly of the same genus. Our attempts to study species behavior in a community setting have been frustrated by the same operational difficulties affecting our attempts to study spatial relationships. Undoubtedly, behavior influences spatial relationships and the invasion process. Unfortunately, gathering this key information may have to await the development of new technology and insights.

CONCLUSIONS

Freshwater protozoan communities appear to have one or more internal or endogenous regulatory systems that influence the number of species present at a particular time. There are two fairly strong sets of data supporting this statement: (1) the sudden increases in numbers of species that appear to be correlated with events associated intimately with the colonization of the substrate instead of with external environment events (Cairns et al. 1969; Cairns and Ruthven 1970); and (2) the sampling of "undisturbed" substrates that was characterized by an increase in species following the first sampling but was not repeated in subsequent sampling. Oscillation of numbers of species on these undisturbed substrates was within a range significantly below that of "disturbed" substrates, suggesting that undisturbed communities excluded invading new species more effectively than disturbed communities (Cairns et al. 1971b). However, factors other than the protozoan community seem to affect the number of species present or the range within which oscillation in number of species will occur. There are several types of evidence supporting this statement: (1) the range within which the number of species oscillated was significantly higher on "disturbed" substrates than on "undisturbed" substrates; (2) the volume or size of substrate available for colonization as well as its physical structure influences the number of species present; (3) sudden application of environmental stress (pH, temperature, zinc, copper) reduces the number of species present; the number of species usually returns to approximately its previous level when the stress is removed (Cairns 1969; and Cairns and Dickson 1970).

ACKNOWLEDGMENTS

This manuscript was prepared at the University of Michigan Biological Station during the summer of 1974. We are grateful both to students (particularly James L. Plafkin) and to colleagues for discussion of these and other topics that have influenced our views.

REFERENCES

Alexander, M. 1971. *Microbial Ecology*. Wiley, New York.

Brown, R. M. Jr., Larson, D. A., and Bold, H. C. 1964. Airborne algae: their abundance and heterogeneity. *Science 143*, 583-585.

Cairns, J. Jr. 1964. The chemical environment of common freshwater protozoa. *Not. Nat. Acad. Sci. Phila*. No. 365, pp. 1-16.

Cairns, J. Jr. 1965a. The protozoa of the Conestoga Basin. *Not. Nat. Acad. Nat. Sci. Phila*. No. 375, pp. 1-14.

Cairns, J. Jr. 1965b. The environmental requirements of protozoa. *Biological Problems in Water Pollution*, Third Seminar, 1962, PHS Publ. No. 999-WP-25, pp. 48-52, Abst. pp. 385-386.

Cairns, J. Jr. 1966. The protozoa of the Potomac River from Point of Rocks to White's Ferry. *Not. Nat. Acad. Nat. Sci. Phila*., No. 387, pp. 1-11; plus 43 pp. supporting data deposited as document No. 8902 with the AID Aux. Pub. Proj. Photodupl. Serv. Library of Congress.

Cairns, J. Jr. 1967. Probable existence of synergistic interactions among different species of protozoans. *Revista de Biologia 6*, 103-108.

Cairns, J. Jr. 1969. Rate of species diversity restoration following stress in protozoan communities. *Univ. Kansas Sci. Bull. 48*, 209-224.

Cairns, J. Jr. 1971. Factors affecting the number of species in freshwater protozoan communities. In *The Structure and Function of Freshwater Microbial Communities*, J. Cairns, Jr. (Ed.). Virginia Polytechnic Institute and State University Press, Blacksburg, Va., pp. 219-247.

Cairns, J. Jr., Beamer, T., Churchill, S., and Ruthven, J. 1971a. Response of protozoans to detergent-enzymes. *Hydrobiologia 38*, 193-205.

Cairns, J. Jr., Dahlberg, M. L., Dickson, K. L., Smith, N., and Waller, W. T. 1969. The relationship of freshwater protozoan communities to the MacArthur-Wilson equilibrium model. *Am. Nat. 103*, 439-454.

Cairns, J. Jr., and Dickson, K. L. 1970. Reduction and restoration of the number of freshwater protozoan species following acute temporary exposure to copper and zinc. *Trans. Kansas Acad. Sci. 73*, 1-10.

Cairns, J. Jr., Dickson, K. L., and Yongue, W. H. Jr. 1971b. The consequences of nonselective periodic removal of portions of freshwater protozoan communities. *Trans. Am. Micros. Soc. 90*, 71-80.

Cairns, J. Jr., Heath, A. G., and Parker, B. C. 1975. Temperature influence on the chemical toxicity to aquatic organisms. *Hydrobiologia 47*(1):135-171.

Cairns, J. Jr., and Lanza. G. R. 1972. Effects of heated waste waters upon some microorganisms. Water Resources Research Center, Virginia Polytechnic Institute and State University, Bull. No. 48, 101 pp.

Cairns, J. Jr., Lanza, G. R., and Parker, B. C. 1972. Pollution related structural and functional changes in aquatic communities with emphasis on freshwater algae and protozoa. *Proc. Acad. Nat. Sci. Phila. 124*, 79-127.

Cairns, J. Jr., and Plafkin, J. L. 1975. Response of protozoan communities exposed to chlorine stress. *Arch. Protistenk. 117*, 47-53.

Cairns, J. Jr., and Ruthven, J. 1970. The relation between artificial substrate area and the number of freshwater protozoan species. *Trans. Am. Micros. Soc. 89*, 100-109.

Cairns, J. Jr., and Ruthven, J. 1972. A test of the cosmopolitan distribution of freshwater protozoans. *Hydrobiologia 39*, 405-427.

Cairns, J. Jr., Ruthven, J. A., and Kaesler, R. L. 1974. Distribution of protozoa in a small stream. *Am. Midl. Nat. 92*(2), 406-414.

Cairns, J. Jr., and Yongue, W. H. Jr. 1968. The distribution of freshwater protozoa on a relatively homogeneous substrate. *Hydrobiologia 31*, 65-72.

Cairns, J. Jr., and Yongue, W. H. Jr. 1973. The effect of an influx of new species on the diversity of protozoan communities. *Revista de Biologia 9*, 187-206.

Cairns, J. Jr., and Yongue, W. H. Jr. 1974. Protozoan colonization rates on artificial substrates suspended at different depths. *Trans. Am. Micros. Soc. 93*, 206-210.

Cairns, J. Jr., Yongue, W. H. Jr., and Boatin, H. Jr. 1973. Relationship between number of protozoan species and duration of habitat immersion. *Revista de Biologia 9*, 35-42.

Cairns, J. Jr., Yongue, W. H. Jr., and Smith, N. In press. The influence of substrate quality upon colonization by freshwater protozoans. *Revista de Biologia*.

Chu, S. P. 1942. The influence of the mineral composition of the medium on the growth of planktonic algae. *J. Ecol. 30*, 284-325.

Fogg, G. E. 1965. *Algal Cultures and Phytoplankton Ecology*. University of Wisconsin Press, Madison, Wisconsin.

Fogg, G. E., and Boalch, G. T. 1958. Extracellular products in pure culture of a brown alga. *Nature 181*, 789.

Hohn, M. H., and Hellerman, J. 1963. The taxonomy and structure of diatom populations from three Eastern North American rivers using three sampling methods. *Trans. Am. Micros. Soc. 82*, 250-329.

Hutner, S. H., Provasoli, L., Stokstad, E. L. R., Hoffman, C. E., Belt, M., Franklin, A. L., and Jukes, T. H. 1949. Assay of antipernicious anemia factor with Euglean. *Proc. Soc. Exp. Biol. Med. 70*, 118-120.

Kidder, G. W. 1941. Growth studies on ciliates. The acceleration and inhibition of ciliates grown in biochemically conditioned medium. *Physiol. Zool. 14*, 209-226.

Lily, D. M., and Stillwell, R. H. 1965. Probiotics: Growth-promoting factors produced by microorganisms. *Science 147*, 747-748.

MacArthur, R., and Wilson, E. O. 1963. An equilibrium theory of insular zoogeography. *Evolution 17*, 373-387.

MacArthur, R., and Wilson, E. O. 1967. *The Theory of Island Biogeography*. Princeton University Press, Princeton, New Jersey, 203 p.

Maguire, B. Jr., and Belk, D. 1967. *Paramecium* transport by land snails. *J. Protozool.* *14*, 445-447.

Mast, E. C., and Pace, D. M. 1938. The effect of substances produced by *Chilomonas paramecium* on the rate of reproduction. *Physiol. Zool.* *11*, 359-382.

Mitchell, R. 1974. *Introduction to Environmental Microbiology*. Prentice-Hall, Englewood Cliffs, New Jersey, 355 p.

Noland, L. E., and Gojdics, M. 1967. Ecology of free-living protozoa. In *Research in Protozoology*, Vol. 2, T. Chen (Ed.). Pergamon, Oxford, pp. 217-266.

Oppenheimer, C. H. 1974. Book Reviews: Freshwater Microbial Communities. *Bioscience 24*, 419.

Parker, B. 1968. The vitamin B_{12} content of rainwater. *Nature 219*, 617-618.

Patrick, R. 1949. A proposed biological measure of stream conditions based on a survey of the Conestoga Basin, Lancaster County, Pennsylvani. *Proc. Acad. Nat. Sci. Phila. 101*, 277-341.

Patrick, R. 1961. A study of the numbers and kinds of species found in rivers in eastern United States. *Proc. Acad. Nat. Sci. Phila. 113*, 215-258.

Patrick, R. 1967. The effect of invasion rate, species pool, and size of area on the structure of the diatom community. *Proc. Nat. Acad. Sci. 58*, 1335-1342.

Patrick, R., Cairns, J. Jr., and Roback, S. S. 1967. An ecosystematic study of the fauna and flora of the Savannah River. *Proc. Acad. Nat. Sci. Phila. 118*, 109-407.

Picken, L. E. R. 1937. The structure of some protozoan communities. *J. Ecol. 25*, 368-384.

Robertson, T. B. 1921. The influence of mutual contiguity upon reproductive rate and the part played therein by the "x-factor" in bacterized infusions which stimulate the multiplication of infusoria. *Biochem. J. 15*, 1240-1247.

Robertson, T. B. 1924. Allelocatalytic effect in cultures of *Colpidium* in hay-infusion and in synthetic media. *Biochem. J. 18*, 612-619.

Rodhe, W. 1948. Environmental requirements of freshwater plankton algae. *Symbolae Bot. Upsaliensis 10*, 1-149.

Ruthven, J. A., and Cairns, J. Jr. 1973. The response of freshwater protozoan communities to concentrations of various toxicants, particularly the heavy metals, zinc, and copper. *J. Protozool. 20*, 127-135.

Salt, G. W. 1967. Predation in an experimental protozoa population: (*Woodruffia-Paramecium*). *Ecol. Monogr. 37*, 113-144.

Schlichting, H. E. Jr. 1961. Viable species of algae and protozoa in the atmosphere. *Lloydia 24*, 81-88.

Schlichting, H. E. Jr. 1964. Meteorological conditions affecting the dispersal of airborne algae and protozoa. *Lloydia 27*, 64-78.

Spoon, D. M., and Burbanck, W. D. 1967. A new method for collecting sessile ciliates in plastic petri dishes with tight-fitting lids. *J. Protozool. 14*, 735-744.

Stillwell, R. H. 1967. Colpidium-produced RNA as a growth stimulant for Tetrahymena. *J. Protozool. 14*, 19-22.

Wihlm, J., and Dorris, T. C. 1973. Biological parameters for water quality criteria. *Bioscience 18*, 477-481.

Yongue, W. H. Jr., and Cairns, J. Jr. 1971. Microhabitat pH differences from those of the surrounding water. *Hydrobiologia 38*, 453-461.

Yongue, W. H. Jr., and Cairns, J. Jr. 1973. Long-term exposure of artificial substrates to colonization by protozoans. *J. Elisha Mitchell Sci. Soc. 89*, 115-119.

Yongue, W. H. Jr., Cairns, J. Jr., and Boatin, H. C. Jr. 1973. Freshwater protozoan species in geographically proximate but chemically dissimilar bodies of water. *Arch. Protistenk. 115*, 154-161.

Zimmerman, P. 1961. Experimental untersuchungen uber die okologische Wirkung der Stromungsgeschwindigkeit auf die Lebensgemeinschaften des fliessenden wassers. *Schweiz. Zeit. Hydrologie 23*, 1-81.

Chapter 10

COMPETITION AND PREDATION IN RELATION TO SPECIES
COMPOSITION OF FRESHWATER ZOOPLANKTON, MAINLY CLADOCERA

Jaraslov Hrbáček

Hydrobiological Laboratory
Botanical Institute, Czechoslovakia Academy of Science
Praha, Czechoslovakia

CONTENTS

Motto: S. Brody (1945): "Science like life in general is
continuously forgetting and re-
discovering old truths."

INTRODUCTION

More than 100 years ago, Darwin recognized that biotic relations are more important than abiotic ones. Nevertheless, the limnological literature has strongly emphasized the study of the influence of the abiotic factors
(e.g., pH, Ca, humic substances; Pacaud 1939) and has virtually neglected the biotic factors.   The reason for this
situation was probably the findings of Thienemann (review,
1939) and his school on the close relation between the
limnological features of a lake (expressed by the oxygen
concentration in the hypolimnion at the end of the stagnation) and the species composition of the bottom fauna
(*Chironomus* and *Tanytarsus* lakes).   From the biotic factors of zooplankton, only algae have been considered as
a food resource, and the biotic relations have been simplified to a one-sided influence of nutrients on algae,
of algae on filter feeders, and of filter feeders on predators of the first and the second level.   This scheme
was so firmly established that Welch (1952) made the statement that "if all of the nonplankton animals were removed
from a lake. . .the plankton, with possibly some minor
modification, would continue to exist."   Schäperclaus
(1961) described the influence of various fish stocks on
the quantitative composition of fish food, and he disre-
garded the possible influence of fish stock on its qualitative composition due to selective predation.   The fact
that a fish can influence the qualitative composition of

its food organisms has been shown in a number of papers that were neglected in the textbooks (Hurlbert *et al.* 1972). Frič and Vávra (1897) considered the possibility that the planting of a new fish species in Schwarzee induced the change of zooplankton species. Birge and Juday (1932) related the high transparency of some lakes to the ability of large *Daphnia* species in keeping down the phytoplankton biomass. Bowkiewicz (1935) described the relation between the presence of some species of crustaceoplankton and the presence of some species of whitefish. Banta (1939; in a footnote) claims that the predators, fish living on cladocerans, act as a limiting factor of the cladoceran population. Hrbáček and his co-workers (e.g., Hrbáček 1969a) have shown changes not only in the zooplankton composition but also in the concentration of seston, the number of phytoplankton cells, and even in some chemical parameters (e.g., pH and oxygen concentration) after the fish stock has been drastically changed. Grygierek (1967), Hall *et al.* (1970), and Losos and Heteša (1973) have found similar effects of low and high fish stock in their experiments. Hall *et al.* (1970) showed that an increased stock of invertebrate predators has no effect on the species composition of zooplankton comparable to that of an increased fish stock. An exact interpretation of the cause-and-effect relation in these field experiments is difficult because changes in the fish stock influence all trophic levels and there exists considerable variation of plankton among bodies of water from year to year. Brooks and Dodson (1965) have proposed the size-selection hypothesis that has considerably increased the interest in this field. The purpose of this chapter is to review in more detail the possible mechanisms of the interaction of various taxa in different bodies of water resulting from different food levels and fish predation.

DISCUSSION

Principal Taxa of Zooplankton and the Components of Their Strategies in the Struggle for Existence

In the following paragraphs zooplankton are defined as those organisms that are found in the open water and are not associated with the littoral or benthic zones.

Ciliates

Ciliates play a minor role in the composition of zooplankton in larger bodies of water, although Gajevskaja (1933) reports a rich fauna of pelagic ciliates in Lake Baikal in winter. Some genera are known to live in the anaerobic layers of the hypolimnion. Most of the epilimnetic species are particle feeders. The locomotion of the animal and the production of feeding currents along the oral groove result from the activity of cilia. These cilia are more or less differentiated both in form and function in various taxonomic groups. Feeding in *Paramecium*, which is not a typical plankton genus, is probably the most intensively investigated. Mast (1947) was unable to observe the mechanism for the selection of particles proposed by Bragg (1936). Jahn and Bovee (1967) demonstrated photographically that the particles are agglutinated as they are driven along the oral groove but not along the ciliary tracts on the other parts of the organism. To my knowledge, data on the size of the ingested particles by various genera are lacking. Pitelka (1963) reviewed the fibrillar apparatus at the base of the cytostome. Predatory ciliates (e.g., *Didinium*) are reported to distend considerably not only the cytostome but also the whole body (Dragesco 1948) so that the prey can be nearly as large as the predator.

Ciliates do not have a pigmented part of the protoplasm analogous to the eye. Propagation proceeds by division, i.e., a short generation time, but it results in
the least number of offspring from one individual per generation.

Rotatorians

Rotatorians are present in most waters and are an
important constituent of zooplankton in seston-rich ponds
that have high fish stock (Hrbáček 1962; Hillbricht-Ilowska
1964). Conversely, rotatorians, along with the ciliates,
also are the only representatives of zooplankton in extreme oligotrophic lakes (because of ice cover most of
the year) or in lakes with low pH. The nannosestonic particles are collected by the movement of the ciliae. There
is considerable variation in the feeding habits of various
genera (Ruttner-Kolisko 1972). The pelagic species of
rotifers have six types of coronal ciliation and five types
of mastax, a peculiar part of the digestive system. Gossler
(1950) has found in various genera that the relationship
between locomotion and the production of the feeding currents is different. In *Conochilus unicornis* the feeding
currents are also responsible for the movement of the animal. In *Anurea cochlearis* and *Notholca longispina*, the
cirgulum is principally responsible for motion while the
trochus is used for feeding. In *Polyarthra* and *Synchaeta*,
the ciliated trochus also produces the feeding currents.
The size of the ingested particles is below 12 μm. In
Asplanchna the size of the ingested particles is considerably larger (above 40 μm), although some authors report
a high proportion of the smaller particles (Galkovskaya
1963). Green and Lan (1974) have documented the ability
of rotifera to capture copepodids.

Some genera have special features for a short, strong displacement (skip), e.g., *Polyarthra* and *Hexarthra*. Szlauer (1965) found that 22% of individuals of *Polyarthra* and 18% of *Notholca* escaped from the mouth of a tube 5.6 mm in diameter that was lowered at a speed of 1 cm/sec. *Keratella cochlearis*, *K. quadrata*, and *Synchaeta* sp. showed lower percentages of escape; respectively, 10, 2, and 0%. Green and Lan (1974) observed that a considerably lower percentage (7 against 21) of *Hexarthra* sp. were found in the stomach of *Asplanchna brightwelli* than in the other zooplankton of a Javanese sewage pond.

Propagation proceeds by eggs that are carried by or develop in the female. The number of eggs carried by one female varies from one to three, and occasionally is higher. At the end of embryonic development, the number of nuclei does not increase and the relative growth increment during the postembryonic life is 50 to 100% in planktonic species. The generation time (from egg to egg-carrying female) is usually short, two days in *Brachionus urceolaris* at 20 C (Ruttner-Kolisko 1972). Pourriot and Deluzarches (1971) have found that the length of embryonic and post-embryonic development differs in various strains of *B. urceolaris* (from 40.4 to 65.0 hr at 20 C). The length of generation of other examined planktonic species (such as *Keratella quadrata*, *K. valga*, *Polyarthra dolichoptera*, *Synchaeta pectinata*) was within this range. The examined species of the genus *Notommata* has a considerably longer generation period of from 122 to 143 hours.

Cladocerans

Cladocerans are an important component in waters with high primary production; seasonally one genus may represent 90% of the biomass of zooplankton (Korinek 1972). They are less important in oligotrophic lakes with plankton-feeding fish, e.g., Lake Tanganyika (Sars 1909; Harding 1957), Lake Baikal (Koshov 1963), and the North American Great Lakes (Palatas 1975). However, in oligotrophic lakes without plankton-feeding fish, cladocerans might be a very important component of the zooplankton at certain seasons, e,g., in Lake Tahoe (Frantz and Cordone 1970), Lake Ohrid (Serafimova-Hadžišče 1957), and Hincovo pleso (Litynski 1913; Ertl *et al.* 1965). Most of the species are particle feeders. They filter the particles by plumose bristles on their thoracic limbs. In pairs, these limbs may act separately (e.g., *Diaphanosoma*) or together to compose an effective filtering apparatus (e.g., *Daphnia*, *Bosmina*; Storch 1929; Cannon 1933). Coker and Hayes (1940) have found that the setae in *Daphnia pulex* and *Diaphanosoma brachyurum* are about 1 μm apart. Burns (1968a) has shown that the maximum size of plastic spherical beads ingested by Cladocera varies from 10 to 75 μm and is roughly one-fiftieth of the carapace length. Banta (1921b) has successfully grown *Daphnia* on undefined bacterial cells and Rodina (1950) has grown *Daphnia* on pure cultures of bacterial cells (size below 1 μm), which are less than one-thousandth of the carapace length. Bogdan and McNaught (1975) found that *Daphnia galeata mendotae* filters nannoplankton and net phytoplankton with similar efficiency. Predatory *Leptodora* were found by Mordukhai-Boltovskaya (1958) to prefer a prey of 0.3 to 0.7 mm. Butorina (1971) has grown *Polyphemus pediculus* on bacteria and protozoans.

Locomotion results from the activity of the second pair of the antennae that beat either in short, nearly constant (e.g., *Daphnia, Bosmina*) or long, irregular intervals (e.g., *Diaphanosoma*). From the limited number of cladocerans examined by Szlauer (1965), it was shown that *Diaphanosoma brachyurum* (71%) best escaped the mouth of a tube 5.6 mm in diameter that was lowered at a speed of

1 cm/sec. A lower percentage of escape was found with
Daphnia cucullata (48% in females and 36% in males), *Bosmina coregoni* (35%), and *Chydorus sphaericus* (24%). *Bosmina longirostris* did not escape.

Genera found in the open water have one large compound eye and many species also have a small simple eye.
The number of ommatidia in the compound eye varies from
several tens in filter feeders to several hundreds in predatory species (Wesenberg-Lund 1939). The pigmentation
of the eye protects the sensitive receptors against excessive illumination and the pigments show photomechanical
movements in response to changes in light conditions
(Kleinholz 1961).

Reproduction is parthenogenetic and the eggs are carried by the female in the brood chamber. At 20 C the
length of time for embryonic development in *Daphnia* is
three days (Esslová 1959; Hall 1964). This apparently
is true for other Cladocera and Copepoda irrespective of
their size (Schindler 1972). This is rather surprising
because one would expect the large newborn individuals to
have a larger number of cells than the smaller ones and
therefore a longer embryonic development. Weglenska (1971)
found for four species of Cladocera that at 17 C low food
concentrations prolonged the length of embryonic development. The length of the embryonic development of *Daphnia*
at high food levels was 3.0 days; at low food levels it
increased to 3.4 days in *D. cucullata*. In *Chydorus sphaericus* the shortest embryonic development period was 2.8
days; under low food concentrations it lengthened to 3.0
days. The effect might be partly due to a difference in
methodology. Weglenska observed the length of the embryonic development in intact animals while other workers
observed embryos cultivated outside the mother. The release of the embryos is related to molting of the female

(Banta 1939). A prolonged period between molting at low food concentrations may also influence the length of the embryonic development. The differences observed by Weglenska are well in the range of the results obtained by various authors (summarized by Schindler 1972). The newborn individual (neonata) is smaller than the mother, but its structure is nearly identical. There is a general tendency for the ratio of the size of a primipara to that of a neonata to be smaller in smaller species. In larger species of *Daphnia* juvenile and adolescent instars increase about one-third in length after each molt (Hrbáček and Hrbáčkova-Esslova 1960).

The number of eggs carried by one individual is from one to several, not exceeding ten, in pelagic populations of lakes; whereas in ponds, especially in temporal ones, it may reach several tens and may even exceed 100. At 20 C the generation time (from neonate to first-egg production) is 2-3 days in *Bosmina* and 5-15 days in *Daphnia*; it is influenced both by the food concentration and the size of the species.

Copepods

Copepods are present in most waters and have one or more reproductive periods per year. They constitute a major portion of the zooplankton biomass in many oligotrophic lakes, and in some seasons they may be important components of zooplankton in eutrophic waters. In contrast to the previous groups, predatory feeding is common in many species of both cyclopoid and calanoid copepods during the later larval stages and in adults. The younger naupliar larvae are herbivorous and feed on individual pieces of algal material. No filtration mechanisms are developed in older copepodite and adult stages of cyclopoid copepods; they feed either raptorially or by scraping off plants or other material from the substrate (Fryer 1957). The carnivorous copepods feed on a wide spectrum of prey with larval stages of copepods, including their own larvae, often being preferred (McQueen 1969; Confer 1971). Brandl (1973) was able to cultivate copepods on *Paramecium* the length of which ranged from 60 μm to 150 μm. For ceriodaphnia in the range of 0.5 to 1.0 mm, a definite preference was shown for smaller individuals. Calanoid copepods have an elaborate filtration system using water moved by the second antennae. Particles carried in this water are filtered by the bristles of the mouthpart appendages (Lowndes 1935). Nauwerck (1963) has found bacteria an important food. The range of particle sizes filtered varies from 5 to 11 μm (Hargrave and Geen 1970; McQueen 1970) to 7 to 70 μm (Wilson 1973).

All copepods have small, simple eyes. The cyclopoids and calanoids move differently. The cyclopoid copepods move in relatively short, continuous, but erratic jerks and encounter their prey as the result of random swimming movements (Fryer 1957). The calanoid copepods make long, rapid jumps but usually they glide, using the first antennae to balance. Szlauer (1965) found that copepods were the most effective of all examined zooplankton in escaping from the mouth of a tube. Eighty-nine percent of the adult *Diaptomus graciloides* escaped, whereas only 70% of the copepodite stages were able to escape. While 84% of adult female *Thermocyclops oithonoides* escaped, only 79% of the males did. Fifty-seven percent of copepodites (instar IV and V) of *T. oithonoides* escaped. Copepodites (instar IV and V) of *Mesocyclops* escaped at the rate of 71%.

Reproduction is always bisexual with eggs carried by the female. The clutch size varies from few to several tens of eggs, with a smaller number of larger eggs in the

summer generations (Wesenberg-Lund 1904; Czeczuga 1960).
Schindler (1972) found the length of the embryonic development to be comparable to that of cladocerans. Wegleńska
(1971) found a somewhat longer development time of copepod
embryos (4.0 days in *Eudiaptomus graciloides* at 17 C) without a prolonged effect of low food concentrations as was
typical of cladocerans. The youngest larvae are considerably smaller than the adult female. The postembryonic
development is rather complicated, having six naupliar
and five copepodid larval stages. This development lasts
from a few weeks in the summer generation of some species
to a whole year in others; in general, there are no more
than three generations per year. Cyclopoid copepods can
spend some time in diapause as resting eggs or larvae in
the bottom deposits either over winter or summer. For
the same species, the number of generations and the presence or absence of the diapause can be correlated with
latitude (Smyly 1961).

Phantom Midge Larvae

Phantom midge larvae are important predators in small
invertebrates. *Chaoborus* larvae are the only freshwater
invertebrates with well-developed hydrostatic organs. The
larvae lie horizontally in the water when at rest, and
they can suddenly change their position by lashing movements of the posterior part of the body. Allan (1973)
found *Bosmina* (size 0.3 mm) was a preferred prey to *Daphnia*
(0.8 mm) and *Cyclops* (0.4 mm). Dodson (1974b) found the
third and the fourth instar of the *Chaoborus* larvae (10-
15 mm long) to prefer a prey in the size range of 0.7 to
1.2 mm.

During the day in fish stocked lakes and ponds, the
larvae of *Chaoborus* live either in the metalimnion, the
hypolimnion, or in bottom deposits, whereas at night they

appear in the epilimnion.   In ponds without fish, they are
known to live not far from the surface even during the day.

The larvae have a pair each of large composite and
simple eyes.

In Central Europe, there are usually two periods of
emergence (Saeter 1972), from April to May and from July
to September.   They overwinter as larvae.   The disc-shaped
eggmass deposited by the female consists of about 350
eggs (Parma 1969).

Strategies in the Competition for Food

General Tendencies in Particle Size Allocation

In some protozoa the prey can be nearly as large as
the predator.   The prey of *Chaoborus* is one-tenth to one-
fiftieth of its body size.   Rotifers and copepods are in
between.   This great relative decrease of the prey size
in *Chaoborus* is at least partly due to the elongate body
shape.   Also, in filter feeders the relative size of the
largest ingested particles decreases with the increasing
organizm size.   In *Keratella* it is about one-tenth of the
total body length and in *Daphnia* it is about one-fiftieth.
There is a considerable range in size of filtered particles
in individual taxonomic groups, but particles of 1-10 µm
can be filtered by representative species of all taxonomic
groups present in the zooplankton.   Contrary to Dodson
(1974a), I do not find 50 µm to be a division between the
size of the filtered particles (both *Daphnia* and *Diaptomus*
are reported to filter particles above this size) and the
size of the prey (e.g., *Didinium, Asplanchna,* and also
Cyclops).   The efficiency of filtration and predation within certain limits is related to size, but field observations do not necessarily show an optimum size because the
most desired size is likely to disappear first.   Morover,
a particle has three dimensions of which any two may be

317

very different. Galkovskaya (1963) is one of the few authors who give two dimensions to food particles. For the
rotifer *Asplanchna priodonta*, she found food particles such
as the diatom *Fragillaria crotonensis* to measure 140 x 40
µm, and particles such as *Keratella cochlearis* to measure
145 x 62 µm. In my opinion, it is not certain which of
these two parameters is decisive for selection.

Burns (1969) was unable to find a clear size selection
of spherical plastic beads by different-sized *Daphnia*. In
cladocerans coexisting in a laboratory microcosm, Neil
(1971) found that smaller species tended to filter out
smaller particles than did the larger ones. This is probably because of the mechanism of action of the filtering
apparatus. The filtering setae are inserted in one row
on each of the filtering thoracic limbs, and during feeding, the distance between neighboring bristles increases
distally. As the body size of the cladocerans increases,
the proportion of the area covered by densely spaced bristles decreases relatively. Consequently, the proportion
of smaller particles in the food also decreases in larger
animals.

In general, the metabolism of the weight unit is indirectly proportional to the weight of the animal. This
has been shown in detail for crustaceans by Winberg (1950)
and is true for different sizes of the same species.
Richmann (1958) has shown that the oxygen consumption in
Daphnia is a function of the body weight (b = 0.881).
Buikema (1972) found considerable variation in this expo-
nent when light conditions were changed (b varied from 0.37
to 1.48). Nearly all values above 1.0 were found in experiments with animals that were not adapted to experimental conditions. The average b value for unacclimated
animals was 0.930; it was 0.780 for acclimated animals.

Therefore, groups of small individuals are at a disadvantage if food is limited. From this, one could predict

that in oligotrophic waters the small ciliates and rotifers would be absent. This is only partly true, as representatives of these two groups are always present at least in low numbers. At this time, it is difficult to determine whether this results from the influence of a low concentration of particles excluding competitors (Luckinbill 1974) or whether this results from some kind of partitioning of the resources based on quality rather than on the size of particles. There are numerous indications that blue-green algae are less readily eaten by Crustacea (Lefevre 1950; Arnold 1971). Burns (1968b) observed that *Daphnia rosea* rejected food when a filament of *Anabaena* reached the labral region. Conversely, Kořínek (1972) has found a high efficiency in the transfer of energy from phytoplankton to *Daphnia* and to fish in ponds, where *Aphanizomenon flos-aquae* was an important constituent of the phytoplankton. This suggests a direct utilization of this blue-green alga by the Cladocera that were the main component of the zooplankton in the ponds. Gophen *et al.* (1974) demonstrated, by differential labeling, that with increased food concentration *Ceriodaphnia* assimilated a greater proportion of the bacterium *Chlorobicum* and a lesser proportion of the alga *Chlorella vulgaris*. This was probably because bacterial cells are more readily lysed.

There is a definite tendency for smaller species to have a shorter generation time than larger individuals, which have more offspring per brood. In small species of ciliates, rotatorians, and cladocerans there is less difference between the size of the mother and the daughter than in larger species and, therefore, a stronger competition for food among individuals of various generations or cohorts. Among the filter feeders, the copepods have the longest generation time and the least competition between mother and daughter. From this it follows that in bodies of water with a short-term fluctuation of particulate

319

food, it is the small organisms such as ciliates, rotatorians, and small cladocerans that, because of their
shorter generation time, have the advantage over the larger cladocerans and especially over the copepods. In addition, the smaller organisms are less vulnerable to fish
predation (see discussion below).

In all groups (except phantom midges), there are representatives of filter feeders and predators. It is rather strange that only in the Cladocera do the predators
have more efficient eyes than the filter feeders. Conversely, the cyclopoid copepods, which are often the most
numerous predators in zooplankton, have small and simple
eyes, whereas the small Cladocera, their potential prey,
have composite and large eyes. This difference, I think,
may be explained only by the assumption that the basic
features of the copepod body organization developed in a
turbid environment, whereas that of Cladocera developed
in a transparent one. This situation is not unique as the
predatory larvae of Chironomidae living on the bottom have
simple eyes, whereas the filtering mosquito larvae have
large, composite eyes.

Effect of Food Concentration on the Size and Growth Rate of Filter Feeders

There are numerous papers describing the effect of
food concentration and temperature on the filtration rate
of cladocerans and calanoid copepods. Chisholm and Stross
(1975) have shown that under optimal temperatures *Daphnia
middendorffiana*, living in arctic ponds with low primary
production, has the largest filtering capability of any
species of *Daphnia* investigated. The difficulty in fully
evaluating the competitive ability of various species is
that we do not have sufficient data on the metabolic cost
of collecting food because there are no comparable data on

the expenditures under various feeding rates. Additional details on this topic are given above. In the following discussion, more emphasis has been placed on the effects of food concentration on growth rate that should summarize long-term gains and losses under various environmental conditions.

Generally, in habitats with less-concentrated food, the size of individuals within comparable taxonomic groups is larger than in habitats with more-concentrated food. This is in accordance with the relatively higher metabolic rates (per unit weight) of small individuals that were mentioned above. Therefore, zooplankton individuals should be larger in big lakes because lake morphology strongly influences the productivity level of lakes (Rawson 1942). Findenegg (1948) has shown that within a set of Karinthian lakes the specimens of *Daphnia cucullata* are definitely larger in larger lakes. The fact that the largest Cladocera and calanoid copepods are found in small, very productive, and often temporary ponds can be explained by the assumption that predation is another factor that influences the size of the species present in a body of water. This effect will be discussed later.

Hrbáčková-Esslová (1962) has compared the growth rate of three species of *Daphnia* under the natural food conditions of Slapy Reservoir and under artificially increased food conditions. Contrary to the expectations derived above, she has found that the individuals that have the first brood of eggs (primiparae) are definitely smaller when grown under natural food conditions (about 15 J/liter in particles) than under enriched food conditions (more than 300 J/liter). Nothing is known about the underlying mechanism that determines the distribution of the produced biomass to the body increment of the mother and the biomass and numbers of the eggs. This seems ecologically unsound because under conditions of scarce food, development is prolonged and the metabolic losses per increase of weight unit are therefore higher. Thus, the biomass produced by the female is preferentially used in the production of new individuals that have a relatively higher metabolic rate per unit weight instead of for individual growth that would lower the weight-specific metabolism of biomass. Conversely, a population composed of smaller individuals is better adapted to survive the stronger fish predation in transparent waters. It can be concluded that the food-limited *Daphnia* populations maximize numbers and not individual biomass.

Hrbáčková (1971) has shown in *Daphnia hyalina* that there is a positive relation between the size of the female and its offspring. The size of neonates of the females from natural food conditions is the same as the size of offspring of the females from enriched cultures. However, the comparable instars of the females under natural food conditions are smaller, so the size of their offspring females from natural conditions of Slapy Reservoir are relatively larger. This supports the previous statement that larger individuals are produced under low-food conditions. Consequently, the decrease in the size difference between mothers and daughters decreases the metabolic disadvantage between daughters and their mothers when competing for the same low concentrations of food.

Weglenska (1971) has observed in Mikolajskie Lake that the growth rate of several species of Cladocera was not increased by the increase of the food concentration, and therefore, it seems that in this lake the populations of Cladocera are not food-limited. Burns (1968b) has found that a considerable decrease of natural food concentration was needed to increase the filtering rate, suggesting that the natural concentration of food particles was

322

well above the incipient limiting concentration for the
filtering rate. On the other hand, the data of Nauwerck
(1963) and the effect of temperature on the metabolism
of Cladocera suggest (Blažka 1966) that the food concentration in the lakes and ponds they studied was food-
limited and below the incipient limiting concentration.

Weglenska (1971) was, so far as I know, the first
author to show that the concentration of food adequate for
Cladocera was inadequate for Copepoda. *Eudiaptomus graciloides* increased its growth rate significantly when the
food concentration was doubled, whereas Cladocera did not
change its growth rate. On the other hand, this finding
contradicts the general tendency for Cladocera to become
more important with increasing eutrophy of the water
(McNaught 1975). Bogdan and McNaught (1975) found that
Diaptomus minutus filtered cells greater than 22 µm much
less efficiently than cells smaller than 22 µm, whereas
Daphnia galeata mendotae filtered both size groups with
the same efficiency. Filtration rate per unit weight of
the organism was nearly three times greater in *Daphnia*
than in *Diaptomus*. In my opinion, this is consistent with
the longer development rate of *Diaptomus* than *Daphnia*, and
it does not necessarily mean that *Daphnia galeata mendotae*
should outcompete *Diaptomus minutus*, because only a part
of their energy budget (consumption) is compared.

r and K Strategists

It may appear that the Cladocera are so adapted to
the intrinsic rate (r) strategy by parthenogenesis that
the K strategy is meaningless. To me it is meaningful if
a K strategist is defined as a population that under a
given concentration of food particles produces the largest
biomass. It follows from this that an r strategist is at
an advantage in a temporary pond because the food there

is plentiful, at least at the beginning, and the species
that can produce the most numerous population during a
short period is at an advantage irrespective of its efficiency in the utilization of the food resources. At first
this might point to a predominance of ciliates and rotatorians as they have a short generation time. However,
it is the average number of offspring produced per day by
one female that is important. Under good food conditions
at 20 C, large female cladocerans can produce about ten
offspring after eight days and then several tens of offspring every three days. This is more than the several
offspring produced in two days by rotatorians and the
three individuals produced per day in ciliates. The production of relatively smaller eggs in *Daphnia* when food
is plentiful is also advantageous in this situation.

It can be assumed, but to my knowledge it has not
been directly examined, that the r strategists of the same
size have a higher maintenance cost (weight-specific metabolism) than the K strategists. To examine this question
indirectly, Hrbáčková-Esslová (1963) has compared two
species of *Daphnia* that are morphologically so similar that
they have not been recognized as different species by the
majority of authors. One of these species (*D. pulex*) is
typical of small pools and ponds rich in nonnosestic algae;
the other (*D. pulicaria*) is typical of seston-poor waters
(high mountain lakes, ponds with *Aphanizomenon flos-aquae*)
in Central Europe and oligotrophic lakes (e.g., Yellowstone
Lake, the type locality, and Lake Tahoe in North America).
At high food concentrations *D. pulex* reaches maturity 8%
faster than *D. pulicaria*. At the concentration of seston
in Slapy Reservoir newborn individuals of *D. pulex* died
after several days under signs of exhaustion without reaching reproductive maturity. When food concentration was
increased to five times greater than natural levels or

when artificially grown algae were added, *D. pulex* reached
maturity after five days, showing that the concentration
of food, not its quality, was critical. Under natural
low-food conditions *D. pulicaria* took about twice as long
to reach maturity as it did at high food concentrations.

This comparison, in my opinion, shows very clearly
the difference between an r and a K specialist: the short
generation time under high food levels in an r specialist
is related to the inability to concentrate an adequate
amount of food under the same conditions that a K specialist is able to do so. Whether this ability of the K specialist results from slower action of the relatively larger
filtration area or from another type of adaptation is unknown at present. Egloff and Palmer (1971) have found that
D. rosea, a typical lake species, has a relatively larger
filtration area than *D. magna*, which is typical of small
eutrophic ponds. These species, which belong to different
subgenera, also differ in many other morphological aspects.
It is not known whether this relatively larger filtration
area results from a greater length of the row of filtration setae or from a greater length of the setae. Of
these two species, *D. magna* can be considered as an r specialist and *D. rosea* as a K specialist.

To achieve a more effective use of low-food concentrations, the K specialist must have a much closer coord-
ination among physiological functions than an r specialist.
Since above-optimum temperatures affect different functions
somewhat differently, it might be expected that the K specialist from the same geographical region (adapted to the
same temperature conditions) should stop some basic functions (e.g., reproduction) at a lower temperature than an
r specialist. Hrbáčková-Esslová (1963) has shown that
Daphnia pulicaria stopped propagation at 28 C, whereas
Daphnia pulex from the same area stopped it at 32 C.

Brandlova *et al.* (1972) found the same difference in the
Canadian populations of these two species. The temperature
at which reproduction ceases is also influenced by the
temperature conditions of the particular habitat. For example, populations of *Daphnia pulicaria* from high-altitude
lakes, which probably never reach 10 C, stopped reproduction at 18 C (Hrbáčková-Esslová 1966).

In addition to the long-term adaptation to temperature
there exists a short-term adaptation. Kibby (1971) has
found a significant difference in the filtering rates at
comparable temperatures between the populations of *Daphnia
rosea* cultured at 12 C and 20 C, respectively. The maximum
rate was close to the acclimation temperature. Thus, for
comparative studies on populations of different origin, it
is necessary to cultivate species or strains under investigation for one or two generations at identical temperatures.

I consider *D. pulex* and *D. pulicaria*, respectively,
as extreme r and K specialists among the species of the
subgenus *Daphnia*, and all the other species of this subgenus are somewhere in between. Hrbáčková (unpublished
manuscript) has compared various *Daphnia* species, and
strains of *D. hyalina*, from oligotrophic and eutrophic
conditions (e.g., *D. galeata mendotae* from Windermere Lake
and Loch Leven). In all the species, except *D. pulex*, she
found that they were able to reproduce at the food conditions found in Slapy Reservoir (average energy content is
between 10 and 20 J/liter). The species from more oligotrophic conditions were relatively more productive (a
relatively shorter generation time and a relatively greater
number of eggs) than those from more eutrophic conditions
when fed the Slapy food concentration. At higher food
concentrations the same species from more eutrophic conditions showed a relatively greater production than those
from the oligotrophic conditions. The difference in results was smaller than between *D. pulex* and *D. pulicaria*.

From this it follows that in all r and K pairs of
species or strains, a concentration of food must exist at
which the growth rate in both species or strains is identical. I do not understand why this food concentration,
in all experiments performed to date by Hrbáčková, has
been higher than that in Slapy Reservoir.

Long-term Physiological Adaptation for Low-Food-Level
Conditions

Hrbáčková (1971) and Kořínek (1970) observed that the
population of *D. hyalina* in Lago Maggiore had a longer
generation time (about 16 days) than the population of *D.
hyalina* from Slapy Reservoir when cultured in Slapy water.
This could be explained by a lower concentration of food
in that lake, but the generation time did not change when
10^6 cells/ml of *Chlorella* and *Scenedesmus* were added to
the culture. At this algal concentration, the generation
time of several other species and strains is five or six
days. After several generations in the laboratory at high
food concentrations the Lago Maggiore strain of *D. hyalina*
shortened its generation time to nearly seven days (about
6% longer than the *D. hyalina* group native to Slapy Reservoir). This decrease in generation time was not caused
by a selection effect, because in June 1972 Blažka brought
specimens from Lago Maggiore to Prague, and the offspring
of specimens that were cultured at high food levels had
a developmental rate identical with that of the earlier
cultivated animals. Kořínek (1970) has confirmed the
longer generation time not only in the Lago Maggiore populations but also in three other species (including *D.
pulicaria*) living in oligotrophic Canadian waters under
high food conditions (Kořínek, unpublished manuscript).
Moreover, Kořínek (1970) has found that the length of the
embryonic development (cultivated outside of the female's

body) was unaffected by the low food adaptation of the mother.

A speculative explanation why the increased food level does not increase the rate of development is that in oligo-trophic water bodies, the concentration of food during some periods is so low that it cannot support the *Daphnia* population unless the animal has a lowered metabolic rate. Under these conditions metabolism is controlled by the level of the enzyme activity and not by the food concen-tration. Hebert (1973) has found that in a clone of *D. magna*, prolonged starvation induced a loss of lactate de-hydrogenase activity in two of the three zones of activity obtained by electrophoresis from well-fed individuals. Brief starvation produced additional bands below each original zone. Blažka (1966) has found a different tem-perature dependence of metabolism in *Daphnia* from natural populations and in *Daphnia* from well-fed laboratory cul-tures.

It is frequently observed that in large lakes the sexual propagation of *Daphnia* species is very low or nearly nonexistent (Wesenberg-Lund 1939). One can speculate that a population with a slower replacement rate is more nu-merous at the end of a low-food period than a population with a higher replacement rate that would considerably decrease population abundance.

Strategies to Lower the Impact by Predation

Effect of Fish Predation on the Size of Prevalent Zoo-planton Species

It may be assumed that populations under high pre-dation pressure can be recognized by a strong predominance of younger instars. Ravera (1954) compared the seasonal changes in the frequency of the smaller *Mesocyclops leuckarti* and the larger *Cyclops abyssorum maiorus* in

328

Lago Maggiore. The abundance of nauplii is considerably greater than that of adults, and this difference is greater in larger species. There are several other papers that show numbers of nauplii to be considerably larger than those of adults in Copepoda. For Cladocera, the predominance of younger instars is much less apparent (Lyakhnovich *et al.* 1969). This would point to a greater predation pressure on Copepoda than on Cladocera. On the other, Hrbáček (1962) and Hrbáček and Novotná-Dvořaková (1965) found that the influence of an increased fish stock on the decrease of the size of prevalent species was more pronounced in Cladocera than in Copepoda. The total biomass of zooplankton was not affected. This discrepancy occurs at least partly because in addition to predation, the length of the preadult and adult instars is another important factor. In Cladocera the adult period, where the instars have small increments of change, is considerably longer than the preadult period.

Hrbáček and Hrbáčková-Esslová (1960) studied the relative increment change in length of juvenile and adolescent instars of various species and strains of *Daphnia* from various habitats with different fish stock in laboratory cultures. These experiments were conducted at a high food concentration and at 20 C. The *Daphnia* populations from habitats with no (high mountain lakes and temporary ponds) or low (fishponds) fish predation pressure showed about a one-third increase in body length irrespective of the species. This relative increase in increment of growth has been observed in natural populations of Stomatopoda (Brooks 1886). This relative increase in growth increment is likely to represent the lowest energy losses due to molting. Species (e.g., *D. cucullata*) and strains (e.g., *D. hyalina*) found in bodies of water with a high fish stock showed lower growth increments not only per instar but also from neonata to primipara. Smaller females produce smaller eggs than larger ones. The smaller relative growth increment of individuals hatched from these smaller eggs also results in small females. Galbraith (1967) and Hutchinson (1971) have shown that rainbow trout and yellow perch exhibit a definite size-selective predation on *Daphnia*. Thus, the advantage of reproduction at a smaller size is apparent. What is less understandable is why the *Daphnia* living under high predation pressures do not decrease the number of instars instead of their relative growth increments. In the preceding paragraphs, it has been shown that a low food level also decreases the relative growth increment. All these observations can be reasonably explained by the assumption that the extra energy expenditures resulting from extra molts are less than the gains from the lower maintenance cost or more effective grazing of larger individuals.

In eutrophic ponds it has been found (Hrbáček and Hrbáčková-Esslová 1966) that some species (e.g., the *D. pulex* group) are restricted to habitats with a low fish stock while others (e.g., *D. cucullata*) are restricted to habitats with a high fish stock; still others (e.g., the *D. longispina* group) have been found in habitats with both high and low fish stock. Nilsson and Pejler (1973) have found different members of the *D. longispina* group in Norwegian lakes that are stocked differently with fish. It may be that the outcome of competition and predation under oligotrophic conditions is different from that of eutrophic conditions. The possibility that other factors are responsible for the observed difference should not be overlooked (e.g., the different ability to withstand intensive flushing of zooplankton in lakes by throughflowing water resulting from different behavioral patterns).

Brooks and Dodson (1965) have formulated the size

selection hypothesis to explain the decrease in size of
dominant zooplankton species when the fish stock is more
numerous. This hypothesis does not take into account the
possibility of considerable differences in resource utilization of species of comparable size that has been shown
by the comparison of *D. pulicaria* and *D. pulex*. The hypothesis holds only if comparable species are strategists
in their size rank. On the contrary, Neil (1975) has
found in a laboratory microcosm that in predictions based
on the size efficiency hypothesis, *Ceriodaphnia* (a smaller
species) outcompeted *Daphnia* (a larger species). *Daphnia*
were able to colonize the microcosm only when *Ceriodaphnia*
were partly suppressed by a temporal predation by a small
fish. It is very likely that the population of *Ceriodaphnia* used in the colonization of the microcosm was a
K strategist, whereas the population of *Daphnia* was an
r strategist.

Effect of Predation by Invertebrates

Dodson (1974a) suggested that large filter feeders
are likely to outcompete small ones over long-term periods,
whereas the observations show a short-term effect. He
proposed the hypothesis that the predation of invertebrates
intensified the relative decrease in numbers of smaller
species. In Copepoda, there are several observations that
the adults prefer smaller individuals of prey. Moreover,
Brandl and Fernando (1975) have found that the part of
Ceriodaphnia ingested by *Cyclops* increased as the size of
the prey decreased. Wong and Ward (1972) have observed
that the very small fry of perch prefer the smaller individuals of *D. pulicaria*. Hall *et al.* (1970) intention-
ally increased the population of larger invertebrate predators (Notonectidae and Odonata) in experimental ponds
but were unable to see a change in the species composition

331

of zooplankton comparable to that of an increased fish
population. Hrbáček and Novotná-Dvořaková (1965) have
destroyed the whole plankton association by copper sulfate
in two experimental ponds. During the recovery period
three consecutive peaks of phytoplankton, Rotatoria, and
large Cladocera occurred. When rotenone was used in other
ponds, phytoplankton, protozoa, and Rotatoria were nearly
unaffected and no peaks developed as in previous experiments. At the end of the recovery period, large *Daphnia*
predominated. There was no prolonged persistence of more
numerous protozoa and Rotatoria populations even though
the invertebrate predators were destroyed by rotenone.
Because of their longer generation time, they were unable
to develop large populations before the development of
populations of larger Cladocera.

The previously discussed strong predominance of younger instars of Copepoda than of Cladocera may be related
to the greater predatory impact of adult Copepods on nauplii than on young cladocerans.

An indirect sign of the importance of invertebrate
predation is the fact that in all taxonomical groups of
zooplankton there are predators and filter feeders with
a strong ability for short, rapid displacement (skip).
Green and Lan (1974) have found that this ability is an
important factor in decreasing the predation of *Hexarthra*
by *Asplanchna*. The decrease of the relative growth increment in populations of *Daphnia* under high predation pressure by fish (Hrbáček and Hrbáčková-Esslová 1960) reduces
the impact of invertebrate predators selecting smaller individuals, because it shortens the period of smallest instars.

Porter (1972) and Fott (1975) have observed that when
the exploitation of phytoplankton populations by zooplankton is increased, the relative proportion of flagellates

increases. An active avoidance of the current produced by filter feeders can be considered as a possible factor responsible for this change.

Very limited data exist on how intensive predation must be in order to induce a change in the species composition. Dodson (1972) has found that a 95% mortality of *D. rosea*, mainly caused by *Chaoborus*, has not induced its replacement by a species more resistant to predation. It is very likely that the predation affecting mainly mature instars typical of fish predation is more effective in the change of species than the predation that affects mainly immature stages. Lyakhnovich *et al*. (1973) have found that increased impact by small fish on *D. pulex* and *D. longispina* not only eliminates larger classes of mature individuals but also increases the number of eggs in the remaining smaller class of mature individuals.

Effect of the Visibility of Prey on Intensity of Predation

Brooks (1965) formulated the hypothesis that it is not the total size of *Daphnia* but the visible part of the body that is the decisive factor in the size-selection mechanism. In his opinion, even though the helmet increases the size of *Daphnia*, the helmet is not visible to fish, and it does not increase the size of *Daphnia* that affects the intensity of predation. This hypothesis does not take into account the discussed decrease of the relative growth of the body, without the helmet, in those species living under high predation pressure. Some of these species have a helmet, whereas others do not, and even in those with a helmet the growth increment per instar is lower than one-third. Nor does this hypothesis explain why turbulence is the effective agent determining the helmet size (Brooks 1947; Hrbáček 1959; Hazelwood 1962; Jacobs 1962, 1970). The argument leads to the question whether the helmet has to be considered as an integral part of the body or whether it is an outgrowth on the head of the animal. The number of cells forming a claw of a certain size is very likely similar in different species and forms. If the helmet is an integral part of the body, the carapace of the helmeted forms necessarily must be smaller in relation to the claw than the carapace of the unhelmeted forms. Hrbáček (1969b) found a very close correlation between the size of the carapace and the size of the claw irrespective of whether the species was helmeted or not.

Zaret (1969) has made an important contribution to the argument of visibility, claiming that it is the eyes of the cladocerans that are visible to the fish. He has shown that intensive predation in the Gatun Lake, Panama, of *Ceriodaphnia* maintains the coexistence of two genetically isolated populations. One population is propagated more successfully, and it is more intensively exploited by fish due to its larger eye. Wawrik (1954, 1966) observed that the size of the eye of *D. longispina* increases with the increasing altitude of the body of water in which the population lives, and this can be considered as supporting evidence. Generally, the predation pressure from both vertebrates (gradual prevalence of salmonid fish and lack of fish and amphibians in higher elevated lakes) and invertebrates decreases with increasing altitude. Fishponds are an exception in this respect because they have artificially decreased fish stock. Therefore, the presence of populations of *D. longispina* with larger eyes in lowland fishponds is in agreement with the hypothesis of the predation effect but not with the light effect supposed by Wawrik.

The critical evaluation of the importance of the size of the eye in the selection by predation has a drawback, namely, that we do not really understand the advantage to

a filter feeder of having a larger eye (only the size is
increased, not the number of *omnatida*). McNaught (1971)
has found that the sensitivity of the *Daphnia* eye to various colors is changed when the photoenvironment is changed.
In my opinion, this supports the conclusion that the primary function of the eye is not in detecting green clouds
of algae; this could counteract the disadvantage of a
greater predation because of a large eye. The other difficulty in the evaluation of the effect of visibility is
that some species of pelagic crustaceans are more or less
definitely colored (brownish, yellow, pink, and bluish)
and this clearly increases their visibility. Nilsson and
Pejler (1973) concluded from their investigation of North
Swedish lakes that the variability in their zooplankton
composition is at least partly caused by the differences
of their fish stock as the planktivorous fish eliminate
colored species.

Mellors (1975) has shown that *Lepomis gibbosus* and
Perca flavescens preferred ephippial females of *Daphnia
galeata mendotae* to similarly sized nonephippial females.
Moreover, females with darker ephippia were consumed at
a greater frequency than those with less pigmented ephippia.

To my knowledge there is no investigation showing the
effect of the food in the gut on the visibility of hyaline
planktonic animals.

Interaction of Feeding, Predation, and Wind-Induced Tur-
bulence on the Vertical and Horizontal Distribution of
Zooplankton

It is a well-established fact that during the daytime
most of the larger zooplankton species are concentrated
in deeper lake layers. This is especially true in lakes
with transparent water. Weismann (1874) explained this
by the crepuscular habits of these animals. Although there

is no lack of papers describing the interaction between
light intensity and its avoidance by planktonic animals
(Hutchinson 1967), the underlying cause-and-effect relations are not yet understood. The maximum density of
phytoplankton and the maximum intensity of phytosynthesis
may not occur at the surface either, but their level of
maximum biomass is usually above that of the zooplankton
biomass. Moreover, in the same lake various species do
not have their maxima at the same depth. Elster (1954)
has found, in the same lake studied by Weismann, that during the day the nauplia of copepods remain in the well-
illuminated upper 5-m layer. Tappa (1965) has shown a
different vertical pattern of six limnetic species of
Daphnia in one lake. Lane and McNaught (1970) used the
difference in the vertical distribution for a quantification of the niche size of some zooplankton species in Lake
Michigan. Lackey (1973) has changed the vertical gradient
of several parameters by artificial destratification, but
the change in zooplankton species composition was slight
(e.g., a lower number of *D. schøedlri* but no change in *D.
galeata mendotae*). In the more transparent lakes, food
is a limiting factor in the development of zooplankton
populations. The question arises as to what are the advantages of zooplankton living during daytime in the deeper
layers remote from the food concentration. During the
night, most of the species living in deeper layers move
upwards where there is the maximum concentration of food.
McLaren (1963) proposed the hypothesis that by accumulating
and digesting food at higher temperatures and by metabolizing it at lower temperatures, a metabolic bonus was
obtained. Skadovsky (1939) observed that under laboratory
conditions the well-fed individuals of *D. pulex* were photonegative, whereas starving individuals from the same culture were photopositive. It is uncertain whether this

change was related to the periodic changes of phototropism
reviewed by Hutchinson (1967). This finding of Skadovsky
(1939) is compatible with McLaren's (1963) hypothesis with
the restriction that its optimal use has to be related to
the digestion period, and therefore to temperature, not
to the light conditions. For marine zooplankton, Kerfoot
(1970) has studied the bioenergetics of vertical migration
and found that although productivity decreases regularly
with decreases in light intensity, the energy available
to consumers adapted to different light levels exhibits
a distinct biomodality. All these field observations, even
when using the most sophisticated approach, have contributed little to the understanding of the reasons of cre-
puscularity in various plankton species.

Forel (1878) noticed that the crepuscular behavior
of the pelagic Entomostraca decreased the impact of fish.
Banta (1921a) has shown that from the same basic stock of
Daphnia it is possible, even under parthenogenetic propagation, to select either a photopositive or a photonegative
strain. Fish predation can function as a selective agent
because of the greater survival of photonegative individuals. There are not adequate field observations from which
the vertical interaction of plankton-feeding fish and zooplankton during a period of 24 hr could be evaluated. As
supporting evidence for the assumption that light is not
the prime factor inducing vertical migration, cases of
reverse diurnal migrations can be mentioned. Bayly (1963)
observed the reversed diurnal migration of *Diaptomus ban-
toranus* in Lake Rudolf and of *Bocckerella propigna* and
Daphnia carinata in Lake Aroarotamanihe; he attributes
this unusual behavior to the unusually high pH values in
these lakes (9.5 and 9.0, respectively). In laboratory
experiments, the photonegative response was changed to a
photopositive response by adding carbon dioxide, and

therefore decreasing pH (Loeb 1906). Dumont (1972) explained that the reverse diurnal migration of *Asplanchna priodonta* in a eutrophic pond is induced by competition for space.

Wind induces water currents close to the surface of water. Stavn (1971) has shown in laboratory experiments that in the Langamuir circulation *Daphnia magna* tends to swim against the current. In a slow-current system (velocity range 0.6-2.4 cm/sec), *Daphnia* accumulated during daytime in the downwellings by swimming upwards. In the fast-current system (velocity range 2.0-8.8 cm/sec) the *Daphnia* accumulated during daytime in the upwellings by swimming downwards. George and Edwards (1973) have observed at a windspeed of 600-650 cm/sec that there is an accumulation of *Daphnia hyalina* var. *lacustris* in the upwelling zone at a depth of 1 m (illumination intensity was about 60% of that at the surface); this is in close agreement with laboratory observations. Hrbáček (1959) has considered that the helmet in *Daphnia cucullata* is a morphological feature that increases the effectiveness of swimming against a current and so increases the ability of *Daphnia* to maintain its position within the vertical light gradient to avoid higher predation close to the surface. Jacobs (1962) and Egloff (1968) have found further morphological changes in turbulent environments that can be interpreted in the same way.

In cases where several species of *Daphnia* coexist in one lake, only those living close to the surface have helmets, whereas those living in deeper layers have rounded heads and very often are larger in size (Woltereck 1930). The only exception to this rule is *D. ambigua*, which is a small species with a small helmet. It is found in deeper layers of some North American lakes and reservoirs (Stavn 1975). This suggests that there is a lesser impact of

fish in the deeper layers. Green (1967) has found a similar situation in Lake Albert in the tropics in the horizontal, but not the vertical gradient. The population near
the shore exhibited a pointed head, whereas the open-lake
population was rounded headed. Green interpreted this
finding as evidence for a different fish predation intensity between the inshore and the open water. Zaret (1969)
has observed two genetically isolated forms of *Ceriodaphnia
dubia* in the open and the inshore water of Lake Gatun, and
he explains this by a differential impact by fish.

CONCLUSIONS

The position of a species in the trophic hierarchy
rests upon a delicate equilibrium between intake and ex-
penditure of energy at the level of an individual and between natality and mortality at the population level. It
is not possible to correlate the seasonal change of the
phenotype with one environmental factor. This has been
exemplified by the effect of predation and food concentrations, both of which can decrease the body size of
mature individuals in nature. With laboratory cultures,
it is possible to get information on the genotype and to
discriminate between the effects of these two factors.
Another complicating factor is that the changing environment has a different effect than a stable environment on
basic population parameters (Dunham 1938). Physical
(temperature, light), chemical (pH, salinity, calcium,
dissolved nutrients), and biological (concentration of
food organisms, competition, predation, parasitism) factors change considerably in time and space, and a successful species must necessarily be able to partially adapt
both phenotypically and genotypically to a wide range of
specific parameters. In the previous discussion this response has been exemplified by changes in organism size.

Organism growth and egg production are easily recognized parameters that summarize the effect of many environmental factors and the adaptive response of animals to environmental change.

The strategies of taxonomically closely related species necessarily must be similar. The necessity for plankton species to make very slight adaptations to their environment is not easily measured because of the crudeness of the methods available for measurement of vital parameters (e.g., concentration of food, predatory impact). Strategies of taxonomically distant species differ in a greater set of parameters, but their relative importance is not immediately apparent. For example, in the previous discussion it has been shown that in every zooplankton genera there are species which vary in their ability to skip. We do not know how effective this saltatory ability is in avoiding predation by various predators. Also, as the concentration of particles from a circular current produced by the cilia of rotifers may lead to a local decrease of the concentration of particles, this ability to skip may be helpful in more effective feeding.

From the viewpoint of the investigator's strategy, more sophisticated methods can be used and a more defined set of environmental parameters can be produced in a laboratory study. However, these studies might be inapplicable to the conditions in nature. In the well-known fundamental laboratory study on energy transformation of *Daphnia pulex* (Richmann 1958), the lowest food concentration may be well above the concentration encountered by the species in nature. This depends on whether the species is *D. pulex* or *D. pulicaria*. Another difficulty in extrapolating laboratory studies to field conditions is that the effect of significant parameters cannot be considered to be nearly linear. This has been exemplified by the comparison of *D. pulex* and *D. pulicaria*, which show only a small difference in the growth rate under high food conditions and great difference at low food conditions (Hrbáčková-Esslová 1963). This would suggest that the most effective strategy for the investigator is to interchange field and laboratory investigations. This approach does not necessarily lead to unequivocal results that can be exemplified by studies of the environmental factors that are involved in the helmet formation of *Daphnia*. As reported in the aforementioned laboratory studies, several investigators found that temperature and turbulence have a positive effect on the length of the helmet during postnatal development. Jacobs (1970) has thoroughly investigated the interaction of temperature, food, and wind-induced turbulence on the size of the helmet in *D. galeata mendotae* in Klamath Lake, Oregon. He has shown satisfactorily by multiple correlations that 78% of the prenatal and 86% of the postnatal variability of the relative size of the helmet can be explained by environmental parameters. The difficulty is that the partial correlation between turbulence and the relative size of the helmet was negative, and temperature had no effect on the postnatal development.

Irrespective of these difficulties, I think that the only research strategy that can avoid circular reasoning, whether expressed by words or by equations, is to compare the results developed by one methodological approach with the results developed by a different approach. It is the researcher's individual decision whether to use a broad or a narrow spectrum of the species under investigation.

Allan, J. D. 1973. Competition and the relative abundance of two cladocerans. *Ecology 54*, 484-498.

Arnold, D. E. 1971. Ingestion, assimilation, survival and reproduction of *Daphnia pulex* fed with seven species of blue-green algae. *Limnol. Oceanog. 16*, 906-920.

Banta, A. M. 1921a. Selection of Cladocera on the basis of a physiological character. *Carnegie Inst. Wash. Publ. No. 305*, 170 p.

Banta, A. M. 1921b. A convenient culture medium for daphnids. *Science 53*, 557-558.

Banta, A. M. 1939. Studies on the physiology, genetics and evolution of some Cladocera. *Carnegie Inst. Wash. Dep. genetics p. Nr. 39*, 285 pp.

Bayly, J. A. E. 1963. Reversed diurnal vertical migration of planktonic Crustacea in inland waters of low hydrogen ion concentration. *Nature 200*(4907), 704-705.

Birge, E. A., and Juday, C. 1932. Dissolved oxygen and oxygen consumed in the lake waters of northeastern Wisconsin. *Trans. Wisc. Acad. Sci. 27*, 315-486.

Blažka, P. 1966. Metabolism of natural and cultured populations of *Daphnia* related to secondary production. *Int. Ver. Theor. Angew. Limnol. Verh. 16*, 380-385.

Bogdan, K. G., and McNaught, D. C. 1975. Selective feeding by *Diaptomus* and *Daphnia*. *Verh. Int. Verein. Limnol. 19*, 2935-2942.

Bowkiewicz, J. 1935. Komplexy entomostraca yako wskazniki wystepowania sielawy. *Fragm. Faun. Mus. Zool. Polon. 22*, 229-241.

Bragg, A. N. 1936. Selection of food in *Paramecium trichium*. *Physiol. Zool. 9*, 433-442.

Brandl, Z. 1973. Laboratory cultures of cyclopoid copepods on a definite food. *Věst. čs. spol. zool. 38*(4), 81-88.

Brandl, Z., and Fernando, C. 1975. Food consumption and utilization by two cyclopoid copepods, *Mesocyclops edax* and *Cyclops vicinus*. *Int. Rev. Ges. Hydrobiol. 60*, 471-494.

Brandlova, J., Brandl, Z., and Fernando, C. H. 1972. The Cladocera of Ontario with remarks on some species and distribution. *Can. J. Zool.* *50*(11), 1373-1403.

Brody, S. 1945. *Bioenergetics and Growth*. Reinhold, New York, 1023 pp.

Brooks, J. L. 1947. Turbulence as an environmental determinant of relative growth in *Daphnia*. *Proc. Nat. Acad. Sci.* *33*, 141-148.

Brooks, J. L. 1965. Predation and relative helmet size in cyclomorphic *Daphnia*. *Proc. Nat. Acad. Sci.* *53*(1), 119-126.

Brooks, J. L., and Dodson, S. T. 1965. Predation, body size and composition of plankton. *Science* *150*(3692), 28-35.

Brooks, W. K. 1886. Report on the Stomatopoda collected by H. M. S. Challenger during the years 1873-1878. In Report of the scientific results of the voyage H. M. S. Challenger during the years 1873-1876. *Zoolog.* *16*, 44-46.

Buikema, A. L. 1972. Oxygen consumption of the cladoceran, *Daphnia pulex*, as a function of body size light and light acclimation. *Comp. Biochem. Physiol.* *42*(A), 877-888.

Burns, C. W. 1968a. The relationship between body size of filter feeding Cladocera and the maximum size of particle ingested. *Limnol. Oceanog.* *13*(4), 675-678.

Burns, C. W. 1968b. Direct observation of mechanisms regulating feeding behavior of *Daphnia* in lakewater. *Int. Rev. Ges. Hydrobiol.* *53*(1), 83-100.

Burns, C. W. 1969. Particle size and sedimentation in the feeding behavior of two species of *Daphnia*. *Limnol. Oceanog.* *14*(3), 392-402.

Butorina, L. G. 1971. O sposobenosti Polyphemus pediculus (L.) pitatsja bakterijami i prostejšimi. *Biol. vnutr. vod, Inform. bjul.* *11*, 47-48.

Cannon, H. G. 1933. On the feeding mechanism of Branchiopoda. *Phil. Trans. Roy. Soc.* *222*(B), 267-352.

Chisholm, S. W., and Stross, G. R. 1975. Environmental and intrinsic control of filtering and feeding rates in arctic *Daphnia*. *J. Fish. Res. B. Can.* *32*, 219-226.

Coker, R. E., and Hayes, W. J. 1940. Biological observations in Mountain Lake, Virginia. *Ecology 21*, 192-
198.

Confer, J. L. 1971. Intrazooplankton predation by *Mesocyclops edax* at natural prey densities. *Limnol.
Oceanog. 16*, 663-666.

Czeczuga, B. 1960. Zmiany plodnosci niektorych przesta-
vicieli zooplankton i Crustacea jezior Raygradskich.
Polski arch. hydrobiol. 7, 61-91.

Dodson, S. J. 1972. Mortality in a population of *Daphnia
rosea*. *Ecology 53*(6), 1011-1023.

Dodson, S. J. 1974a. Zooplankton competition and predation: an experimental test of the size-efficiency hypothesis. *Ecology 55*, 605-613.

Dodson, S. J. 1974b. Adaptive change in plankton morphology in response to size selective predation: a new
hypothesis of cyclomorphosis. *Limnol. Oceanog. 19*(5),
721-729.

Dragesco, J. 1948. Étude microcinématographique de la
capture et de l'ingestion de proses chez les cilies
holotriches gymnostomes. *C. R. Int. Congr. Zool.
Paris 13*, 226-227.

Dumont, A. J. 1972. A competition-based approach of the
reverse vertical migration in zooplankton and its
implications, chiefly based on a study of the interactions of the rotifer, *Asplanchna priodonte* (Gosse)
with several Crustacea Entomostraca. *Int. Rev. Ges.
Hydrobiol. 57*(1), 1-38.

Dunham, H. H. 1938. Abundant feeding followed by restricted and longevity in *Daphnia*. *Physiol. Zool. 11*,
339-407.

Egloff, D. A. 1968. The relative growth and seasonal
variation of several cyclomorphic structures of *Daphnia
catawba* Coker in natural populations. *Arch. Hydrobiol.
65*, 325-359.

Egloff, D. A., and Palmer, D. S. 1971. Size relations
of the filtering area of two *Daphnia* species. *Limnol.
Oceanog. 16*, 900-905.

Elster, H. J. 1954. Uber die Populationsdynamik von
Eudiaptomus gracilia Sars and *Heterocope borealis*
Fisher in Bodensee-Obsersee. *Arch. Hydrobiol.* (Suppl.)
20, 546-614.

Ertl, M., Juriš, Š., and Vranovský, M. 1965. Some remarks about the plankton of the great and little Hincovo Pleso in the High Tatra. *Sborník prác o Tatransko národním parku 8*, 57-69.

Esslová, M. 1959. Embryonic development of parthenogenetic eggs of *Daphnia pulex*. *Věst. čs. spol. zool. 23*, 80-88.

Findenegg, J. 1948. Die *Daphnia*-Arten der Kärtner Gewässer und ihre Beziehung zur Grösse des Lebensraumes. *Öst. zool. Zsch. 1*(6), 519-532.

Forel, F. A. 1878. Faunistische Studien in den Süsswasserseen der Schweiz. *Zsch. f. wiss. Zool. 30 Suppl. Bd. 1878*, 383-391.

Fott, J. 1975. Seasonal succession of phytoplankton in the fishpond Smyslov near Blatna, Czechoslovakia. *Arch. Hydrobiol.* (Suppl. 46 Bd. Algological Studies) *12*, 259-279.

Frantz, T. C., and Cordone, A. J. 1970. Food of Lake Trout in Lake Tahoe, California. *Fish and Game 56*(1), 21-35.

Frič and Vávra. 1897. Untersuchungen über Fauna der Gewässer Böhmens III. Untersuchung zweier Böhmerwaldseen des Schwarzen u. des Teufelsees. *Arch. Naturn. Landesforschung von Böhmen 10*, 74 pp.

Fryer, G. 1957. The feeding mechanisms of some freshwater cyclopoid copepods. *Proc. Zool. Soc. London 129*, 1-25.

Gajevskaja, N. 1933. Zur Ökologie, Morphologie und Systematik der Infussion des Baikalsees. *Zoologica Stuttg. 83*, 298 pp.

Galbraith, M. G. Jr. 1967. Size-selection predation on *Daphnia* by rainbow trout and yellow perch. *Trans. Am. Fish. Soc. 96*, 1-10.

Galkovskaya, A. G. 1963. About the feeding of planktonic rotifers. *Dokl. An BSSR 7*, 202-205.

George, D. G., and Edwards, R. W. 1973. *Daphnia* distribution within Langmuir circulation. *Limnol. Oceanog. 18*(5), 798-800.

Gophen, M., Cavari, B. Z., and Berman, T. 1974. Zooplankton feeding on differentially labeled algae and bacteria. *Nature 247*, 393-394.

Gossler, O. 1950. Funktionsanalysen des Räderorgans von Rotatorien durch optische Verlangsamung. *Österr. zool. z.* 2(5-6), 568-584.

Green, J. 1967. The distribution and variation of *Daphnia umholtzi* in relation to fish population in Lake Albert, East Africa. *J. Zool. London 151*, 181-197.

Green, J., and Lan, O. B. 1974. *Asplanchna* and the spines of *Brachionus calyciflorus* in two Japanese Sewage Ponds. *Freshwater Biol. 4*, 223-226.

Grygierek, E. 1967. Formation of fish pond biocenosis exemplified by planktonic crustaceans. *Ecol. Polska A 15*(8), 155-181.

Hall, D. J. 1964. An experimental approach to the dynamics of a natural population of *Daphnia galeata mendotae*. *Ecology 45*, 94-112.

Hall, D. J., Cooper, W. E., and Werner, E. E. 1970. An experimental approach to the production dynamics and structure of freshwater animal communities. *Limnol. Oceanog. 15*(6), 839-928.

Harding, J. P. 1957. Crustacea: Cladocera. *Expl. Hydrol. Lac. Tanganica 1946-1947 III 6*, 52-89.

Hargrave, B. T., and Geen, G. H. 1970. Effect of copepod grazing on two natural phytoplankton populations. *J. Fish. Res. Bd. Can. 27*, 1395-1403.

Hazelwood, D. H. 1962. Temperature and photoperiod effects on cyclomorphosis in *Daphnia*. *Limnol. Oceanog. 7*, 230-232.

Hebert, P. D. N. 1973. Phenotypic variability of lactate dehydrogenase in *Daphnia magna*. *J. Exp. Zool. 186*(1), 33-38.

Hillbricht-Ilkowska, A. 1964. The influence of the fish population on the biocenosis of a pond, using Rotifera fauna as an illustration. *Ecol. Polska A 12*, 453-503.

Hrbáček, J. 1959. Circulation of water as a main factor influencing the development of helmets in *Daphnia cucullata* Sars. *Hydrobiologia 13*, 170-185.

Hrbáček, J. 1962. Species composition and the amount of the zooplankton in relation to the fish stock. *Rozpr. ČSAV, řada matem. a přír. věd 72*(10), 116 pp.

Hrbáček, J. 1969a. Relation of productivity phenomena to the water quality criteria in ponds and reservoirs. *Adv. Water Poll. Res., Proceedings of the 4th Int. Conference*, Pergamon, Oxford and New York, 717-724.

Hrbáček, J. 1969b. On the possibility of estimating predation pressure and nutrition level of populations of *Daphnia* (Crustacea, Cladocera) from their remains in sediments. *Mitt. Int. Verein. Limnol. 17*, 269-274.

Hrbáček, J., and Hrbáčková-Esslová, M. 1960. Fish stock as a protective agent in the occurrence of slow-developing dwarf species and strain of the genus, *Daphnia. Int. Rev. ges. Hydrobiol. 45*(3), 355-358.

Hrbáček, J., and Hrbáčková-Esslová, M. 1966. The taxonomy of the genus *Daphnia* and the problem of "Biological Indication." *Verh. Int. Verein. Limnol. 16*(3), 1661-1667.

Hrbáček, J., and Novotná-Dvořáková, M. 1965. Plankton of four backwaters related to their size and fish stock. *Rozpr. CSAV, řada matem. a přír. věd 75*(13), 64 pp.

Hrbáčková, M. 1971. The size distribution of neonates and growth of *Daphnia hyalina* Leydig (Crustacea, Cladocera) from Lago Maggiore under laboratory conditions. *Mem. Inst. Ital. Idrobiol. 27*, 357-367.

Hrbáčková-Esslová, M. 1962. Postembryonic development of Cladocerans I. *Daphnia pulex* group. *Věst. čs. spol. zool. 16*(3), 212-233.

Hrbáčková-Esslová, M. 1963. The development of three species of *Daphnia* in the surface water of the Slapy Reservoir. *Int. Rev. ges. Hydrobiol. 48*(2), 325-333.

Hrbáčková-Esslová, M. 1966. The differences in the growth and reproduction at 8 C and 20 C of *Daphnia pulicaria* Forbes (Crustacea, Cladocera). *Věst. čs. spol. zool. 30*(1), 30-38.

Hurlbert, S. H., Zedler, J., and Fairbanks, D. 1972. Ecosystem alteration by Mosquitofish (*Gambusia affinis*) predation. *Science 175*(4022), 639-641.

Hutchinson, B. P. 1971. The effect of fish predation on the zooplankton of ten Adirondack Lakes, with particular reference to the alewife Alosa pseudoharengus. *Trans. Am. Fish. Soc. 100*, 325-335.

Hutchinson, E. 1967. *Introduction to the Lake Biology in Hutchinson E: Treatise on Limnology II.* Wiley, New York, 1115 pp.

Jacobs, J. 1962. Light and turbulence as co-determinants and relative growth rates in cyclomorphic *Daphnia. Int. Rev. ges. Hydrobiol. 47*, 146-156.

Jacobs, J. 1970. Multiple determination der Zyklomorphose durch Umweltfaktoren. *Okologia 5*, 96-126.

Jahn, T. L., and Bovee, E. C. 1967. Motile behavior of Protozoa. In Tze-Tuan Chen (Ed.), *Research in Protozoology, Vol. 1.* Pergamon, Oxford, pp. 40-200.

Kerfoot, W. B. 1970. Bioenergetics of vertical migration. *Nature 104*(940), 529-546.

Kibby, H. V. 1971. Effect of temperature on feeding behavior of *Daphnia rosea. Limnol. Oceanog. 16*, 580.

Kleinholz, L. H. 1961. Pigmentary effectors. In T. H. Waterman (Ed.), *Physiology of Crustacea, Vol. II.* Academic Press, New York and London.

Kořínek, V. 1970. The embryonic and postembryonic development of *Daphnia hyalina* Leydig (Crustacea, Cladocera) from Lago Maggiore. *Mem. Inst. Ital. Idrobiol. 26*, 85-95.

Kořínek, V. 1972. Results of the study of some links of the food chain in a carp pond in Czechoslovakia. In Z. Kajak and A. Hillbricht-Ilkowska (Eds.), *Productivity Problems of Freshwaters Warszawa-Krakow.* pp. 541-553.

Koshov, M. 1963. Lake Baikal and its life. *Monogr. Biol., Vol. II.* The Hague.

Lackey, R. T. 1973. Effect of artificial destratification on zooplankton in Parvin Lake. *Trans. Am. Fish. Soc. 102*(2), 450-452.

Lane, P. A., and McNaught, D. C. 1970. A mathematical analysis of the niches of Lake Michigan zooplankton. *Proc. 13th Conf. Great Lakes Res. 47-57,* Int. Assoc. Great Lakes Res.

Lefèvre, M. 1950. *Aphanizomenon gracile* Lemm. cyanophyte défavorable au zooplankton. *Ann. Stat. Centr. Hydrobiol. app. 3*, 205-208.

Litynski, A. 1913. Revision der Cladocerenfauna der Tatra-Seen, 1. Teil Daphnidae. *Bull. Acad. Sci. Cracovie, Series B,* 566-623.

Loeb, J. 1906. Über die Erregung von positiven Heliotropismus durch Saure, insbesondere Kohlensaure und negativen Heliotropismus durch utraviolette Strahlen. *Pflügers Arch. Ges. Physiol. 115,* 151-181.

Losos, B., and Heteša, J. 1973. The effect of mineral fertilization and of carp fly on the composition and dynamics of plankton. *Hydrobiol. Studies 3, Academia Prague,* 173-217.

Lowndes, A. G. 1935. The swimming and feeding of certain calanoid copepods. *Proc. Zool. Soc. London,* 687-715.

Luckinbill, L. S. 1974. The effect of space and enrichment on a predator-prey system. *Ecology 55*(5), 1142-1147.

Lyakhnovich, V. P. 1973. The biotic matter circulation and energy flow in the ecosystems of fishponds. *Verh. Int. Verein. Limnol. 18*(3), 1809-1817.

Lyakhnovich, V. P., Galkovskaya, G. A., and Kazyuchic. 1969. Vzrastnoy sostav i plodovitost populyacii Dafniy v rybovodnykh prudakh. *Voprosy rybnogo chozjajstva Belorussii, Moskva,* pp. 33-38.

Mast, S. O. 1947. The food vacuole in *Paramecium. Biol. Bull. 97,* 31-72.

McLaren, J. A. 1963. Effects of temperature on growth of zooplankton and the adaptive value of vertical migration. *J. Fish. Res. Bd. Can. 20*(3), 685-727.

McNaught, D. C. 1971. Plasticity of cladoceran visual systems to environmental changes. *Trans. Am. Micros. Soc. 90*(1), 113-114.

McNaught, D. C. 1975. A hypothesis to explain the succession from cladocerans during eutrophication. *Verh. Int. Verein. Limnol. 19*(1), 724-731.

McQueen, D. J. 1969. Reduction of zooplankton standing stocks by predaceous *Cyclops bicuspidatus thomasi* in Marion Lake, British Columbia. *J. Fish. Res. Bd. Can. 26,* 1605-1618.

McQueen, D. J. 1970. Grazing rates and food selection in *Diaptomus oregonensis* (Copepoda) from Marion Lake, British Columbia. *J. Fish. Res. Bd. Can. 27*, 13-20.

Mellors, W. K. 1975. Selective predation of ephippial *Daphnia* and the resistance of ephippial eggs to digestion. *Ecology 56(4)*, 974-980.

Mordukhai-Boltovskaya, E. D. 1958. Predvaritelnie dannie po pitaniu khishnykh kladocer *Leptodora kindti* i *bytotrephes* (Preliminary data concerning the diet of the Cladocera, *Leptodora kindti* and *bytotrephes*).

Nauwerck, A. 1963. Die Beziehungen zwischen zooplankton und Phytoplankton im See Erken. *Symbolae Bot. Upsalienses 17(5)*, 163 pp.

Neil, W. E. 1971. Resource partitioning by competing microcrustaceans in stable laboratory microecosystems. *Verh. Int. Verein. Limnol. 19*, 2885-2890.

Neil, W. E. 1975. Experimental studies of microcrustacean competition, community composition and efficiency of resource utilization. *Ecology 54(4)*, 809-826.

Nilsson, N. A., and Pejler, B. 1973. On the relation between fish fauna and zooplankton composition in North Swedish Lakes. *Inst. Freshwater Res. Drottnigholm, Report No. 53*, 51-77.

Pacaud, L. A. 1939. Contribution à l'ecologie des cladocers. *Bull. biol. France et Belgique, Suppl. 21*, 260 pp.

Palatas, K. 1975. The crustacean plankton communities of fourteen North American Great Lakes. *Verh. Int. Verein. Limnol. 19(1)*, 504-511.

Parma, S. 1969. The life cycle of *Chaoborus crystallinus* (de Geen) (Diptera, Chaoboridae) in a Dutch pond. *Int. Ver. Theor. Angew. Limnol. Verh. 17*, 888-894.

Pitelka, D. R. 1963. *Electron-Microscopic Structure of Protozoa*. Pergamon, New York, 269 pp.

Porter, K. G. 1972. A method for the in situ study of zooplankton grazing effects on algal species composition and standing crop. *Limnol. Oceanog. 17(6)*, 913-917.

Pourriot, R., and Deluzarches, M. 1971. Recherches sur la biologie des Rotifères. II. Influence de la temperature sur la durée du developmental embryonnaire et post-embryonnaire. *Ann. Limnol. 7*, fasc. 1, 25-52.

Ravera, O. 1954. La struttura demografica dei Copepodi del Lago Maggiore. *Mem. Inst. Ital. Idrobiol. 8*, 109-150.

Rawson, D. S. 1942. A comparison of some large alpine lakes in western Canada. *Ecology 23*, 143-161.

Richmann, S. 1958. Transformation of energy by *Daphnia pulex*. *Ecol. Monogr. 28*(3), 273-291.

Rodina, A. G. 1950. Experimentalnoe issledovanie pitanija dafnii. *Tr. Vses, Gidrob, Obchestva 2*, 169-193.

Ruttner-Kolisko, A. 1972. Rotatoria. In *Das Zooplankton der Binnengewässer, 1. Teil*. E. Schweizerbartsche Verlagsbuchhandlung, Stuttgart, pp. 99-234.

Saeter, O. A. 1972. Chaoborinae. In *Das Zooplankton der Binnengewasser, 1. Teil*. E. Schweizerbartsche Verlagsbuchhandlung, Stuttgart, pp. 257-280.

Sars, G. O. 1909. Zoological results of the third Tanganyika expedition report on the Copepoda. *Proc. Zool. Soc. London*, pp. 31-76.

Schäperclaus. 1961. *Lehrbuch der Teichwirtschaft*. Paul Parey, Berlin und Hamburg, 582 pp.

Schindler, D. W. 1972. Production of phytoplankton and zooplankton in Canadian Schield Lakes. In Z. Kayak and A. Hillbricht-Ilkowska (Eds.), *Productivity Problems of Freshwaters*. Warszawa-Kraków, pp. 311-331.

Serafimova-Hadžišče. 1957. Le plancton du Lac d'Ohrid au cours des années 1952, 1953 et 1954. Station hydrobiologique Ohrid, *Editions Speciales 1*, 65 pp.

Skadovsky, S. N. 1939. Fiziologichesky analiz fototaksisa u. dafnij (*D. pulex*). *Uch. Zapiski MGU 33*(3), Fyziologija ryb, 236-237.

Smyly, W. J. P. 1961. The life-cycle of the freshwater copepod *Cyclops leuckarti* Claus in Esthwaite Water. *J. Anim. Ecol. 30*, 153-169.

Stavn, R. H. 1971. The horizontal-vertical distribution hypothesis: Langmuir circulations and *Daphnia* distributions. *Limnol. Oceanog. 16*(2), 453-466.

Stavn, R. H. 1975. The effects of predator pressure on species composition and vertical distribution of *Daphnia* in Piedmont Lakes of North Carolina, U.S.A. *Verh. Int. Verein. Limnol. 19*, 2891-2897.

Storch, O. 1929. Analyse der Fangapparate niederee Krebse auf Grund von Mikro-Zeit Lupen Aufnahmen analysiert. *Biol. Genezalis 5*, 1-62.

Szlauer, L. 1965. The refuge ability of plankton animals before models of plankton eating animals. *Polsk. Arch. Hydrobiol. 13*(26)(1), 89-95.

Tappa, W. 1965. The dynamics of the association of six limnetic species of *Daphnia* in Aziscoos Lake, Maine. *Ecol. Monogr. 35*, 395-423.

Thienemann, A. 1939. Die Chiromidenforschung in ihrer Bedeutung fur Limnologie und Biologie. *Biologish Yaarboek Dodonaea 10*, 96 pp.

Wawrik, F. 1954. Limnologische Studien an Hochgebirgs-Kleingewässern im Arlberggebiet I. *Sitzb. Öst. Akad. Wiss., Abt. I 163*, 277-296.

Wawrik, F. 1966. Zur Kenntnis Alpiner Hochgebirgsklein-gewässer. *Verh. Int. Verein. Limnol. 16*(1), 543-553.

Weglénska, T. 1971. The influence of various concentrations of natural food on the development, fecundity and production of planktonic crustacean filtrations. *Ecol. Polska 19*(30), 427-473.

Weismann, A. 1874. Über Bau und Lebenserscheinungen von *Leptodora hyalina*. *Lillyeborg Zschr. wiss. Zool. 24* (3), 349-418.

Welch, P. S. 1952. *Limnology,* 2nd ed. McGraw-Hill, New York, Toronto, London, 538 pp.

Wesenberg-Lund, C. 1904. Plankton investigations of the Danish Lakes. In *Special Part*. Copenhagen, 223 pp.

Wesenberg-Lund, C. 1939. Biologie der Süsswasser-Tiere. In *Wirbellose Tiere*. Springer, Wien, 817 pp.

Wilson, D. S. 1973. Food size selection among copepods. *Ecology 54*, 909-914.

Winberg, G. G.   1950.   Intenzivnost obmena i razmery
    rakoobraznych.   *Zh. obsch. biol. 11*, 367-438.

Woltereck, R.   1930.   Alte und neue Beobachtungen über
    die geographische und zonare Verteilung der helmlosen
    und helmtragenden Biotyp von *Daphnia.   Int. Rev. Ges.
    Hydrobiol. 24*, 358-380.

Wong, B., and Ward, F. J.   1972.   Size selection of *Daphnia
    pulicaria* by yellow perch (*Perca flavescens*) fry in
    West Blue Lake, Manitoba.   *J. Fish. Res. Bd. Can. 29*,
    1761-1764.

Zaret, T. M.   1969.   Predation-balanced polymorphism of
    Ceriodaphnia cornuta Sars.   *Limnol. Oceanog. 14*(2),
    301-303.

Chapter 11

COMMUNITY STRUCTURE OF PROTOZOANS AND ALGAE WITH PARTICULAR EMPHASIS ON RECENTLY COLONIZED BODIES OF WATER

Bassett Maguire, Jr.

Department of Zoology
University of Texas
Austin, Texas

CONTENTS

ABSTRACT

Twenty beakers, which originally were sterile, were filled with artificial lake water and then sampled periodically. Originally sterile polyethylene tubs were placed on the newly formed island Surtsey. Colonization curves are regular, rising at first rapidly and then progressively more slowly with time. It is possible, however, that the apparent (near) equilibrium will be only "temporary" and that colonization curves are not always monotonic. Correlation coefficient of variation of number of taxa per beaker with mean number of taxa and mean number of total individuals per milliliter was positive and significant. Diversity [$H' = -\Sigma(n_i/N)(\log_2 n_i/N)$] had essentially zero correlation with coefficients of variation of number of taxa and number of individuals. The autotroph-heterotroph ratio changed through seven orders of magnitude in the beakers; except in the beginning, autotrophs were dominant. Ecological activity rose with the increase in number of taxa and individuals but did not decline when the growth rate of either of these slowed considerably.

INTRODUCTION

The rates and patterns of colonization and of formation and subsequent change of community structure in initially sterile, really isolated habitats have been studied very little. Such studies, especially in very small, isolated aquatic habitats, can be very rewarding because their full biological and physicochemical history is known, the developing communities are relatively simple, adequate replication and experimental manipulation are possible, and sampling problems are less severe than in other natural environments. Small, isolated habitats, therefore, may be used as especially effective tools for the elucidation of many ecological processes of fundamental importance.

Study of the related processes of colonization and development of community structure on clean surfaces immersed in lakes, streams, or the ocean has produced considerable information. Ruth Patrick has, over a number of years, done some of the best and most interesting of this kind of work; her paper (1971) reviews some of the results she has obtained through her study of diatoms. John Cairns (1971) writes on what is essentially the same process, although his ecological tools are protozoans in or on sponges.

The first extensive experimental study of the colonization of small originally sterile bodies of water was carried out in 1959 when it was determined that a surprisingly complete aquatic biota developed in 265 ml of water in isolated bottles within several weeks (Maguire 1963a). Raccoons were of major importance in transport of aquatic organisms in this experiment. Colonization rates initially were high and then decreased regularly; the number of species per bottle increased rapidly at first and then progressively more slowly toward a maximum determined by the available species pool, the considerably restrictive resources of the small water masses, and the relatively small number of niches provided by a water-filled bottle. A very striking example of interaction was observed in which the ciliate *Colpoda* was eliminated from the active community, apparently by biotic mechanisms (Maguire 1963b).

A very stimulating little book by MacArthur and Wilson (1967) gives important theoretical development to our understanding of some aspects of island colonization and community structure and provides an excellent review and synthesis of recent theoretical work. Some of this theoretical development has been recently tested by experimental defaunation and observation of subsequent recolonization of small mangrove islands of the Florida keys by Wilson and Simberloff (1969) and Simberloff and Wilson (1969). Their experimental results generally support the theoretical developments of MacArthur and Wilson.

This chapter considers the rates and patterns of colonization and development of community structure in small bodies of freshwater variously isolated from possible sources of aquatic organisms.

COLONIZATION

Pattern of Increase of Species Number

In late 1963, 33 km off the south coast of Iceland, a volcano arose from the sea to produce the island Surtsey. Eruptions continued sporadically on Surtsey and on several small islands that formed nearby until the spring of 1967. The last sterilization ash fall occurred in June, 1965 (Fridriksson 1968). Surtsey is about 170 m high and has an area of about 2.3 km^2. Its southern portion has been covered by flowing lava so that, unless this lava layer is undercut by the sea, the island probably will last for a long time. The other small islands formed near Surtsey were not subsequently covered by lava and as a result were quickly eroded away.

The Icelandic government has realized Surtsey's value to geological and biological scientists and has placed it under the aegis of the Surtsey Research Society. This Society, which is composed of scientists, restricts visitation of the island to those who have good scientific reason for going. Those permitted on the island are made aware of the need to keep human transport of plant and animal dissemules to the island at a minimum. The Society publishes a valuable annual report concerning work accomplished on Surtsey (edited by Steingrimur Hermannsson, the Society's chairman).

In early June, 1967 (as early in the year as it is practical to try to get to Surtsey), 11 aquatic "traps" were placed on Surtsey. They were eliptical polyethylene laundry tubs (80 cm long, 50 cm wide, and 30 cm deep), and they were placed on various parts of the island and held down by lava blocks to prevent them from blowing away. The expectation was that these traps would fill with rainwater and that communities of freshwater organisms would then develop in them. Since then the traps have been sampled twice annually, once in early June and again in late August. Aseptic techniques have been used. The traps initially were sterilized by rubbing them with alcohol, and all sampling apparatus and jars were autoclaved and kept wrapped or sealed until use. Cultures of sweepings from the cabin floor (where presumably man-imported dissemules would be found in the greatest concentration) have been negative for algae, protozoa, and multicellular organisms.

The summer of 1967 was much drier than usual and the result was that only four of the traps contained any water when the island was visited in the fall. Since then some of the traps have been broken, some filled with blowing ash, one filled with salt water, and one destroyed when the winter storms of 1968-1969 eroded about 100 m from the south shore. (These have been replaced or emptied of ash and moved.) Because of these difficulties, the sample size is smaller than had been hoped. Nevertheless, when those traps that held adequate amounts of freshwater were considered, the curve produced by the data looks like that I obtained in Texas in 1959 (see below) and is similar to the theoretical curves drawn by MacArthur and Wilson (1967). The average number of taxa per trap grew from 6.75 in the

fall of 1967 to 15.8 in the spring of 1968, 15.4 in the
fall of 1968 and 16.5 in the spring of 1969 (Fig. 11.1).

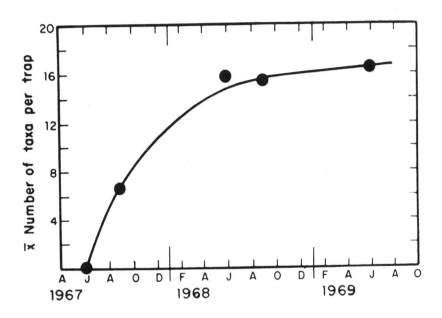

FIG. 11.1. Colonization curve: Surtsey. (400 X)

It might appear that already those traps were approaching
equilibrium value. This may be true and, if so, shows a
remarkably rapid development toward this level. However,
it is possible that the curve will continue to rise for
a long time. The species now represented in the traps are
probably only the most vagile of the species that live on
Iceland and that could survive and grow in the water in
the traps. For example, only three or four species of
rotifers have reached the traps; all other micrometozoa,

such as copepods, ostracods, cladocera, gastrotrichs, small
annelids, and so on, still are absent. Many algae and
protozoans that presumably could live in the traps are
still absent; for example, there are no desmids or peritrichs. It is also possible, as will be discussed below,
that species number cannot increase beyond some low maximum until more efficient predators (including herbivores)
invade the traps.

Colonization of apparently unfilled niches will proceed at a decreasing rate and will be accompanied by some
niche reshuffling and readjustment. The result will be
that of the number of species will continue to grow (slowly
in the absence of predators — see below) for many years,
as will the extinction rate, until an equilibrium level
is reached. Then there will be a significantly higher
number of species than are now observed.

In any event, the detailed shape of this (and any)
colonization curve is influenced, sometimes to a considerable degree, by the frequency distribution of vagilities
among species of potential colonizers. This point needs
greater attention.

The colonization curve for 32 bottles initially filled
with sterile water and placed at different distances from
a pond and different heights above the ground near San
Marcos, Texas, in 1959, is given in Figure 11.2. This
curve is for the number of taxa that could be distinguished
with a dissecting microscope and a magnification of 45X.
The circles and the eye fit curve show the colonization
rate. Note that this colonization curve (which is on a
different time scale from that of Surtsey) is relatively
flat. This flatness is in part an artifact using data
obtained with low magnification; presumably if all species
of algae and protozoan were differentiated, the curve would
show a more rapid rise during the first few weeks, in part

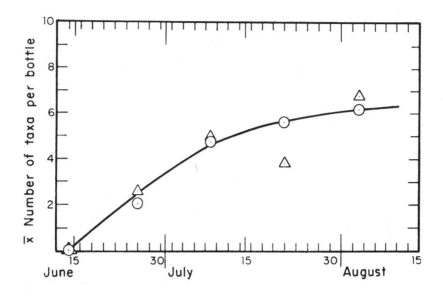

FIG. 11.2. Colonization curve: San Marcos, Texas, 1959. (○),
Bottles stayed in place; (△), bottles replaced at each sampling
time.

the result of different frequency distributions of vagility
and availability of protists and of metazoans. The triangles show the relative rates of dispersal and colonization during the periods immediately preceding them. They
give evidence that the flatness of the colonization curve
is also partly the result of differential input (of the
larger species) with respect to time, with the greatest
input occurring late in the experiment.

There is a good correlation between the rate of colonization during the various intersample periods of this
experiment and the number of muddy raccoon footprints on
and around the bottles. This, plus comparison with a
somewhat similar but raccoon-free colonization experiment
to be described later, shows that these raccoons provided
the dominant fraction of the input to these bottles. See
Maguire (1963a) for further details.

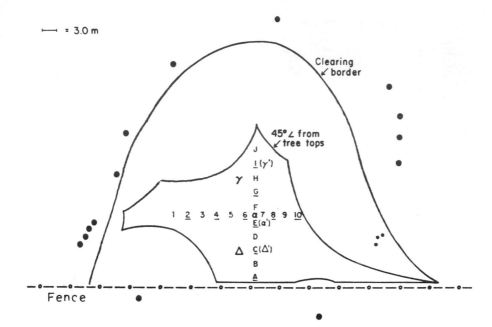

FIG. 11.3. Map of Austin colonization experiment. (●), Trees; underscore, 400 ml; others, 800 ml.

In the spring and summer of 1969 a number of hard-working students and I carried out another colonization experiment in which we gathered information in an attempt to show, in some detail, the changing community structure throughout the early phases of colonization.

On March 12, 1969, 10 beakers of 800-ml capacity and 10 beakers of 400-ml capacity were placed 3 m apart in a cross-shaped pattern in a grassy field near the Colorado River in Austin, Texas. Figure 11.3 shows their arrangement and the location of trees, a wire fence, and the

river. An electric shock fence (not shown on the map) kept
raccoons and other larger animals from affecting the experiments. Note that the beakers in line perpendicular
to the river are lettered consecutively; A is closest to
and J farthest from the river. The centralmost (scalloped)
line indicates where in the field the tops of the nearest
and/or tallest trees were at 45° altitudes. Large and
small beakers were alternated in the "cross"; the small
ones were suspended with stainless steel wire harnesses
in 800-ml, distilled water-filled "overbeakers," so that
the total volume and surface/volume relationships of the
400- and 800-ml systems were the same. Their temperatures,
therefore, should have been the same. All beakers were
fastened to steel posts, the tops of all beakers were 60
cm above the ground and all were shaded by a 25 x 25 cm
sheet of galvanized iron held 10 cm above them. Each
shade held a funnel that emptied into the beaker below
and had the same diameter as the beaker into which it emptied. Beaker and funnel diameters were 75 and 100 mm for
small and large, respectively. This produced rain collecting and air-water surfaces that were the same per unit
volume for large and small beakers.

All beakers, funnels, and shades were sterilized and
the beakers were filled with artificial lake water (see
Maguire 1963a). Collections were made at 2-day intervals
for another month (when analysis time increased the pro-
hibitive point), then at 4-day intervals for another month,
and later at irregular intervals as time permitted.

Aseptic techniques were used. At each sample time
the beakers were first filled with sterile, distilled water
(to compensate for the evaporation that had occurred),
stirred very vigorously, and then sampled with effort to
take parts of the sample from each part of the beaker.
Sampled volume was replaced with sterile artificial lake

water. (During the last two months some collections were
made prior to filling of the beakers.) Samples were kept
cool until they were examined, always on the same day.

Initially, when organisms densities were somewhat
low, species estimates were made from counts from five
transects of a Sedgwick Rafter cell using an ocular micro-
meter and a magnification of 100. When population densities increased to the point that counts became very high,
we changed to a chamber made by supporting a No. 1-1/2
coverslip on flanking No. 1 coverslips (thickness 0.14-
0.17 mm, assumed to be 0.15 mm). The coverslips were held
down by capillary forces, as care was taken never to completely fill the chamber. (Occasionally small metal
weights were used, apparently without effect.) Transect
length was 15 mm and indicated by two lines scratched on
a microscope slide parallel to its edges. Five transects
were counted. Filamentous algae were counted only when
they were intersected by a given line of the ocular micro-
meter and filament length per milliliter is calculated:

$$mm/ml = \frac{(\text{total intersections for 5 transects})(1)(1000)}{5 \cos 45° (15)(.15)}$$

Results of change in counting technique are not detectable
in the data.

As filamentous, sheath-producing algae (esp. Oscilla-
torales) became common, it was necessary to count them,
and many of the Chlorococcales and other small algae that
became imbedded in or stuck to their gelatinous products,
after blending the sample for 30 sec in a microcup attachment of a Waring Blendor. Fragile algal species and most
protozoa could not withstand this treatment and so were
counted before the sample was blended (experiments demonstrated that the number of the more robust algae was not
affected).

PROTOZOAN AND ALGAE COMMUNITY STRUCTURE

Effect of Distance from the River

Analysis by the rank difference correlation coefficient, r_d (Tate and Clelland 1957) of each sample series
through May 19 did not demonstrate a significant difference in number of recognized taxa as a function of distance
from the river. Average r_d from March 24 through May 5
was 0.24 [r_d is significant (n = 10, p = 0.05) at 0.65].
On May 11, r_d was 0.66, on May 15 it was 0.84, and then
on May 19 it had dropped, without obvious cause, to -0.12.
On the whole, there is a suggestion that the river may have
preferentially contributed a few more taxa to the beakers
near to it but the effect, if it exists, is small.

Comparisons of Large and Small Beakers

The average number of taxa observed (excluding bacteria and fungi) per 800-ml beaker (from March 24 through
June 23) was 8.87 and the mean for the 400-ml beakers was
8.51. If the period of early colonization is excluded by
choosing the interval April 9 through July 19, the means
are 9.39 taxa for the large beakers and 9.55 for the small
ones. These figures suggest that there is no difference
between number of taxa as a function of beaker size when
sample size totals 20 and size difference between beaker
is 1:2. Perhaps, however, the colonization rate of the
larger beakers may have exceeded (slightly) that of the
smaller ones during the initial phases.

The average number of organisms (excluding bacteria
and fungi) per milliliter between April 9 and July 19 was
3.0×10^6 for the large and 2.4×10^6 for the small beakers, but the difference does not approach statistical
significance. (The April 9-July 19 period was one in
which increase in number of individuals was approximately

exponential but at a rate considerably less than that which occurred as the first colonists of the beaker underwent the initial population "explosion" during which they "filled" the habitat — see below.)

Input Pattern

Three beaker stands (α, Δ, γ,) similar to those described above were used to measure dissemule input to the experimental beakers as a function of time. Figure 11.3 shows their location with respect to the experimental stands. Sterile, water-filled beakers were placed on these stands at each time of collection and the water, which had been sterile at the preceding sample time and therefore contained 2- or 4-day input, was used to provide innoculum for flasks of Bold's Basal Medium (Bold 1942), which is designed to provide nutrients necessary for the growth of many species of algae. These were cultured under a mixture of fluorescent and tungsten light for daily 12-hr periods and examined at periods of several weeks. These data gave important information on the rate of input of dissemules into the beakers, showing that it was somewhat irregular but that there was no discernible pattern through time except for the possibility that input may have been slightly greater during the first months of the experiment when rain fell most frequently. Of the algae that grew from cultures taken from March 16 through May 3, *Chlorella* had by far the highest vagility and/or availability, as it was found in 16 of these cultures. *Chlorococcum* was second with seven occurrences, *Oocystis* and Bodonids had three, and *Palmellococcus*, diatoms, *Oscillatoria, Kirchneriella, Rhizochlonium, Scytonema,* and *Coccochloris* were seen once or twice.

Colonization Curve

Figure 11.4 illustrates colonization pattern in terms of average number of taxa observed per beaker as a function of time. The theoretically expected colonization curve

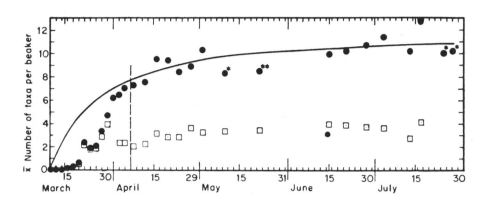

FIG. 11.4. Colonization curve: Austin, Texas, 1969. (400 X)

is approximated by the line. This expected colonization curve fits points after April 4 nicely, but there is considerable deviation between the observed pattern and the expected pattern at the beginning of the experiment. This is the result of the methods used in making the estimates of the number of taxa.

Knowledge of the time of each successful colonization is necessary to demonstrate the classical colonization curve. During the early parts of the beaker study this curve shape, which most probably followed the expected form, was obscured. This was because, after different numbers of dissemules of the various species arrived in

the beakers at different times, they then required different times for excystment and reproduced at different rates.
Therefore, different amounts of time elapsed (after arrival) before they reached densities at which they could be
observed. Each variable (number of dissemules, time of
arrival, excystment time, and reproductive rate) tends to
have a normal distribution; cumulative normal curves are
of sigmoid shape.

Although in some ways it is a disadvantage that there
must be at least several cycles of reproduction of a species in the new habitat before that species will, on the
average, be detected, it may turn out that this disadvantage will be outweighed by the advantage resulting from
better representation of the biological realities of colonization than otherwise would be possible. The definition
of colonization as being the arrival in the new area of
a single potential reproductive individual or pair (or
more) is convenient and lends itself well to theoretical
manipulation as it has been carried out by MacArthur and
Wilson (1967) and by Wilson and Simberloff (1969). In
biological terms, however, a definition of colonization
as the arrival and subsequent reproduction (through several generations?) of a new species may be more useful.
The problem, which the definition of MacArthur and Wilson
avoided, is that it is difficult to decide how much reproduction is enough to qualify the new species as a colonist.
In some species it might reasonably be more than several
generations. For example, it is well known that some species of algae require vitamin B_{12} and can store enough to
permit growth and reproduction through several generations.
Should one of these B_{12}-requiring cells obtain a large
amount of the vitamin and then be transported to an environment devoid of B_{12}, all of the cells will die, but
only after several generations of reproduction. It seems

370

to me that this species could only be a considered transient or a noncolonist (failure as a colonist) but never
a colonizer of this B_{12}-free habitat. Colonization as a
biologically meaningful activity can only occur when the
species potentially can survive the conditions of the new
environment for many generations. Under these conditions
a colonizing species can also be thought of as having an
appreciable ecological effect on the environment and/or
the populations already living in the area. The difficulty
in delimitation of the proposed definition should be more
than compensated for by the more realistic biological
meaning that the term would have.

As far as one can tell from an examination of the
data, the number of taxa colonization curve for the 1969
Austin experiment is continuing to climb slowly. It may
be that this will continue for a long time for reasons
outlined above.

The difference between the increase in number of taxa
of heterotrophs (circles) and that of the autotrophs
(squares) as seen in Figure 11.4 is interesting. The heterotrophs made up nearly all of the total during the early
part of the colonization (until March 28); then after
March 30 the curves separated rapidly and the number of
kinds of heterotrophs remained at about three to four per
beaker, while the number of taxa of autotrophs continued
to increase to about 11. The heterotrophs were primarily
monads, bodonids, a few small holotrichs, and a few small
amoebae. Before the growth of algal populations they apparently fed on bacteria growing on the pollen grains,
leaf bud scales, anthers, and small unidentifiable pieces
of organic debris, which were fairly common almost immediately after the beakers were put in the field. The autotroph-heterotroph ratio will be discussed at greater length
below.

Equilibrium from Above

In an attempt to determine equilibrium level (see MacArthur and Wilson 1967, for development of this concept) for these beakers in a different way, a large number of taxa were added to four of the beakers on July 20. These were obtained by collecting samples (somewhat concentrated with a plankton net) from two rivers, a permanent pond, and water in which mud forms the bottom of some ephemeral ponds had been cultured for several days. These samples were then stirred together and the resultant taxon-rich mix was used to replace about 50 ml of beakers 1, 4, 6, and 7. These beakers therefore contained the communities that had developed in them plus a large number of new taxa. The number of taxa that had made up the communities in these beakers was about 12 on the average; on July 23, 3 days after the mix was added, the beakers contained an average of 25.5 taxa. This number did not change appreciably by September 2, the last day that they were sampled (when they also contained 25.5 taxa, although there had been a little fluctuation between these dates).

One might theorize that an input of a large number of taxa into beakers in which colonization had proceeded for four months would be followed by violent ecological activity resulting in rapid modification of the structure of the community that had developed through the operation of natural mechanisms of dispersal, colonization, and community development. This activity presumably would quickly result in some reduction in the number of taxa present. With time, as the more long-term extinction mechanisms operate, the number of taxa will gradually decrease and approach the equilibrium value from above.

The lack of change in the number of taxa from July 23 to September 2 suggests either that 25 is at or below equilibrium number for these beakers or that extinction rates had no effect in six weeks.

It would not have been surprising if the addition of new taxa, including several kinds of microcrustacea, would have resulted in a rapid reduction of number of taxa. I observed natural events of this kind of rainwater caught and held by the flower bracts of the wild banana, *Heliconia*, in the forest at El Verde Experiment Station, Luquillo Natural Forest, Puerto Rico. Protozoan communities develop very high densities in the water of some of these flower bracts, and then when mosquito larvae appear their feeding activities are so efficient that the protozoan populations are reduced to very low densities, frequently with extinction of many of the taxa that had been abundant (Maguire *et al.* 1968).

It is possible that, given enough time, the ostracods and copepods added to the beaker communities will have a similar (although less extreme) effect to that of mosquito larvae on *Heliconia* communities. However, there is a very interesting alternative hypothesis that may have great implications with respect to both theoretical and practical considerations of island colonization. It is possible that the equilibrium value for the beakers is 25 or more taxa but that this cannot be even approached until one or more relatively efficient (but not too efficient) herbivores which feed on the originally dominant alga taxa, are present. It seems probable that a community in a beaker would frequently act to prevent colonization by new immigrants through mechanisms discussed by Elton (1958). This is especially probable where, for example, several species of algae are greatly dominant because they are most rapid and efficient in utilization of algal nutrients that become available. A rather simple algal community could be indefinitely stable under these conditions. The communities that developed in the beakers that relied only

on natural mechanisms for dissemule input may have reached
close to an equilibrium of this kind. (The protozoa and
rotifers that colonized the beakers consumed some algae
but were observed to have eaten algal cells sufficiently
infrequently to suggest that they were not effective in
limiting the algal populations.) This kind of intermediate
equilibrium could be maintained for a long time and might
well include a much lower number of species than could
exist in the beakers (or on islands) at highest equilibrium.

A major discontinuity in the colonization curve could
then be produced by the colonization of the beaker (or
island) by one or more efficient herbivore(s). The results
from the addition of many taxa to the beakers as described
above suggest that the additional number of taxa that might
be permitted by the activities of the herbivores could be
as great as the number at intermediate equilibrium. It
is conceivable that a second and higher intermediate level
occasionally occurs in the absence of carnivores. Then
their addition would be necessary before highest equilibrium could develop. In either event the colonization
curve is not necessarily monotonic. Instead, it could be
of the shape shown in Figure 11.5. In this curve the
dotted line illustrates the average course of colonization
when efficient herbivores and carnivores enter the community appropriately soon after the beginning of the colonization process. The solid line illustrates the expected
average (idealized) course of colonization of beakers (and
islands) where efficient herbivores have significantly
poorer dispersal mechanisms, and therefore considerably
later times of arrival than the autotrophs, and the carnivores have lower dispersal rates than the herbivores, and
therefore arrive still later.

This colonization pattern is consistent with the finding of Paine (1966) in which the number of species on

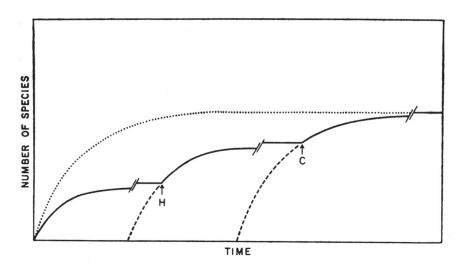

FIG. 11.5. Nonmonotonic colonization curve. H and C indicate time of arrival of adequately efficient herbivores and carnivores, respectively. See text for other details.

intertidal rocks was considerably increased by the presence of a moderately efficient predator.

If, however, the predation (including herbivores feeding on autotrophs) is exceedingly efficient and involves a large enough number of prey species, a "final" equilibrium lower than the intermediate equilibrium (and much lower than highest equilibrium) can be established. The effect of mosquito larvae on communities held by *Heliconia* bracts (given above) provides a good example.

In sum, it is the interaction between the different colonizing potentials of some groups of the herbivore and carnivore trophic levels, and the competition-reducing effects of these groups on the trophic level below them, that can produce immigration, extinction, and colonization curves that are not monotonic.

Change in Number of Individuals

Figure 11.6 illustrates the average number of individuals per milliliter of all taxa as a function of time. The number increases at a rapid rate, which is approximately exponential for about the first month. During this time it passes through six orders of magnitude. Then it abruptly changes to a smaller slope but continues to increase exponentially. Two straight lines, both passing through the April 8 data point, could be drawn and would closely fit all points. This April 8 point also could be the one that might be chosen if it were desirable to pick a point representing the ending of the most rapidly rising part of the taxa colonization curve. It would appear that the high initial rate was the result of rapid growth during the "filling" of the environment under conditions of little or no competition. The later, slower growth probably was accompanied by considerable reshuffling of resource utilization through differential growth of the various taxa. This reshuffling would, other things being equal, provide for more efficient use of the total resources and permit increase in number of organisms (and number of taxa).

Comparison of the mean number of taxa per beaker and the mean number of individual organisms per milliliter during the April 9–July 19 period gives an r_d of 0.47 that is significant at 0.05. Both increase in the number of taxa and appropriate adjustment of the number of organisms within each taxon operate toward a maximization of number of organisms that can share the resources and coexist in the beakers.

Community Stability

The suggestion was originally made by MacArthur (1955) that the stability of a community may (among other things) be directly related to the number of taxa in that

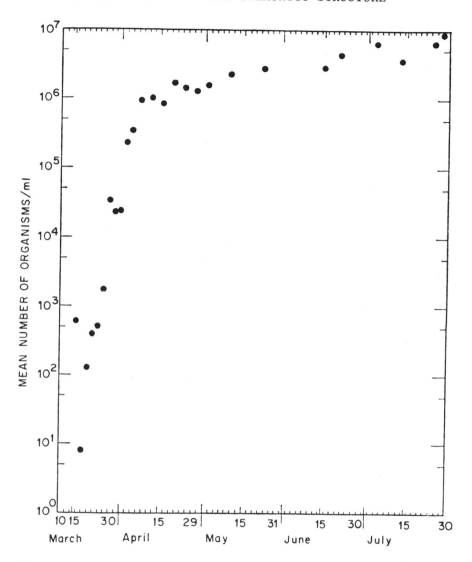

FIG. 11.6. Colonization, Austin, Texas, 1969. Mean number of organisms per beaker through time.

community. Hairston *et al.* (1968) described an experiment, although not as satisfactory as might be wished, which tended to show that in some instances assemblages containing fewer taxa are more stable than assemblages with greater number of taxa.

Community stability criteria include constancy of level of number of taxa and number of individuals, the degree to which constancy of number of taxa is represented by continuation of a particular assemblage of taxa, and the degree to which the number of individuals within each taxon continues to represent the same fraction of the total number of individuals. Normal seasonal fluctuations must, of course, frequently be taken into account.

To determine the degree to which the communities in the beakers remained stable with respect to constancy of number of taxa and number of individuals, coefficients of variation (CV) were calculated for each beaker through time. The advantage of CV is that it not only gives the relative variation of number of individuals or taxa through time but that it also provides an estimate that is independent of the magnitude of that number (Snedecor 1956). Rank difference correlation coefficients (r_d) were run to compare the stability (CV) of number of individuals and of number of taxa of the communities within the beakers with the average number of individuals and taxa in the beakers for the time interval of April 9–July 19 (after the first spurt of growth): r_d for CV of number of taxa and mean number of taxa was -0.43, and r_d for CV of number of taxa and mean number of organisms was -0.53. Significance at 0.05 for r_d (n = 20) is 0.45. Therefore both variation of number of taxa and variation of number of organisms in a beaker appear to have been reduced by the presence of greater number of taxa. In other words, in these systems (and at this stage of succession-community

development), stability increases with increase in number
of taxa present. The CV of number or organisms, however,
did not show significant correlation coefficients with
average number of organisms, average number of taxa, or
the CV of number of taxa (r_d was 0.30, 0.26, and 0.12,
respectively).

Diversity

Diversities, estimated by

$$H' = -\Sigma - \frac{n_i}{N} \log_2 \frac{n_i}{N}$$

were computed for each sample. The means of the diversities for all beakers are plotted as a function of time in
Figure 11.7. These increased during the early period of
rapid colonization in a form very similar to the change
of number of taxa curve. The diversities, however, increased even more rapidly than the number of taxa and
peaked about a week before the taxa data points caught up
with the expected colonization curve. This peak was also

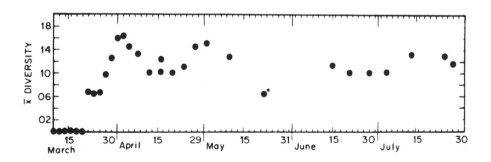

FIG. 11.7. Mean diversity of beaker communities through time.

a week earlier than the break in the number of individuals curve. The diversity curve then may have undergone something of a dampened oscillation before it settled down to a rather constant value.

The initial high peak is primarily the result of simultaneous growth of populations of several species to somewhat comparable high levels. Then, some of the species were able to continue their population growth and reach very high densities, while other species, presumably at least partly because of reduction of resources (nutrient?) to below their growth requirement level, were unable to maintain their rate of increase. This produced a community in which one species became greatly dominant, with resultant reduction of diversity.

Figure 11.8 shows the change of number of the most numerically dominant taxa (which also make up the bulk of the biovolume) for beaker F, which had a majority of the beakers. In this developing community, diversity increased very rapidly between March 26 and March 28; on the latter date it was higher than for any but one other beaker of the whole experimental series. The reasons for this were that four taxa (two ciliates, a monad, and a bodonid), an unusually high number for so early a date, were present and represented by approximately equal numbers of individuals. This equality of number of individuals, of course, resulted in a H' that was nearly as high as it could have been (near H_{max}) considering the number of taxa present.

Immediately after this, however, *Chlorella* increased much more rapidly than any other taxon in the beaker, and soon made up almost all of the community, which as a result had lower diversity. This situation continued until the last of May or early June when the *Kirchneriella* population began to grow explosively. When *Kirchneriella's* numbers (1.7×10^6) approached those of *Chlorella*

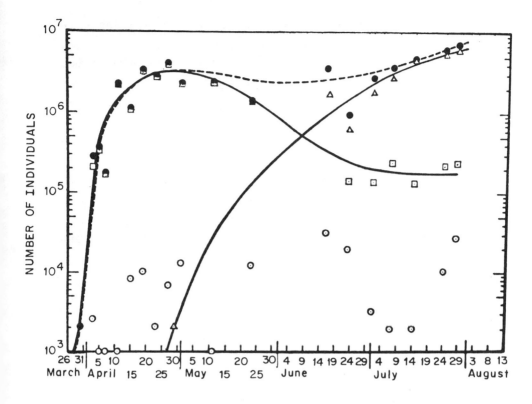

FIG. 11.8. Beaker F species growth curves. (●), Total; (○), *Monas*; (△), *Kirchnariella*; (□), *Chlorella*.

(1.8×10^6) slightly after the middle of June, community diversity rose strikingly, with the components contributed by these two taxa making up over 80% of the total. After that, the *Chlorella* population slowly declined while that of *Kirchneriella* rose with the result that the diversity of the beaker's community dropped precipitously. The data show clearly that a large fraction of the diversity (H') of the community in the beakers frequently results from equitability of two or three abundant species and then

381

may show negligible response to the number of taxa present.

Oscillatoria began to become important in most beakers toward the end of June or early July; beaker F was no exception. Dominance with respect to volume by the middle of August was held indisputably by *Oscillatoria*, although numbers of *Chlorella* and/or *Kirchneriella* frequently were also high.

If one of the ecological results of a high diversity is stability, then it seems reasonable that there would be an inverse correlation between the diversity and the amount of variation of the community through time. The coefficient of variation (CV) of number of organisms per milliliter in each beaker through time was compared with the mean of that beaker's diversity by the rank difference correlation coefficient; r_d was -0.05. This correlation is so low that it fails to support any relationship. If mean diversity is compared with CV of the number of taxa, r_d is 0.18, which also is too low to be meaningful. This suggests that if either diversity or coefficient of variation is a good measure of stability, the other is not. Since the coefficient of variation as used here is a good and direct measure of degree of fluctuation of numbers of organisms and number of taxa through time, evidence is good that the relationship between diversity and stability is not close. This is because of the confounding effect of the equitability and number of taxa components of the diversity index. It suggests that difficulties may follow an attempt to use a simple numerical index that summarizes (and confounds) data of dissimilar kinds.

Several less interesting relationships were suggested by the data. When mean total biomass of all organisms and mean diversity are compared, r_d is -0.28; when the comparison is between mean total number of organisms and mean diversity, r_d is -0.44; and when it is between mean number of taxa and mean diversity, r_d is 0.21. An r_d of 0.45 is significant at the traditional 0.05 level (N = 20). The only one of these r_d's likely to be meaningful is that between the mean number of organisms and mean diversity, and this is negative, which suggests that those beakers that on the average have the greatest number of organisms also have the lowest diversities. This resulted from the tendency for a single species, usually *Chlorella*, to completely dominate the community (both in number of individuals and in biomass).

Autotroph-Heterotroph Ratio

Figure 11.9 shows the change in the autotroph to heterotroph ratio in the beakers from the middle of March to the end of July. The early part of the curves are not surprising because they are consistent with the early invasion by the small heterotrophs followed by the invasion of autotrophs that was illustrated in the taxa colonization curve of Figure 11.4. The surprise was that the ratio then continued to rise and went so far (through over seven orders of magnitude, from about 10^{-2} to over 10^5), the combined result of the growth of the populations of algal taxa and a diminution in the densities of the heterotrophs. Most of the heterotrophs apparently were bacterial feeders, although some were seen to ingest algal cells.

If heterotrophs capable of ingesting the algae reached the beakers, and it appears that some did, then there are several alternatives to consider. Perhaps only some algal taxa were edible and they were soon eliminated. Alternatively, at least several of the algal species might have released substances that very effectively inhibited the heterotrophic populations. Likewise, the algae (and the bacterial populations that they support), at least in

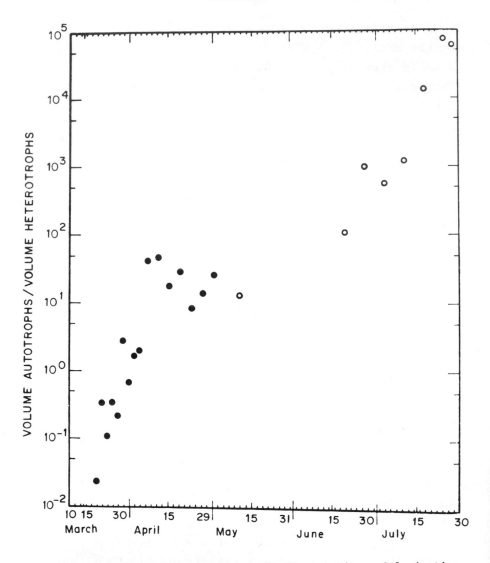

FIG. 11.9. Autotroph-heterotroph ratio through time. Colonization, Austin, Texas, 1969.

beaker communities that have been developing for three or
more months, do not contain enough usable food (or sufficient localized concentrations of food) to provide for the
growth of those heterotrophs that have been able to invade
the water in the beakers. It may be that critical changes
occurred in chemical content of many algal taxa as the
populations aged and, as Fogg's (1965) data suggest, nutritional value declined. There was a general change in
color of the total algal population from green to yellow
or yellow-brown during the study. This kind of color
change coincides with algal nutrient starvation, which is
accompanied by reduction in nutritional value of the algae
(for example, there is a lowered protein content).

Chlorella is known to produce an antibiotic that has
inhibitory effects on other species (Fogg 1965), and
Silvey and Wyatt (1971) has discussed these kinds of interactions. Oscillatoria also is known to have inhibitory
effects, and its biomass began to become appreciable by
mid- to late May.

Even though the effects of inhibitory metabolites may
be great and could explain the great increase in the autotroph-heterotroph ratio that has been observed, it still
may be useful to examine another possible major explanation. Ecologists frequently assume that whenever food
appears (to them) to be present there must be plenty that
the heterotrophs can eat. Hairston et al. (1960) and
Slobodkin et al. (1967) have suggested that all trophic
levels except that of the herbivores are, as a whole, food
limited. The argument is that since the world is green,
obviously there is plenty of food for the herbivore trophic level, and therefore that food limitation cannot control this trophic level.

The major difficulty with this hypothesis is that it
considers food quantity and does not take quality into

account. Required substances may be absent or may be present but not available. In the Slobodkin *et al*. (1967)
paper, a table shows that various species, of different
trophic levels, utilize different proportions of the net
production of the food that is assumed to be available to
them. The herbivores have much lower percent utilization
rates than do members of other trophic levels that they
list.

Even if available energy content is adequate, other
necessary nutrients may not be available or sufficiently
abundant or concentrated. For example, both vitamin and
protein levels may be inadequate; they differ from one part
of a plant to another, from one plant to another, or within
a single plant at different times. In addition, plant
physical characteristics change with time; for example,
cell walls become heavier, with the result that breaking
open the wall to get to the cell contents (which may be
reduced) becomes more difficult. Anyone with a flower
garden knows that the small herbivores that attack plants
frequently preferentially eat the young, growing plant
parts. Of the vegetative parts, they have the highest
protein and water content and the thinnest cell walls.
Flower and fruit, which also are frequently eaten preferentially, have especially high concentrations of easily
digestable carbohydrate, protein, and other valuable nutrients. Perhaps the reason that the world is green is
that herbivores, as a group, have difficulty in finding
enough food with high enough nutritional quality to enable
them to survive, grow, and produce strong offspring. Population control by food limitation must not be thought of
as operating only through the production of animals that
are obviously starving. It is more likely to operate most
of the time through reduction in vigor, producing an increase in mortality and a decrease in reproductive rates.

These changes in mortality and reproductive rates do not need to be large to be very effective.

Taub (1971, 1969) outlines some of the problems related to constructing simple self-sustaining gnotobiotic communities. One major problem, as her valuable and persistent work demonstrates, is providing the top heterotrophs (protozoa) with food. They need organisms that they will eat and that have sufficient, usable nutrient of adequate quality to permit them to reproduce enough to at least maintain their population. Taub and Dollar (1968) make this point even more forcefully in their paper describing the nutritional inadequacy of *Chlorella* and *Chlamydomonas* for *Daphnia*.

An accurate summary of trophic level limitation is that: All trophic levels are, in general, food limited.

Interspecific Association Patterns

In the study of colonization of bottles in San Marcos, Texas, when raccoons were active in the transport of a large number of organisms from the pond to the bottles, there were a large number of positive association coefficients (Cole 1949) between the species involved. These association coefficients reflected common occurrence of the species in the pond and their simultaneous transport to some bottles and not to other bottles. The only striking exception to this pattern was that of the ciliate, *Colpoda*, which was transported to many of the bottles by rapid, nonraccoon mechanisms. Shortly after the assemblage of species carried by a raccoon was added to a bottle that held active *Colpoda*, the ciliate either encysted or was eliminated (Maguire 1963a,b).

The association patterns between the species that have colonized the beakers in the Austin study is very different. They were calculated for March 28, April 3, July 2

and 7, and July 14 and 18 communities. Out of the 184
association coefficients calculated for the taxa that
occurred with sufficient frequency (and not with too great
a frequency), only four were "significant" at the 0.05
level. On the basis of chance alone, one would expect to
find nine "significant" associations in a sample of this
size. This is evidence that there was, at most, little
very strong interaction between taxa that were intermediate
in frequency of occurrence. This does not mean, however,
and this point should be emphasized, that there is evidence
for lack of any interaction. It is very possible that
there is much interaction that is not mandatory to some
species occurrence in a given beaker or not so strong
that it produced the elimination (exclusion) of species
from the beakers. It also is very possible that there is
interaction, either positive or negative, but that it
occurs between two taxa, at least one of which has great
enough vagility and colonizing ability to develop populations in at least most of the beakers. This species could
easily have had a major effect on other species, and the
association coefficients would be either very high or very
low, but they would not be meaningful because of the extreme breadth of the confidence band. Another kind of
experiment will be necessary to adequately define these
interactions.

Small, isolated bodies of water do not always have
the very low level of associations measurable by use of
the beaker data. Figure 11.10 shows part of the community
structure of organisms living in rainwater caught in the
leaf axils of the bromeliad, *Guzmania*, in the forest of
El Verde Experiment Station, Puerto Rico. The solid lines
indicate positive association and the broken lines show
negative association. The thickness of the line indicates
the general magnitude of the association. (The thinnest

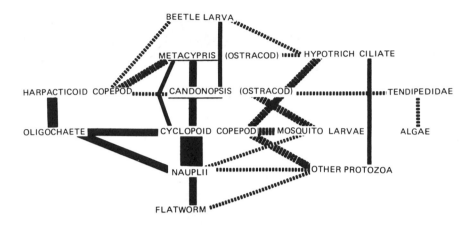

FIG. 11.10. *Guzmania* community structure; association pattern.

lines show associations that are relatively large, but a
little below significance level of 0.05.) In addition to
illustrating the sign and magnitude of the strands of the
web of interaction between these species, this diagram
also tells us that, at least when these data were collected, cyclopoid copepods were having much greater reproductive success than were harpacticoids. There was, for that
matt, a negative coefficient between harpacticoids and
nauplii, but it was small enough to have little meaning
in itself (for more detail see Maguire, 1970).

Ecological Activity

The effect that each species has on the community is
a function of the amount and kind of resources acquired,
held, and released per unit time. In a stable community,
the long-term rate of acquisition of resources will be
balanced by the rate at which they are released; estimates
of rates of uptake or release give measures of one kind

of a species activity. The amount of each resource stored in or occupied by a species is also of importance because it represents resource that either cannot be used by a competitor and/or is held in a form that is more or less available to predators (including herbivores).

Rates of production or of metabolism were not measured directly in this study; the only pertinent data available concern changes in the number of individuals and biovolume of each taxon. Obviously considerable ecological activity can take place when the number of organisms within each species is constant, not only through simple turnover of matter and energy by the species, but also through change in the amounts and kinds of materials making up the individuals in the species populations. Fogg (1965), for example, reports on some of the kinds of changes that occur in the size and chemical composition of algal cells as a species population ages. Decrease in photosynthetic and metabolic rates also can occur in aging (nutrient-starved) algal populations.

A more obvious ecological activity is expressed by change of community biomass and/or by a change in the relative fraction of the community represented by each species. In the beakers there was a rapid initial exponential increase in biovolume (biomass), followed by a continued but much slower exponential rate of increase. The solid circles of Figure 11.11 illustrate this pattern, a pattern that expectedly follows that of number of organisms (Fig. 11.6). Presumably what happened was that there was a rapid initial "beaker-filling" spurt of growth that was followed by slower but continued increase as interspecific adjustments occurred, which resulted in a rather slow but continuous (through the time of this experiment) increase in efficiency of resource utilization (and therefore of biovolume) by the community. The open circles and triangles of Figure 11.11 illustrate the average summed change

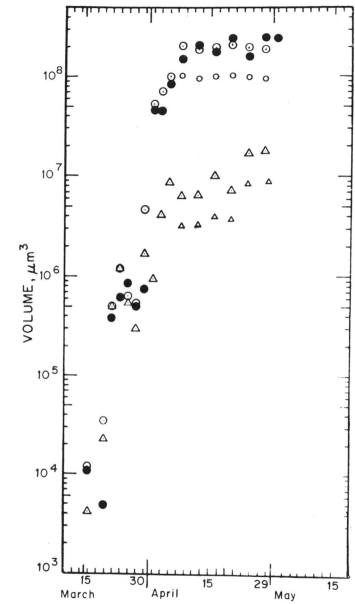

FIG. 11.11. Mean total biovolume per beaker, and rate of change (per
2 days) through time. Rate of change is one-half the difference in
volume between time t and t + 1 plus one-half that between t and
t - 1, summed for all species. The small figures represent change
rates which occurred if division of the measured change by 2 was
appropriate after intersample periods were lengthened from 2 to 4
days. (This procedure is appropriate only to the degree that the
average species population fluctuated with a "period" of 8 days or
longer.) (●), Total; (○), total change; (△), heterotroph change.

in biovolume of all species within the beaker communities (increase and decrease of each species biovolume were added without regard to sign). The striking feature of these curves is that the rate of change remained so high after April 9 when the initial beaker-filling growth spurt was over. This is evidence for intense ecological activity during which biomass was, in effect, exchanged between species. These kinds of activity presumably lead toward a more stable community in the long run; in the short-term duration of this experiment they certainly seem to lead to more efficient utilization of the beaker's resources.

SUMMARY

1. Colonization experiments with initially sterile water on the newly formed volcanic island, Surtsey, and in bottles in Austin, Texas, demonstrate, with expected deviations, normal and expected colonization curves. Species number increase rapidly at first and then with decreasing rate through time.

2. In both instances there are reasons to suspect, although species number had apparently reached a nearly constant level, that the apparent species equilibrium number was considerably lower than the final value eventually will be. The "temporary" intermediate equilibrium, however, may last for considerable time.

3. Evidence is available to suggest that immigration, extinction, and colonization curves may not always be monotonic.

4. In some systems, colonizing potentials of efficient herbivores and predators are considerably lower than those of the autotrophs. This, in conjunction with the competition-reducing effects of grazing and predation, explains the lack of monotonicity of the curves listed above.

5. In the Austin study:

    a. Distance from the river had little influence on number of taxa that became established in a beaker.

    b. No difference in number of taxa was observed as a function of beaker volume (400 and 800 ml).

    c. Input pattern did not vary appreciably with time.

    d. Mean number of taxa per beaker through time was postively and significantly correlated with mean number of total individuals per milliliter.

    e. Coefficient of variation of number of taxa per beaker through time was negatively and significantly correlated with both mean number of taxa and mean number of total individuals.

    f. Diversity ($H' = -\Sigma\, n_i/N \, \log_2 n_i/N$) was computed, and its correlation with the coefficient of variation (CV) of number of taxa and CV of number of individuals per milliliter was very close to zero. The confounding effects of number of taxa and the equitability component of this index render it useless as a measure of stability under the conditions of this experiment.

    g. The autotroph-heterotroph ratio (both number of individuals and biovolume) changed through over seven orders of magnitude, with the autotrophs always, except at the beginning, being greatly dominant.

    h. This development of autotroph-heterotroph ratio could have been the result of antibiotic production by some of the algae, but more probably was because the algae became progressively less nutritious and less edible. Lack of availability of enough food of adequate quality was probably limiting to the protozoan populations and to all trophic levels in general.

i. Fewer interspecific associations were observed
    (the measuring method was very crude) than would
    have been expected by chance alone (the difference
    was much too small to be significant), although
    this is not always the pattern in similarly small
    natural bodies of water.

j. Ecological activity rose with the increase in num-
    ber of individuals in the beakers. When total
    growth rate decreased, the ecological activity
    did not fall off but remained high and was ex-
    pressed by a reshuffling of species as, in effect,
    biovolume was transferred from one to another.

ACKNOWLEDGMENTS

Thanks go to the National Science Foundation, the
Surtsey Research Society, and the University Research Institute of the University of Texas at Austin for financial
support of parts of this study. I wish to express appreciation to Mr. Steingrimur Hermannsson, Chairman of the
Surtsey Research Society, for his help with logistics of
the Surtsey study. I also am grateful to the following
students whose work made the Austin colonization study
possible: Sue Anderson, Bill Gorman, Bill Houk, Steve
James, Claudia Jones, Russell Mase, Dan Udovic, Jay Wilson,
and Ed Zyzner.

REFERENCES

Bold, H. C. 1942. The cultivation of algae. *Bot. Rev.* *8*, 69-138.

Cairns, J. Jr. 1971. Factors affecting the number of species in freshwater protozoan communities. In *The Structure and Function of Freshwater Microbial Communities*, J. Cairns, Jr. (Ed.). Virginia Polytechnic Institute and State University Press, Blacksburg, Va., pp. 219-247.

Cole, L. C. 1949. The measurement of interspecific association. *Ecology 30*, 411-424.

Elton, C. S. 1958. *The Ecology of Invasions by Animals and Plants*. Methuen, London, 181 pp.

Fogg, G. E. 1965. *Algal Cultures and Phytoplankton Ecology*. University of Wisconsin Press, Madison, 126 pp.

Fridriksson, S. 1968. Life arrives on Surtsey. *New Sci.* *37*(590), 684-687.

Hairston, N. G., Smith, F. E., and Slobodkin, L. B. 1960. Community structure, population control and competition. *Am. Nat. 94*, 421-425.

Hairston, N. G., Smith, F. E., and Slobodkin, L. B. 1968. The relationship between species diversity and stability: An experimental approach with protozoa and bacteria. *Ecology 49*, 1091-1101.

MacArthur, R. H. 1955. Fluctuations of animal populations and a measure of community stability. *Ecology 36*, 533-536.

MacArthur, R. H., and Wilson, E. O. 1967. *The Theory of Island Biogeography*. Princeton University Press, Princeton, 203 pp.

Maguire, B. Jr. 1963a. The passive dispersal of small aquatic organisms and their colonization of isolated bodies of water. *Ecol. Monogr. 33*, 161-185.

Maguire, B. Jr. 1963b. The exclusion of *Colpoda (Ciliati)* from superficially favorable habitats. *Ecology 44*, 781-784.

Maguire, B. Jr. 1970. Aquatic communities in bromeliad leaf axils and the influence of radiation. In *The Tropical Rain Forest*, H. T. Odum (Ed.). Atomic Energy Commission, Washington, D. C., pp. E95-E101.

Maguire, B. Jr., Belk, D., and Wells, G. 1968. Control of community structure of mosquito larvae. *Ecology* *49*, 207-210.

Paine, R. T. 1966. Food web complexity and species diversity. *Am. Nat. 100*, 65-75.

Patrick, R. 1971. Diatom communities. In *The Structure and Function of Freshwater Microbial Communities*, J. Cairns, Jr. (Ed.). Virginia Polytechnic Institute and State University Press, Blacksburg, Va., pp. 151-164.

Silvey, J. K. G., and Wyatt, J. T. 1971. Interactions between freshwater bacteria, algae, and fungi. In *The Structure and Function of Freshwater Microbial Communities*, J. Cairns, Jr. (Ed.). Virginia Polytechnic Institute and State University Press, Blacksburg, Va., pp. 249-275.

Simberloff, D. S., and Wilson, E. O. 1969. Experimental zoogeography of islands: The colonization of empty islands. *Ecology 50*, 267-278.

Slobodkin, L. B., Smith, F. E., and Hairston, N. G. 1967. Regulation in terrestrial ecosystems and the implied balance of nature. *Am. Nat. 101*, 109-124.

Snedecor, G. W. 1956. *Statistical Methods* (5th ed.). Iowa State College Press, Ames, 534 pp.

Tate, M. W., and Clelland, R. C. 1957. *Non Parametric and Shortcut-Statistics*. Interstate, Danville, Ill.

Taub, F. B. 1969. A biological model of a freshwater community. *Limnol. Oceanog. 14*, 136-142.

Taub, F. B. 1971. A continuous gnotobiotic (species defined) ecosystem. In *The Structure and Function of Freshwater Microbial Communities*, J. Cairns, Jr. (Ed.). Virginia Polytechnic Institute and State University, Blacksburg, Va., pp. 101-120.

Taub, F. B., and Dollar, A. M. 1968. The nutritional inadequacy of *Chlorella* and *Chlamydomonas* as food for *Daphnia*. *Limnol. Oceanog. 13*, 607-617.

Wilson, E. O., and Simberloff, D. S. 1969. Experimental zoogeography of islands: Defaunation and monitoring techniques. *Ecology 50*, 278-296.

Chapter 12

LACUSTRINE FUNGAL COMMUNITIES

Robert A. Paterson

Department of Biology
Virginia Polytechnic Institute and State University
Blacksburg, Virginia

CONTENTS

ABSTRACT

Although representatives of all fungi are to be found
in the aquatic habitat, the most abundant in terms of numbers of species are the aquatic Phycomycetes. In planktonic and benthic communities of the lacustrine ecosystem
chytridiaceous and saprolegniaceous fungi have significant
roles as parasites of algae and decomposers of bottom detritus. In the case of epidemic fungal infections of
planktonic algae, activities of fungi may affect the composition of phytoplankton communities by delaying the time
of algal maximum and by reducing the population of certain
algae so that other phytoplankters will replace the infected algal populations. In the case of other infections
that are not epidemic the fungi may not influence populations of algae during periods of maximum algal population.
Instead, the fungi may only infect phytoplankters during
periods of decline in the algal population and thus only
hasten the decomposition of the plants. Some chytridiaceous fungi grow only on algal cells that are obviously dead.
The foregoing indicates that there are probably three situations with regard to the relationship of these fungi to
algae in the plankton: (1) Fungi may be obligately parasitic attacking living algal cells during periods of active
growth; (2) fungi may not be obligately parasitic and only
attack algal cells in a senescent condition during periods
of decline in the algal population; and (3) the fungus
lives only on dead algal cells.

There appears to be a pattern to the distribution of
chytridiaceous planktonic fungi that infect diatoms at
various depths in a lake. For the most part the oldest
structures in the fungal life cycles are found at the

greatest depths, indicating that fungus infections may occur near the surface and that the life cycles of the fungi proceed as the diatom cells drop to the bottom.

Aquatic Phycomycetes are commonly found in the benthic community. Fungi most commonly found during studies of Douglas Lake, Michigan, were members of the Chytridiales that decompose chitin. Thus these organisms may play a significant role in chitin decomposition in this lake. However, studies in Lake Michigan where the bottom depths were greater indicated the oomycetous fungi were the most common.

INTRODUCTION

The role of green plants as primary producers in the aquatic environment has received considerable attention. In addition, the contribution of bacteria to the nutritional cycles in aquatic ecosystems has become better understood in the last few years. Probably least known is the role of the true fungi in the freshwater habitat. Although representatives of all major groups of fungi have been found in aquatic situations, the most abundant in terms of numbers of species are undoubtedly the Phycomycetes. In the lacustrine ecosystem more is probably known about the Phycomycetes that are planktonic and benthic than those found elsewhere in lakes.

DISCUSSION

Planktonic Fungi

The occurrence of phycomycetous thalli and such propagules as zoospores, gemmae, and hyphal fragments in the plankton has been well established. Indeed, many investigators have shown the presence of zoospores and other propagules in lake waters. Noteworthy among these are workers

who have attempted quantitative or distributional studies
of these structures. For example, Suzuki (1960, 1961b,c)
and Suzuki and Nimura (1961) have attempted to enumerate
the phycomycetous spores present in several Japanese lakes
and to determine their distribution. Studies by Willoughby
and Collins (1966; Collins and Willoughby 1962) in lakes
in England utilized laborious plating methods to determine
zoospore numbers. Fuller and Poyton (1964) have isolated
freshwater and marine Phycomycetes by plating techniques
and special methods involving superspeed centrifugation.
Although the latter method has not been fully tested it
appears to be the most promising quantitative technique
for determining numbers of zoospores and other propagules
in lake waters. Such other methods as those used by the
Japanese and English workers have not yielded reliable
quantitative data.

Although the occurrence of phycomycetous propagules
in lake waters is of considerable interest, many of these
structures originate from thalli that are not planktonic.
Indeed, many fungal sporangia that contribute zoospores
to the plankton are from members of the Oomycetes and are
undoubtedly found in the bottom sediments. Exceptions
are species of such genera as *Leptolegnia, Aphanomyces,*
and other fungi that attack planktonic crustaceans and
rotifers (Paterson 1958; Petersen 1910; Prowse 1954; Scott
1961). Other exceptions are species that grow on what
Sparrow (1968) terms adventitious plankters. Examples of
these are dead insects, pollen, and other floating debris.

Of the truly planktonic fungi only those members of
the Chytridiales that attack phytoplankton have been studied quantitatively. Several years ago Canter and Lund
(1948, 1951, 1953) investigated the effects of fungal infection on planktonic algal populations in English lakes.
Their careful quantitative work revealed several interesting situations with regard to the relationship between

403

fungal parasitism and factors acting on diatoms and on other algae. One situation was concerned with the effect of parasitism on algal maxima. Where the percent of parasitism was as high as 40 or more of the frustrules infected with chytridiaceous thalli there was a delay in the time of the algal maxima. In addition, the expected number of diatoms in a maximum were decreased. In another situation an increase in population occurred in uninfected algae when an algal species in the same community was parsitized by chytrids. As described by Canter and Lund from their study of Esthwaite water, the course of events that involved one blue-green alga and three species of diatoms was as follows: During the spring of 1949 the presence of *Oscillatoria agardhii* var. *isothrix* Skuja in great abundance (100-250 filaments/ml) near the surface of the lake had the effect of reducing light penetration. The effect reduced the growth rate of *Asterionella formosa* Hass., *Tabellaria fenestrata* var. *asterionelloides* Grun., and *Fragilaria crotonensis* (Edw.) Kitton. In early May the *Oscillatoria* filaments became heavily infected by *Rhizophydium megarrhizum* Sparrow, causing a rapid decrease in the numbers of *Oscillatoria* filaments. As the latter decreased there was a rapid increase in numbers of *Fragilaria* and *Tabellaria* colonies. During the last part of May *Asterionella formosa* became infected by *Rhizophydium planktonicum* Canter, causing a rapid decline in the numbers of diatom colonies. The species of *Fragilaria* and *Tabellaria* remained uninfected and continued to increase in growth rate until the middle of June, when silica became lacking and thus limited further increase in growth. A third observation by Canter and Lund indicated that the amount of infection of the algal population was related to the relative growth rates of the host and the fungus. Indeed, they reported infections so severe that the number of fungal thalli was greater than the number of algal cells.

In other studies Canter and Lund (1948) observed that occasionally an epidemic occurred when a host population was about to decline from other causes. They suggested that the decrease in numbers of algae in a population was hastened by the parasitism. Koob (1966) and Paterson (1960) observed parasitism of phytoplankters by chytrids to occur only during periods of decline in the infected algal population. Furthermore, there was no apparent hastening of the decline in numbers of algae. Indeed in the study by Koob of *Asterionella formosa* parasitized by *Rhizophydium planktonicum* in two Colorado lakes the fungus infection did not appear to have the effect observed by Canter and Lund. Although the parasitism reached a maximum of 40% during Koob's investigation he could find no evidence that parasitism hastened the decrease in numbers of diatom cells. In the study by Paterson in a Michigan lake the infestation of *Anabaena planktonica* Brunnthaler by *Rhizosiphon anabaenae* (Rodhe and Skuja) Canter also occurred during the decline in the algal population. However, the percent parasitism was very small and never exceeded 6% of the algal filaments. In this case the percent parasitism was too small to have any effect on the algal population. Therefore, Koob's and Paterson's investigations suggest that in some situations chytrids attack algae only during periods of algal decline and probably have little or no effect on the algal population.

It is evident from the foregoing that the seasonal distribution of chytrids infesting live algal cells is determined by the seasonal periodicity of the host. Obviously, the fungus can only be observed to infect the host when the host is present. However, at least one planktonic chytrid has been shown to have a periodicity in seasonal distribution when the host cells are continuously present.

In a study by Paterson (1960) of the chytrid *Amphicypellus elegans* Ingold that infects only dead thecae of *Ceratium hirundinella* (O.F.M.) Schrank the chytrid was present only during June and August. The percent of *Ceratium* cells supporting the growth of the chytrid was high, being 38% in June and 92% in August. The host was present continuously from the first of June to the middle of October. Although *A. elegans* cannot be considered a parasite it is nevertheless planktonic.

The foregoing discussion on planktonic fungi indicates that the relationships between the planktonic chytrids and their algal hosts is not the same in each situation. I suggest that there are at least three relationships. In the first case the percent parasitism is high and the alga is attacked during its growth phase before a maximum in numbers of algae is reached. This suggests that the fungus is truly parasitic in that it attacks only actively growing, healthy algal cells. In the second situation algal cells are attacked only during the declining phase of the population after a maximum in their numbers is reached. This indicates that the fungus is specific in its requirements and different from the first case. It does not infest active living cells but only those becoming moribund or senescent. In the third situation the fungus attacks only dead cells or nonliving parts of an alga. An example is the invasion of dead *C. hirundinella* thecae by *A. elegans*. In the first two situations the affinity of the fungus for the alga is probably specific. Although *A. elegans* has been found on dinoflagellates other than *Ceratium* it is probably specific for that group. It undoubtedly has a requirement for some substance in the thecal wall since the protoplasm is always observed to be absent in infested cells of the host and the rhizoids appear to be only in the wall.

Studies by Koob (1966) and an unpublished investigation by Paterson have shown two interesting features of
chytrid specificity. Koob found in his populations of
Asterionella formosa five distinct classes of colonies
based on frustule length. Furthermore, he found that only
the populations of one of the size classes was infected
by *Rhizophydium planktonicum*. The other four populations
remained uninfected. Paterson also found in his investigation of *Chytridium marylandicum* Paterson a specificity
within a host species. *C. marylandicum* attacks *Botryococ-
cus braunii* Kütz. However, the fungal plant body does not
penetrate the living cells of the colony but is found only
in the mucilaginous sheath. During the investigation *B.
braunii* was isolated in unialgal cultures from each of
seven lakes in Michigan and one in Maryland. The cultures
were inoculated with *C. marylandicum*. In spite of repeated
attempts at culturing and inoculation growth did not occur
in the cultures of *B. braunii* isolated from three of the
Michigan lakes. The fungus grew abundantly on the four
other algal isolates from Michigan and on the one from
Maryland. Therefore, there are some features of specificity that remain unknown at present.

Concerning the role of chytridiaceous fungi that infest phytoplankters in the lacustrine ecosystem Wesenburg-
Lund (1905) stated the following.

I have shown, further, that nearly all the protoplasm of the cells in plankton is eaten by
Phycomycetes before reaching the bottom; my observations prove that an organism in the latter
part of the period of maximum development may
very often be infected by Phycomycetes, which
feed upon the protoplasm and kill it leaving
the skeleton intact.

A study by Paterson (1967) on the vertical distribution

of chytridiaceous fungi tends to support Wesenburg-Lund's contention that phytoplankters are attacked and sink with their parasites to the bottom of lakes. The following table shows the vertical distribution of three chytrids

Vertical Distribution of Some Fungi and Temperature Data from Grand Traverse Bay, Lake Michigan, 1963[a]

Temp (7/16)	Depth (m)	7/16	7/22
19.5	0	*	
	2		
19.0	5		
18.2	10		
16.5	11	≡	
11.5	15	s–	
			≡
8.5	20	<	
			<
7.8	25	–	
		≡	≡
7.2	30	s–	
		<	
			<
	80		

[a]* = *Zygorhizidium melosirae* on *Asterionella formosa;* – = *Zygorhizidium melosirae* on *Synedra* sp.; ≡ = *Rhizophydium fragilariae* on *Fragilaria crotonensis;* < = *Rhizophydium* sp. on *Tabellaria flocculosa;* s = sexual stages.

that were found in greatest abundance and their algal hosts on 2 days a week apart in Lake Michigan. Of the three chytrids, *Zygorhizidium melosirae* Canter occurred most frequently. It was found on July 16 at the surface when it infected *Asterionella formosa* at 15, 25, and 30 m and on July 22 at 2 m when it infected *Synedra* sp. Even though those thalli that occurred at 2 m on July 22 and at the surface on July 16 probably belonged to the same fungus population, it is possible that *Z. melosirae* at the other depths were members of different populations that could have been separated from the former by the thermal stratification of the lake. That thermal stratification can influence the distribution on the host, thereby limiting the invasion by chytrid zoospores, has been suggested by Lund (1957). It seems entirely possible that a thermal stratification may also restrict the movement of zoospores from layer to layer. In this investigation a major discontinuity layer existed at 10 and 11 m. There was thus a thermocline between chytrids at the surface and the populations at 15, 25, and 30 m.

Two populations were probably involved in the distribution of *Rhizophydium fragilariae* Canter infecting *Fragilaria crotonensis* (A. M. Edw.) Kitton. The population of *R. fragilariae* at 10 m on July 16 and that at 15 m on July 22 were probably the same, the latter having sunk from 10 to 15 m during the week. This situation with regard to *Rhizophydium* sp. infecting *Tabellaria flocculosa* (Roth) Kütz is similar. The infested diatoms were found at the lower depths of 20 and 30 m on July 16, and 30 and 80 m on July 22. These distribution patterns did not result from the presence of the host only at these depths since they were generally present at all depths on both dates. It appears that the infected diatoms seem to have sunk in the water from July 16 to July 22.

Evidence that those fungal populations at the lower depths were older than those near the surface was the presence of sexual structures in *Z. melosirae* at 15 and 30 m but not at the surface and 2 m. According to Canter and Lund (1948) the sequence of events in a fungal infection is as follows. The zoospore cysts appear first on the host and are most abundant. Then sporangia with protoplasm are most numerous, followed by empty sporangia, and then sexually formed resting spores. Therefore, it is probable that populations of *Z. melosirae* at 15 and 30 m were relatively old. Perhaps they were formed previously nearer the surface and had sunk to those depths just previous to July 16.

Benthic Fungi

Although the presence of chytridiaceous and oomycetous fungi in lake bottoms has been well established, little else is known other than some information about their distribution and ecology. Most studies have been made using the typical phycomycetous substrates or "baits." Some examples are cellophane, chitin of insects, snake skin, hemp seeds, grass leaves, and pollen. In a study by Paterson (1967) in Douglas Lake, Michigan, "baits" were placed in containers that were put on the bottom of the lake at various depths from 0.5 to 15 m, the most common fungi found were such chitinophilic chytridioceous species as *Asterophylctis sarcoptoides* H. E. Petersen, *Obelidium mucronatum* Nowakowskiella, *Rhizoclosmatium aurantiacum* Sparrow, *Chytridiomyces chyalinus* Karling, and a species of *Siphonaria*. The greatest number of chytrid species were recovered from the bottom of Douglas Lake during the first week or 10 days in July to the end of that month, whereas very few were found in August. Although very few oomycetous fungi were recovered from the bottom of Douglas

Lake they were found only during the first 2 weeks in July.
Concerning distribution by depth, some species of chitrids
were restricted to shallow waters, others to deep waters,
and others wer⌐ ubiquitous. Studies were also conducted
by Paterson ⟨1967) on Grand Traverse Bay, Lake Michigan.
As in the investigation of Douglas Lake, substrates or
"baits" were placed on the bottom. In Lake Michigan the
areas studied were 26 and 31 m deep. Few chytrideaceous
fungi were found and the most numerous fungi recovered
were members of the Oomycetes. No seasonal distribution
was evident in Grand Traverse Bay. Willoughby (1961a,b;
1962, 1965) has studied the distribution of fungi in the
muds of English lakes. In these studies chitinophilic
fungi and chytridiaceous species growing on cellulosic
substrates were found abundantly. Saprolegniaceous propagules in Blelham Tarn were more frequently collected from
soils of the lake margin than from the muds in the center
of the lake. High numbers of mucors were collected from
bottom muds, indicating that fungus propagules sedimented
and remained viable. However, these fungi could make no
contribution to the aquatic ecosystem since they were terrestrial forms. Suzuki (1961a,d,e) has observed seasonal
changes in the distribution of oomycetous fungi. More of
these organisms were present in the bottom of Lake Nakanuma
during the winter than during the summer. Fungi in other
Japanese lakes had different seasonal distribution patterns.
For example, in Lake Yamanakako the greatest numbers of
fungi were found in the fall (Suzuki and Hatakeyama 1961).
The distribution of aquatic fungi in Japanese lake bottoms
shows a relationship to the amount of dissolved oxygen.
During periods of anaerobic conditions fungi in lake muds
were scarce. For more complete information on Suzuki's
work consult Sparrow (1968).

CONCLUSION

In conclusion, planktonic and benthic fungi are commonly found in lacustrine habitats. In some situations fungal parasitism may have an effect on increasing or decreasing populations of phytoplankters. In other situations chytridiaceous fungi may be decomposers that break down the algae as they become senescent or die and fall to the bottom of a lake. Benthic fungi are composed of chytridiaceous or oomycetous fungi and are found with enough frequency to conclude that they contribute to the breakdown of such substances as chitin, cellulose, and perhaps other materials as yet unknown.

REFERENCES

Canter, H. M., and Lund, J. W. G. 1948. Studies on plankton parasites. I. Fluctuations in the numbers of *Asterionella formosa* Hass. in relation to fungal epidemics. *New Phytol. 47*, 238-261.

Canter, H. M., and Lund, J. W. G. 1951. Studies on plankton parasites. III. Examples of the interaction between parasitism and other factors determining the growth of diatoms. *Ann. Bot. (London)*, New Series *15*, 359-371.

Canter, H. M., and Lund, J. W. G. 1953. Studies on plankton parasites. II. The parasitism of diatoms with special reference to lakes in the English lake district. *Brit. Mycol. Soc. Trans. 36*, 13-37.

Collins, V. G., and Willoughby, L. G. 1962. The distribution of bacteria and fungal spores in Blelham Tarn with particular reference to an experimental overturn. *Arch. Mikrobiol. 43*, 244-307.

Fuller, M. S., and Poyton, R. O. 1964. A new technique for the isolation of aquatic fungi. *Bioscience 14*, 45-46.

Koob, D. D. 1966. Parasitism of *Asterionella formosa* Hass. by a chytrid in two lakes of the Rawah wild area of Colorado. *J. Phycol. 2*, 41-45.

Lund, J. W. G. 1957. Fungal diseases of plankton algae. In *Aspects of the Transmission of Disease*, C. Horton-Smith (Ed.). Hafner, New York, pp. 19-23.

Paterson, R. A. 1958. Parasitic and saprophytic Phycomycetes which invade planktonic organisms. II. A new species of *Dangeardia* with notes on other lacustrine fungi. *Mycologia 50*, 453-468.

Paterson, R. A. 1960. Infestation of chytridiaceous fungi on phytoplankton in relation to certain environmental factors. *Ecology 41*, 416-424.

Paterson, R. A. 1967. Benthic and planktonic Phycomycetes from northern Michigan. *Mycologia 59*, 405-416.

Petersen, H. E. 1910. An account of Danish freshwater-Phycomycetes, with biological and systematical remarks. *Ann. Mycol. 8*, 494-560.

Prowse, G. A. 1954. *Aphanomyces daphniae* sp. nov., parasitic on *Daphnia hyalina*. *Brit. Mycol. Soc. Trans.* *37*, 22-28.

Scott, W. W. 1961. A monograph of the genus *Aphanomyces*. *Virginia Agri. Exp. Sta. Tech. Bull.* *151*, 1-95.

Sparrow, F. K. Jr. 1968. Ecology of freshwater fungi. In *The Fungi, An Advanced Treatise*, G. C. Ainsworth and A. S. Sussman (Ed.). Acadmic, New York and London, pp. 41-93.

Suzuki, S. 1960. Microbiological studies of the lakes of Volcano Bandai. I. Ecological studies on aquatic Phycomycetes in the Goshikinuma Lake group. *Jap. J. Ecol.* *10*, 172-176.

Suzuki, S. 1961a. The seasonal changes of aquatic fungi in lake bottom of Lake Nakanuma. *Bot. Mag. (Tokyo)* *74*, 30-33.

Suzuki, S. 1961b. The vertical distribution of the zoospores of aquatic fungi during the circulation and stagnation period. *Bot. Mag. (Tokyo) 74*, 254-258.

Suzuki, S. 1961c. Distribution of aquatic Phycomycetes in some inorganic acidotrophic lakes of Japan. *Bot. Mag. (Tokyo) 74*, 317-320.

Suzuki, S. 1961d. Ecological studies on aquatic fungi in the lakes of Volcano Nikko. *Jap. J. Ecol. 11*, 1-4.

Suzuki, S. 1961e. Some considerations on anaerobic life of aquatic fungi in lake bottom. *Jap. J. Ecol. 11*, 219-221.

Suzuki, S., and Hatakeyama, T. 1961. Ecological studies on the aquatic fungi in Lake Yamanakako. *Jap. J. Ecol. 11*, 173-175.

Suzuki, S., and Nimura, H. 1961. The vertical distributions of fungi and bacteria in lake during the circulation and stagnation period. *Jap. Mycol. Soc. Trans. 7*, 115-117.

Wesenburg-Lund, C. 1905. A comparative study of the lakes of Scotland and Denmark. *Proc. Roy. Soc. Edinburg 25*, 401-480.

Willoughby, L. G. 1961a. The ecology of some lower fungi at Esthwaite Water. *Brit. Mycol. Soc. Trans. 44*, 305-332.

Willoughby, L. G.   1961b.   Chitinophilic chytrids from lake muds.   *Brit. Mycol. Soc. Trans. 44*, 586-592.

Willoughby, L. G.   1962.   New species of *Nephrochytrium* from the English lake district.   *Nova Hedwigia 3*, 439-444.

Willoughby, L. G.   1965.   Some observations of the location of sites of fungal activity at Blelham Tarn. *Hydrobiologia 25*, 352-356.

Willoughby, L. G., and Collins, V. G.   1966.   A study of the distribution of fungal spores and bacteria in Blelham Tarn and its associated streams.   *Nova Hedwigia 12*, 150-171.

Chapter 13

CARBON FLOW IN THE AQUATIC SYSTEM

George W. Saunders

Environmental Science Branch
Division of Biology and Medicine
U.S. Atomic Energy Commission
Washington, D. C.

CONTENTS

ABSTRACT

Available information on the distribution and activities of microorganisms in the sea and in inland waters indicates that these two variables are continually fluctuating during the annual period, i.e., they are never in steady state. It is known for one Michigan lake that the annual cycles of phytoplankton, bacteria, zooplankton, soluble carbohydrate, soluble organic nitrogen, and soluble organic carbon are different and are not in phase. This indicates that both the quality and quantity of the microorganisms vary spatially and temporally within aquatic systems. This is true in the tropics as well as north temperate regions and therefore is probably a worldwide phenomenon.

It is possible to estimate the initial conditions for carbon concentration of all the general forms of carbon and most of the rates of carbon transport in certain aquatic subsystems. It appears that changes in the structure of the carbon flow system are controlled by the microorganisms and not directly by the environment, i.e., microorganisms have considerable capacity to adjust their responses to changes in environmental conditions.

Descriptions of longer term fluctuations in distributions of microorganisms and their activities, as well as descriptions of daily or semidaily rates of carbon transport, cannot reveal the control mechanisms that are operating to determine the density structure or the functional structure of the microbial component of the aquatic system.

Generalized data from field analyses are presented to show that light is a fluctuating forcing variable within

the 24-hr period. The photosynthetic rate responds with
a fluctuation not directly proportional to fluctuating
light intensity, and both secretion rate of soluble organic
matter and bacterial assimilation rate of this secreted
organic material fluctuate during the 24-hr period. These
latter two fluctuations are not in phase with one another
or with the photosynthetic rate fluctuation. This sequence
of forcing variable and nonphased fluctuating responses
can be viewed as a linearly coupled control system. Another
control system that exists is that of light-forced photosynthesis and the light-cued zooplankton migration that
results in a coupled producer-grazer control system. It
is suggested that changes in the composition of algae under
simulated day-night conditions reflect an active internal
control system for these organisms and imply that possibility for other organisms.

An aquatic subsystem, and in fact the whole aquatic
system, can be viewed as a set of oscillating control systems having a period of 24 hr, in addition to other longer
or shorter period fluctuations. If a sequence of 24-hr
oscillating sets is examined, the pathway by which a subsystem moves through a series of states can be described.
If the subsystem is subjected to certain critical minor environmental perturbations and the responses of the system
observed, it may be possible to reveal where the fluctuations are tightly coupled and the nature of the control
mechanisms operating in the system and dictating the manner
in which that system passes through a certain sequence of
states.

DISCUSSION

Structure and function usually are very closely related, but the knowledge of one does not necessarily reveal
the exact nature of the other. Structure in the aquatic

system has been examined in many ways but traditionally it
has been integrated as energy flow on an annual basis in
trophic levels within the system (Juday 1940; Lindeman
1942; Dineen 1953; Odum 1957; Teal 1957; Nauwerck 1963).
Because of the difficulty in treating the bacteria they
have usually been omitted from the budget. Any analysis
to be developed from annual data usually must apply to
some degree of subgeological time and therefore to some
aspect of lake evolution. Since the annual period is
greater than the longevity of most microorganisms, the historical record of the sequence of microbial communities
is lost in the integration or, in fact, by never having
been examined. Therefore it becomes impossible to make
any specific statement about the structure or the role of
microbial communities within shorter time intervals. Many
of the questions we purport to be asking require that analyses be conducted within time intervals of hours or days
rather than years.

I want to describe the results of microbial activities
and the structure of microbial processes in subsystems of
the aquatic system within short intervals of time. I want
also to indicate how short-term measurements might be useful in revealing function and may lead to an analysis of
mechanisms of control and regulation within the system.
I want to do this from the point of view of carbon flow,
recognizing that energy flow and nutrient cycling are absolutely coupled and that to discuss one without the other
is merely for convenience. The cycling of carbon is a
vehicle for the transport of energy along one pathway in
the dissipation of energy from the ecosystem.

When one examines the distribution of nutrients and
the metabolic products of organisms during the period of
one year in lakes and oceans (Duursma 1961; Belser 1959;
Brehm 1967; Allen 1969) it is striking that the

concentrations of such substances vary during the year.
In a detailed study of a Michigan lake (Saunders and Lauff,
unpublished) the annual concentration distributions in the
lake volume of soluble carbohydrate, soluble organic nitrogen, soluble organic carbon, phytoplankton, and bacteria
are very different. Presumably these fluctuations reflect
changes in the quality and quantity of microorganisms as
well as of their metabolic activities. Even in the tropics, where light and temperature are relatively more constant, major fluctuations in concentration of organisms
or their metabolites are observed. The structure, qualitative and quantitative, and the activity of the microbial
community does not appear to be in steady state during
this interval of time. Where data are available on the
time-space distribution of phytoplankton, bacteria, and
zooplankton (Overbeck 1968; Gambaryan 1968; Rasumov 1962;
Saunders 1963) it is possible to generate some general
correlations among these groups of organisms, which possibility suggests that they are interacting and exerting
some kind of control over one another. However, it is
difficult to extract any more specific correlation that
would reveal the true mechanisms of interaction.

In annual studies of two Michigan lakes (Saunders,
unpublished) only three points in time were observed where
phytoplankton and bacteria were specifically correlated
in terms of changes in their populations' densities. In
the single case where a zooplankton increase and a phytoplankton-bacterial minimum were observed, subsequent data
suggest that grazing by zooplankton was not the cause of
the phytoplankton-bacterial minimum. Therefore another
factor or other factors must have been operating to determine the simultaneous changes of the phytoplankton and
bacteria. The simple correlation observed did not reveal
the true control mechanism for the phytoplankton and

bacterial concentrations. During the remainder of the year, which constitutes 90% of the period, no obvious direct correlations among phytoplankton, soluble organic matter, and bacteria were observed. This suggests either that there were no direct relationships or that the analytical procedures were such that they could not reveal any specific relationships. In earlier investigations of the time course of bacterial numbers in natural waters, it has not been possible to predict their distribution during the year with any confidence. The season of maximum development differs among lakes and, within a lake, may vary from year to year (Henrici 1939; Taylor 1940; Rasumov 1962).

It is also known that the vertical distribution of bacteria in stratified lakes can change very rapidly (Rasumov 1962). An example of short-term changes in the vertical distribution of bacteria in Frains Lake is presented in Figure 13.1. The most drastic changes occurred in mid-June following a pulse of phytoplankton. At this time the numbers of bacteria increased sharply and decreased again in a few days. The distribution of bacteria inverted and reverted in this same time interval. This indicates that there are shorter term variables and processes that control the distribution of bacteria and that are not essentially revealed by the more general correlations discussed in the preceding paragraph.

In ecology we have long discussed the law of the minimum. In nature, however, it may be only rarely that a population responds to a single environmental variable. This would occur when the population has been stressed in the extreme and driven toward the boundaries of its possible existence. At this time a specific correlation would easily be observed. At other times, which probably constitute the majority, any population may be responding to a set of variables (McCombie 1953, 1960). It becomes

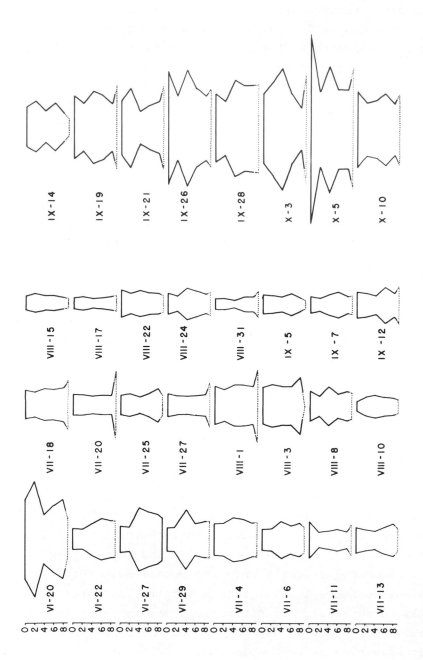

FIG. 13.1. Relative vertical distribution of bacteria (direct counts) by date in Frains Lake, Michigan, 1960. Ordinate, depth in meters; abscissa, relative count.

difficult to assess the impact of the separate variables of the set and also to determine the true response of the population. The analytical procedure needed for meaningful investigation must have sufficient sensitivity to resolve changes in the operating variables and their impact as well as the qualitative and quantitative responses of the organisms. The frequency of sampling must lie well within the longevity of the organisms and, in fact, well within the period of the event, if we are to discern the pathway by which the system moves from its initial to final state and the mechanisms that determine this pathway.

It is necessary to conclude that we have been examining the microbial component of ecosystems in a manner that is inappropriate for revealing microbial function and the mechanisms that control and regulate this function.

I suggest that if we are to answer contemporary questions concerning the microbial component of ecosystems, it is necessary to discontinue casual examination of the system and it is absolutely necessary to examine the fine structure of the system and its perturbations. This is a procedure that has been well used with profit in other fields. It is also necessary to conduct the analysis in a natural system that is minimally perturbed by the manipulation. While laboratory experiments can be used to answer more general questions it is also well documented that extrapolation of results from reconstructed systems is not always reliable.

I would like in the following to indicate that it is feasible to estimate the concentration structure and the metabolic structure of a microbial community. It should also be possible to analyze the perturbations of environmental variables and the response of the community to those perturbations. It is also possible to do this within very short time periods, including parts of a day.

If a set of variables operating on a system changes, the microbiota will respond to the changes in three possible ways: (1) the organisms will respond passively, in which case the organisms are caused to change to some new state where the change is controlled directly by the environment; (2) the organisms will regulate, i.e., they will attempt to maintain constancy and the changes observed will merely reflect the degree to which they are unable to do so; (3) the organisms will control their change of state, i.e., they will dictate the exact path they take in proceeding to the new state. I suggest that it is the last that is in fact the case (there is abundant evidence for regulation and control at the molecular level and the same is intuitively obvious at other levels of organization in the hierarchy of life structures).

It is possible to estimate the initial conditions at sunrise and the rates of carbon flow in the aquatic system on a 24-hr basis for all the general forms of carbon (Fig. 13.2). The analysis was made on a subsystem of lake water

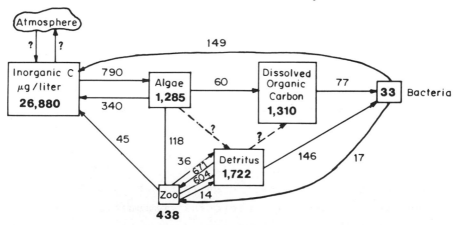

FIG. 13.2. Carbon flow in microbial system at 1.0 m in Frains Lake, Michigan, July 22, 1968. Numbers within blocks show the carbon concentration at sunrise in µg C/liter; numbers on arrows are the transfer rates in µg C/liter/24 hr.

CARBON FLOW

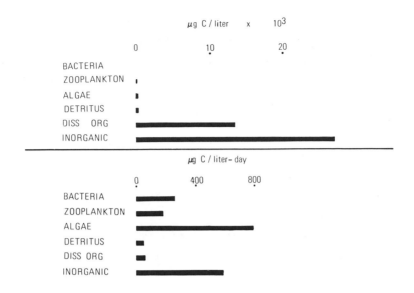

FIG. 13.3. Structure of microbial system at 1.0 m in Frains Lake, Michigan, July 22, 1968. Upper bar graphs show the initial concentration structure at sunrise; lower bar graphs show the metabolic structure as input rate in μg C/liter/day.

from a single depth in the photic zone of Frains Lake. Radioactive tracers were used as an analytical tool and the experiment summarized is the first of a series designed to test the feasibility of such an approach. The figure implies a steady-state condition for the rates during the period. One point is very apparent. The carbon cycling between the algal and inorganic forms exceeds the carbon being cycled through all the other forms of carbon in this subsystem.

The structure of the carbon system is summarized in Figure 13.3. The upper bar graphs represent the concentration structure of the system. There are two very large

427

pools of nonliving carbon, the inorganic and the dissolved organic carbon. Both of these forms are very much in excess of the apparent requirements of the living organisms. The algae and detritus (probably planktogenic) are of intermediate concentration and the bacteria and zooplankton (consumer organisms) have the lowest concentrations. The bacteria amount to about 10% of the zooplankton concentration. This pyramid of carbon distribution probably should not be unexpected for the upper photic zone in lakes. It is unique that all the forms of carbon have been estimated and included in the budget.

The metabolic structure of the subsystem is presented in the lower part of Figure 13.3. These bar graphs represent the input rates or the gross assimilation rates of carbon in micrograms of carbon per liter of lake water per day for each form of carbon. It becomes obvious from the figure that the microbiota control the movement of carbon in the system even though the nonliving forms of carbon are by far the dominant ones in terms of biomass. This implies that although the intensity of sets of environmental variables will set the general level of operation of the system, the chemical processes of life have some freedom of expression and will control the modification of the initial environmental distributions and will dictate their own performance and distributions within that environment.

The statement of initial conditions and the daily rates of transfer of various forms of carbon may be somewhat misleading, because within a 24-hr period the amounts of carbon species are not constant nor are the rates of transfer. The 24-hr period can be partitioned into at least two periods, day and night. The initial conditions and the total transfer rates for the daylight period are presented in Figure 13.4. The concentration structure is

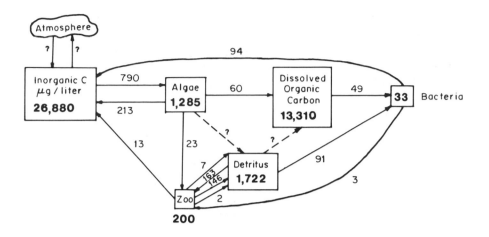

FIG. 13.4. Carbon flow in microbial system at 1.0 m in Frains Lake, Michigan during daylight, July 22, 1968. Numbers within blocks are the carbon concentration at sunrise in µg C/liter; numbers on arrows are the transfer rates in µg C/liter/15 hr.

not changed fundamentally but the metabolic structure is shifted toward a greater relative importance of cycling between the algae and inorganic carbon than in Figure 13.2. This should obviously be true since photosynthesis can occur only during daylight whereas heterotrophic activity occurs throughout the day. The initial conditions and the rates of transfer for the nighttime period are shown in Figure 13.5. The concentration structure is somewhat modified because of changes during the day. The initial conditions for the zooplankton include those organisms that have undergone vertical migration into the depth layer of the subsystem studied as well as growth within the system during the day. The metabolic structure of the subsystem has changed completely. Algal photosynthesis and algal secretion of soluble organic matter have gone to zero rate. The system is completely heterotrophic and the zooplankton

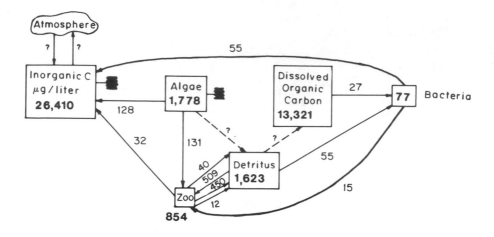

FIG. 13.5. Carbon flow in microbial system at 1.0 m in Frains Lake, Michigan during nighttime, July 22, 1968. Numbers within blocks are the initial carbon concentration at sunset in µg C/liter; numbers on arrows are the transfer rates in µg C/liter/9 hr.

have come to dominate metabolism in the system. This is because there has been immigration of zooplankton into the depth volume studied, which occurs at the onset of darkness. This complete shift in the metabolic structure of the system is quite remarkable. It results, however, only in more minor changes in the concentration structure.

It is, however, more interesting and more realistic to examine the time course of concentrations and rates continuously within a 24-hr period. The data points for the time course of cumulative net particulate photosynthetic carbon fixation and of cumulative carbon secretion for a sample of lake water are given in Figure 13.6. There are two points that need to be made. First, there are inflections in the curves for particulate fixation and secretion. This means that the rates go from zero to some high value to zero, i.e., the rates fluctuate. Second,

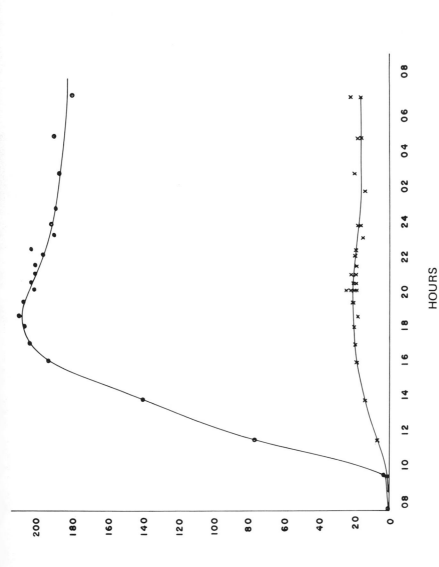

FIG. 13.6. Cumulative carbon in particulate and soluble phases resulting from photosynthesis during 24 hrs in Frains Lake, Michigan, at surface under ice, February 12, 1969. Sunrise, 0730.(0), Photosynthesis; (x), secretion.

the points of inflection are not in phase so that the secretion rate curve lags the photosynthetic rate curve by
about two hours, in this case.

This kind of data is generalized in the set of schematic diagrams given in Figure 13.7 where the time interval
is 24 hrs and the initial time is sunrise. Light intensity
varies with the sun's angle during the day. The phytoplankton respond to changes in light intensity but the response is not directly proportional. The photosynthetic
rate per unit volume is asymmetric about midsolar day, being depressed in the afternoon. This suggests that the
phytoplankton have some control over the manner in which
they respond to light. The phytoplankton secrete photosynthate in which the secretion rate is out of phase with
the rate of photosynthesis. In an independent experiment
I have been able to demonstrate that the assimilation rate
of secreted material by the indigenous bacterial community

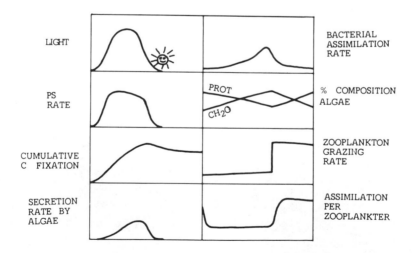

FIG. 13.7. Schematic diagrams of the distribution of concentration
or rates of processes from sunrise to sunrise in the upper photic
zone of an aquatic system.

of a lakewater sample fluctuates and is out of phase with
the secretion rate, in this case by approximately 6 hr.
It is possible to view this sequence of events as representative of a tightly coupled reaction system in which
light intensity, in its ordinary diurnal variation, is a
forcing environmental variable. The biota, phytoplankton
and bacteria, respond to this perturbation in a sequence
of fluctuating causal reactions that presumably are under
the control of the organisms. A part of the soluble organic pool intervening the phytoplankton and bacteria is
a component of this sequence.

Another coupled reaction system is that of the primary producers and the herbivorous zooplankton grazers.
This can, in fact, be viewed as two systems. In one system the biomass of phytoplankton fluctuates during the day
as a function of light intensity, daytime grazing rate,
respiration, nighttime grazing rate, as well as natural
mortality rate. The curve for cumulative phytoplankton
carbon fixation (Fig. 13.6) is representative of this,
where the initial biomass is not zero. The grazing rate
is a function of diurnal migration of the zooplankton that
is cued by changes in light intensity or rate of change
in light intensity. The second system has to do with assimilation of algae by zooplankton. I have some preliminary evidence that the assimilation rate of algae by
migrating *Daphnia* fluctuates during the 24-hr period, being greater during the night than during the day.

In addition to the diel fluctuations of such gross
external and internal variables, it is well known that
cultures of algae undergoing sequential light and dark
exposures exhibit marked changes in the proportions of
their biochemical constituents. A schematic representation
of this is given for the time course of protein and carbohydrate in percentage in Figure 13.7. These and other

fluctuations in biochemical constituents fall generally
within the realm of regulation and control at the molecular
level.

The above general considerations can be extrapolated
broadly to the ecosystem so that it may be viewed as a
network or a set of connected fluctuating reaction systems,
in which the fluctuations vary in shape from quasisinu-
soidal to square waves or paired step functions and in
which the fluctuations are not in phase and may be more
or less damped. The change of state of any system over
a unit time interval may be regarded as the summation of
this network of fluctuations.

If the state of an ecosystem is known at S_1 and S_2
(Fig. 13.8), there is no way of knowing from that information alone how the system moved from state 1 to state 2.
It could have moved along an infinite number of pathways.
If the system is fluctuating in shorter time intervals
with a more or less constant period and if one samples as
frequently as the period of fluctuation, the path taken

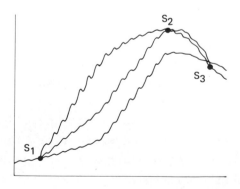

FIG. 13.8. The state of a fluctuating integrated reaction system with
time. Ordinate, S = f (r, s, t, ...z); abscissa, time.

in moving from state 1 to state 2 can be described crude-
ly. If the system is examined more frequently than the
period of fluctuation, the path taken from state 1 to state
2 can be described more precisely. However such information
will not reveal the functional mechanisms that caused the
system to move from state 1 to state 2. The mechanisms
can only be discerned if the causal operators on the system can be revealed (Fig. 13.9). If one is clever enough
to guess the major environmental operators, perturbations
of these operators can be performed and the response of
the biota and the rest of the system determined. In such

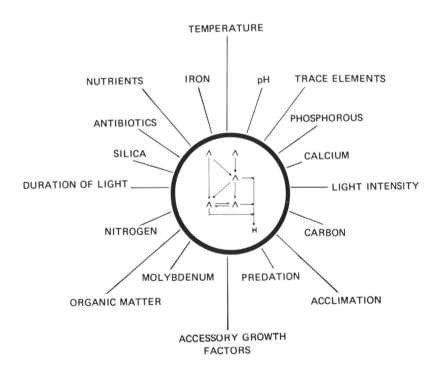

FIG. 13.9. Schematic diagram of a set of environmental factors
operating on an energy flow system at an instant in time.

a manner the major causal reaction systems can be revealed
and presumably also the reaction mechanism.

The pelagic region of a lake is constituted mainly
of microbial or microscopic communities, in which I include
the zooplankton. Such being the case, it is necessary to
examine this system within the lifetime of the populations
and more importantly within the critical response time of
the populations. Otherwise the historical record of the
perturbation-response reaction is lost, never having been
examined. Microorganisms can respond to environmental
changes in a matter of seconds or minutes. This has become obvious in terms of induction-repression mechanisms.
The basis for these mechanisms lies at the molecular level.
If the time constant for the molecular mechanisms is very
short relative to population responses, perhaps the molecular mechanisms can be considered as parameters for longer
term reaction systems. If the 24-hr period is considered
as a pragmatic compromise, it has a periodic fluctuation
in light that is constant and nearly symmetrical for many
latitudes; it is not much longer than the lifetime of some
microbes and is certainly much shorter than the longevity
of most microbial populations. Therefore if we examine
perturbations in the fine structure of the environment and
in fine structure of the ecosystem or a subsystem within
a 24-hr period, it should be possible to analyze function
and control within that system during this period. If a
sequence of 24-hr analyses is made, it should be possible
to perform something analogous to a power spectrum analysis
of observed fluctuations. If a residuum occurs after
short-term interactions have been removed, this should
reveal longer term interactions and suggest mechanisms
of control operating during these longer periods.

The preceding discussion indicates that aquatic ecosystems are dynamic. They are probably never in steady

state, at least within time intervals for which we can
ask questions about the real world and expect to test hypotheses arising from these questions. There are both
long-term and short-term fluctuations in concentrations
as well as in rates of processes. The manner of examining
the microbial subsystem has not been adequate to answer
the questions we have been asking about aquatic ecosystems.
I suggest that we must examine, therefore, the fine structure of the system and we must do this in some detail.
We know that the response of the microbial system to environmental perturbation is extremely rapid. Therefore
we must examine this system within the generation time of
the response. This must lie at least within a 24-hr period
at most latitudes and has the pragmatic advantage that
the light intensity as a forcing environmental variable
is periodic over this interval. Of course at high latitudes this last fact does not hold.

I have shown that it is indeed feasible to examine
the structure of a system and changes in this structure
within such a time interval. I have given examples of
three possible fluctuating coupled reaction systems occurring in aquatic ecosystems. It is inferred that we
should expect to describe many more such systems. Although
it has not yet been done, it should be possible to examine
the response of such reaction systems to small changes in
environmental variables. Such an analysis would reveal
the critical points of coupling and the major control
mechanisms operating to determine the nature of the system at various points in time-space. The preliminary success obtained in describing short-term changes in the
structure of a few microbial reaction systems suggests
that it would be possible to accomplish this last objective
also with success. Thus we can visualize the initial
phases in the analysis of total integration within the
aquatic ecosystem, at least within the limnetic zone.

ACKNOWLEDGMENTS

The work on Frains Lake was supported in part on AEC
Contract No. AT(11-1)-781, Michigan Memorial Phoenix Project Grants No. 192 and 267, and by a grant from the
Louis W. and Maud Hill Family Foundation. I would like
to thank Dr. G. H. Lauff and Dr. D. C. Chandler for their
help and Dr. Lauff for provision of facilities and supplies
while I was in residence at the W. K. Kellogg Biological
Station during the summer 1969. The work was carried out
with the help of S. Sanders, L. S. Chertkov, M. S. Misch,
D. Brubaker, J. C. Roth, and T. A. Storch. I have had the
benefit of discussing the work with all of these persons
as well as detailed and lengthy discussions of certain
aspects of the general problem with Dr. D. Woodring. I
would like to thank Dr. C. L. Schelske and J. C. Roth for
critical readings of the manuscript.

REFERENCES

Allen, H. L. 1969. Chemo-organotrophic utilization of dissolved organic compounds by planktic algae and bacteria in a pond. *Int. Rev. Ges. Hydrobiol. 54*, 1-33.

Belser, W. L. 1959. Bioassay of organic micronutrients in the sea. *Proc. Natl. Acad. Sci. 45*, 1533-1542.

Brehm, J. 1967. Untersuchungen über den Aminosäure-Haushalt holsteinischer Gewässer, insbesondere des Pluss-sees. *Arch. Hydrobiol. Suppl. 32*, 313-435.

Dineen, C. F. 1953. An ecological study of a Minnesota pond. *Am. Midl. Nat. 50*, 349-376.

Duursma, E. K. 1961. Dissolved organic carbon, nitrogen and phosphorous in the sea. *Neth. J. Sea Res. 1*, 1-148.

Gambaryan, M. E. 1968. *Mikrobiologicheskie issledovaniya Ozera Sevan* (Microbiological studies of Lake Sevan). AN SSSR, 165 pp.

Henrici, A. T. 1939. The distribution of bacteria in lakes. In *Problems of Lake Biology*. AAAS Pub. 10, Washington, D.C., pp. 39-64.

Juday, C. 1940. The annual energy budget of an inland lake. *Ecology 21*, 438-450.

Lindeman, R. L. 1942. The trophic dynamic aspect of ecology. *Ecology 23*, 399-418.

McCombie, A. M. 1953. Factors influencing the growth of phytoplankton. *J. Fish. Res. Bd. Can 10*, 253-280.

McCombie, A. M. 1960. Actions and interactions of temperature, light intensity and nutrient concentration on the growth of the green alga, *Chlamydomonas reinhardii* Dangeard. *J. Fish. Res. Bd. Can. 17*, 871-894.

Nauwerck, A. 1963. Die Beziehungen zwischen Zooplankton und Phytoplankton im See Erken. *Symb. Bot. Upsal. 17*, 1-163.

Odum, H. T. 1957. Trophic structure and productivity of Silver Springs, Florida. *Ecol. Monogr. 27*, 55-112.

Overbeck, J. 1968. Prinzipielles zum Vorkommen der Bakterien im See. *Mitt. Int. Verein. Theor. Angew. Limnol. 14*, 134-144.

Rasumov, A. S. 1962. Mikrobial'ny plankton vody (The microbial plankton of water). *Trud. Vsesoy. Gidrobiol. Obshch. 12*, 60-190.

Saunders, G. W. 1963. The biological characteristics of freshwater. *Proc. Great Lakes Res. Conf.*, GLRD Pub. No. 10, 245-257.

Taylor, C. B. 1940. Bacteriology of freshwater, I. Distribution of bacteria in English lakes. *J. Hyg. 40*, 616-640.

Teal, J. M. 1957. Community metabolism in a temperate cold spring. *Ecol. Monogr. 27*, 283-301.

Chapter 14

RADIOISOTOPE STUDIES OF HETEROTROPHIC
BACTERIA IN AQUATIC ECOSYSTEMS

J. E. Hobbie
and
P. Rublee

*Department of Zoology
North Carolina State University
Raleigh, North Carolina*

CONTENTS

INTRODUCTION

In recent years, ecologists interested in aquatic bacteria have developed a number of techniques that have given data on the type, number, biomass, and activites of bacteria in situ. These range from microscopical techniques (fluorescence, scanning electron microscope) to chemical methods (gas chromatographs, infrared gas analyzers). Isotopes have also been useful; among those tested have been ^{35}S, ^{15}N, ^{65}Zn, and ^{3}H. This review will be restricted to the most used, ^{14}C.

The investigator of aquatic bacteria has several formidable obstacles to overcome when he or she leaves the laboratory to find out what the natural bacteria are really doing. First of all, the natural bacteria often are not very active compared with laboratory bacteria. Second, their numbers are usually lower than those of laboratory populations. Therefore, methods that work in the laboratory do not work in the field (e.g., respiration with a Warburg manometer). Third, any treatment of natural water, even merely putting the sample into a glass bottle, immediately sets off changes in the microbial community. Thus, only short-term experiments can be carried out and these must be a maximum of a few hours long.

In the late 1950s, a number of workers added $[^{14}C]$-glucose or other organic forms of ^{14}C to lake water in heterotrophic uptake experiments patterned after measurements of plankton photosynthesis with $[^{14}C]$-bicarbonate. The analogy with the primary production measurements turned out to be false because the natural substrate concentration is high in photosynthesis measurements (always greater than

1 mg inorganic carbon/liter but low in the heterotrophic uptake experiments (for example 5 μg glucose/liter). For this reason, the amount of substrate added in the experiment may be usually ignored in photosynthesis experiments but in heterotrophic uptake experiments the amount added may affect the uptake velocity. Therefore, if uptake of an organic substrate is to be tested at only one concentration, then a tracer level must be used (for glucose, a tracer level may be 0.1 μg/liter).

The proper analogy was with biochemical experiments involving an enzyme and its substrate. This was pointed out by Parsons and Strickland (1962) after they measured uptake velocities of glucose and acetate in the ocean at a variety of substrate concentrations. As the concentration of added glucose increased, the rate of uptake $[^{14}C]$-glucose increased but eventually reached a plateau. With appropriate calculations, this maximum velocity was estimated; they called this potential heterotrophy.

The technique was adapted for freshwater work by Wright and Hobbie (1965, 1966), who showed the importance of keeping the levels of added substrate extremely low and close to the natural concentrations. At these concentrations, bacteria were responsible for most of the uptake. In these studies, no assumptions were made about the natural level of substrate. Instead, the calculations used only the added substrate concentrations to calculate a maximum velocity of uptake. Also, a turnover time for the substrate and a constant, the sum of the natural substrate level and a half-saturation constant, were estimated. At this point, the technique had developed into a useful handle on the activity of heterotrophic bacteria; the rates of uptake spanned six orders of magnitude and correlated nicely with the trophic state of the measured lakes (ultra-oligotrophic to heavily polluted).

444

The next step was to refine the technique to include in the calculations the $^{14}CO_2$ produced by bacterial respiration. The importance of this $^{14}CO_2$ was shown by Hamilton and Austin (1967) and an easily usable technique for the collection and counting of the $^{14}CO_2$ was soon developed (Hobbie and Crawford 1969a). These heterotrophic uptake techniques were next combined with chemical measurements of the natural substrate concentration; the result is the true uptake rate by bacteria of each compound tested. In this way, Crawford et al. (1973, 1974) measured the flux of dissolved free amino acids into the bacteria of an estuary.

These same methods have also been used for uptake studies of sediment bacteria. Unfortunately, the bacteria take up the isotopes so rapidly that mixing and dilution are necessary (Wood 1970). As a result, the sediment system is drastically altered and the heterotrophic activity may be increased by as much as an order of magnitude (Hall et al. 1972). One way to avoid this problem is to use a small addition of isotope, dilute only a small amount, and collect only the respired $^{14}CO_2$ (Harrison et al. 1971). Even with this correction, the sediment activity is at least 100 times higher than that of the water column above.

A totally different approach to measuring heterotrophic activity uses the mineralization of the substrate, i.e., the rate of production of $^{14}CO_2$. This technique was developed particularly for the extremely oligotrophic systems of the ocean by Williams and Askew (1968). This is a tracer technique as the levels of glucose and total amino acids added were 0.4 and 0.1 µg/liter, respectively (Andrews and Williams 1971).

The recent developments have all been refinements of the basic techniques. Various workers have shown that almost all the simple organic compounds tested have been taken up by the bacteria of natural waters (e.g., nine organic acids; Ward and Robinson 1974). There have also been criticisms of errors in the methods, such as possible filtration errors (Thompson and Hamilton 1973) and errors caused by release of $[^{14}C]$-glutamate when the bacteria are treated with acid (Griffiths *et al*. 1974a).

There is an extensive literature dealing with the uptake of organic compounds by algae. Almost all of this literature treats only laboratory studies where high levels of substrate were tested for uptake. Bennett and Hobbie (1972) concluded that previous workers were studying laboratory artifacts and that the studies proved that the algae so far tested could not effectively take up organic compounds at in situ concentrations. For this reason, only bacteria were likely to be actively taking up substrate in nature. Munro and Brock (1968) came to the same conclusion from autoradiographic studies of natural marine populations. All the evidence notwithstanding, however, there is still the possibility that some algae do take up large quantities of organics and do compete with bacteria in nature. Most algae tested have been greens; yet, based on their close relationship to bacteria, the blue-green algae are the most likely to have bacterialike transport systems. Thus we believe that algal use of organic compounds in nature has not been proved by earlier studies but may be proved by future work using better techniques to investigate the blue-green algae.

THEORY AND DESCRIPTION OF METHODS

Bacterial heterotrophy has been studied with tracer and nontracer levels of isotope addition. Both methods follow the basic formula for calculating a velocity of uptake:

$$v_{(S+A)} = (f/t)(S+A) \qquad [1]$$

Here $v_{(S+A)}$ is the measured uptake rate (μg/liter/hr), f is the fraction of the added isotope that is taken up, t (hr) is the incubation time, S (μg/liter) is the natural concentration of the substrate under study, and A (μg/liter) is the labeled and unlabeled substrate added during the experiment.

The basic assumption in tracer-level studies is that the A is practically 0. Therefore, the $v_{(S+A)}$ in equation [1] becomes v_S, the actual uptake rate in nature. This v_S may be calculated only if S is measured. If S is unknown, then a turnover time (T, in hr) may still be calculated, by rearranging equation [1], as:

$$t/f = \frac{S+A}{v_{(S+A)}} \qquad [2]$$

Again, in a tracer experiment the A = 0 so that the equation equals S/v_S or T, the time required for the bacteria to take up all the substrate.

The assumption in non-tracer-level experiments is that the substrate added in the experiment (A) does change the uptake kinetics but that measurement of the uptake at a number of concentrations of A will allow extrapolation back to a v_S at S. Most of the experimental data from natural populations fit a model proposed for bacterial transport systems. That is, the velocity of uptake at a given level of substrate, $v_{(S+A)}$, increases as (S+A) increases but above a certain concentration the transport systems become saturated and the $v_{(S+A)}$ no longer increases (Fig. 14.1, top). This relationship is described by:

$$v_{(S+A)} = V_m \frac{S+A}{K + S + A} \qquad [3]$$

where V_m (μg/liter/hr) is the maximum velocity of uptake and K (μg/liter) is a half-saturation constant. In Figure 14.1, a bacteria culture was used so S was 0. To better estimate V_m, a linear transformation is employed so that (after Wright and Hobbie 1966)

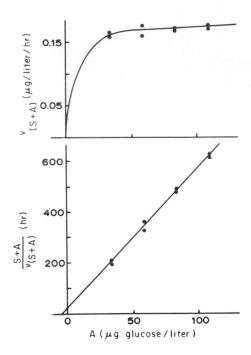

FIG. 14.1. Uptake of glucose by a bacterial culture. In the top graph v is plotted (curve describes a rectangular hyperbola), while in the bottom graph the same data are plotted as (S + A)/v (a straight line). Data from Wright and Hobbie (1966).

$$\frac{S+A}{v_{(S+A)}} = \frac{S + A + K}{V_m} \qquad [4]$$

In this plot (Fig. 14.1, bottom), the V_m is the (slope)$^{-1}$ and S = 0.

Combining equations [2] and [4] it is seen that:

$$t/f = \frac{K+S}{V_m} + \frac{A}{V_m} \qquad [5]$$

From this plot the V_m and (K+S) are easily found when S is unknown (Fig. 14.2). Another value, the (K+S)/V_m is found by extrapolation to the point where A = 0. From equation [4]:

448

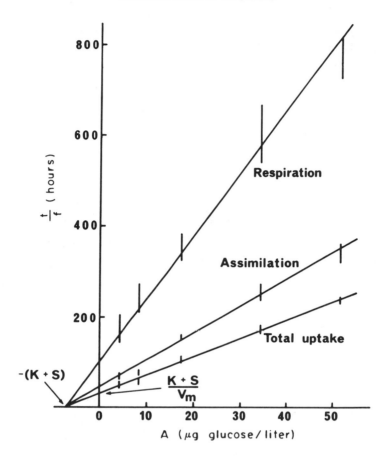

FIG. 14.2. Heterotrophic uptake of glucose in the surface water of Klamath Lake (redrawn from Wright 1973). Each range represents three points. The lower line, total uptake, is the sum of the assimilated and respired glucose. Its slope is the reciprocal of V_m, while its y intercept is $(K + S)/V_m$, and its x intercept is $-(K + S)$.

$$\frac{S}{v_S} = \frac{K+S}{V_m} = T \qquad [6]$$

when A = 0. For this reason, the glucose in Klamath Lake (Fig. 14.2) had a turnover time of 30 hr; a maximum S of 7 µg/liter, from (K+S); and a V_m of 0.2 µg/liter/hr. Note that the v_S can only be calculated if S is known.

The methods for the tracer-level experiments are given in detail in Williams and Askew (1968) and Andrews and Williams (1971). Briefly, 0.2 µg [^{14}C]-glucose is added to 500 ml of sample and the sample incubated in the dark for 3-4 hr. Biological activity is stopped with phosphoric acid and 100 ml of the sample is filtered through membrane filters to collect the bacteria with their incorporated ^{14}C. In the remaining sample, the respired carbon dioxide is converted to barium carbonate and collected for later counting. An amino acid mixture has also been used.

The methods for studies at nontracer levels vary slightly with each investigator. For example, incubation time should be long enough so that adequate amounts of ^{14}C are taken up (>400 CPM/filter) yet not so long that adaptation or growth begins. Twenty minutes to 6 hr is the usual range but each system at each season should be tested for linearity of uptake over time. Volume of the subsample is another variable. Filtration should be rapid yet adequate amounts of ^{14}C should be taken up so that counting is easy. Finally, the levels of added substrate should be kept low; if possible, the level of A should be in the 1-20 µg/liter range. Details are given in Hobbie and Crawford (1969a) and a detailed review of methodological problems in Wright (1973).

In a typical experiment in a eutrophic pond, 5 ml of sample are placed in a 25-ml erlenmeyer flask and the labeled substrate is added with a micropipet. The flask is immediately sealed with a serum stopper that has a suspended plastic cup containing a folded filter paper strip. After incubation for 1 hr the reaction is stopped by injecting 0.2 ml 2N H_2SO_4 through the stopper. Next, still working through the septum of the stopper, 0.2 ml of phenethylamine is added to the filter paper and the flask is shaken for 1 hr. Finally, the sample is filtered through a 0.45 μm pore size membrane filter and the membrane filter later counted with liquid scintillation. The paper filter strip is also counted with liquid scintillation. Usually, four different levels of added substrate are tested, each level is replicated, and several killed samples are also run as blanks. One possible error occurs from loss of some ^{14}C from the cells when strong acids are used to end the reactions (Griffiths *et al*. 1974a). Wood and Chua (personal communication) used 0.2 ml of a mixture of 60 g trichloroacetic acid and 1 mg $HgCl_2$ to reduce this problem.

Method Problems

Tracer-Level Experiments

The first problem in tracer experiments is to prove that the concentrations of isotope added are indeed at true tracer levels. There is usually no problem with this in experiments using $H^{14}CO_3$, $^{32}PO_4$, or $^{35}SO_4$, for example, for the isotope is usually 100 to 1000 times lower in concentration than the natural substrate. This is not always true, however, as both Schindler and Fee (1973) and Allen (1972) found such a great reduction in available inorganic carbon around noon in soft water lakes that nearly 50% of the ^{14}C is quickly taken up. This has also been a problem in tracer-level additions of glucose to sediment samples, as Harrison *et al*. (1971) have found that the lowest levels they have added (a total of seven orders of magnitude were tested) have been rapidly depleted (minutes). In contrast, depletion has not been a problem in plankton experiments but the substrate levels used are only about a tenth of the concentration of S. At these levels, there is some danger that A may effect $v_{(S+A)}$ so tests should be run to find the highest level of A that is still a tracer level. This problem would be solved if the organic radioisotopes

451

were available at very high specific activity but so far only U-[^{14}C]-glucose is really suitable. Wright (1973) points out that glucose is available that is about 80% ^{14}C (320 μCi/μM), while glycollate is available only at 12% ^{14}C. Tritium-labeled compounds are available at very high specific activities and have been successfully used (Azam and Holm-Hansen 1973). However, respiration cannot be measured in this case.

The length of the incubation period is another variable. While a relatively long period is desirable to give the maximum CPM taken up and respired, there are also likely to be changes in the bacterial population during the incubation. The best way is to measure uptake over time and choose the longest time that will still be in the linear portion of the curve. Williams and Askew (1968) used a 4-5 hr incubation for oceanic plankton, while Harrison *et al.* (1971) chose 5 min for lake sediments.

The main reason that the tracer-level measurements have not been popular is the difficulty of recovering the respired ^{14}CO$_2$ from large volumes of water. This could possibly be solved by absorbing the CO$_2$ into a base, such as hyamine or phenethylamine, which can be counted by liquid scintillation.

It is extremely important in this method to make sure that the isotope is free from ^{14}CO$_2$. This is a decomposition product that is always present in solutions of labeled organic compounds. It is usually easy to acidify the solutions and drive off the ^{14}CO$_2$ but Andrews and Williams (1971) found it necessary to use isotope exchange to remove the ^{14}CO$_2$ from amino acid solutions.

Non-Tracer-Level Experiments

The major problem in these experiments continues to be that the uptake is by a heterogeneous population of microorganisms. Thus, if the only organisms taking up the substrate were a single species of bacteria, then uptake should follow the model proposed in Eq. [3] and Figure 14.1 and the data could be interpreted on the basis of bacterial transport systems (Pardee 1968). Instead, a number of bacterial species are present and each has, potentially, different uptake parameters (V_m, K). In addition, algae can also take up the isotope, although their uptake parameters are so different from those of bacteria that they will take up only a small percentage of the total CPM. It should be noted that during intense blooms of algae their uptake may overpower that of the bacteria so that it is impossible to analyze the data by equation [2].

The theoretical implications of uptake by a diversity of organisms has been explored by Williams (1973). When V_m and K are varied tenfold for a number of species, the overall or community V_m becomes constant as the substrate concentration increases. However, at the very low substrate levels, the V_m of the organisms with the lowest K dominates the uptake and the plot deviates from linearity and slopes toward the origin. Williams estimates that errors of 25% are likely in T when "best guess" estimates of the range of S and K are used.

For ecological purposes, the question must be, can we obtain useful information on the uptake and turnover of dissolved organic compounds by treating the entire population as a single species? The test is easy; merely compare the T obtained by the tracer-level technique with the T obtained by the non-tracer-level approach. If the two are equal, then the non-tracer-level technique is valid. Unfortunately, only a few comparisons have been made (Hobbie, unpublished) but these did show that the two methods gave similar results. A great variety of systems and substrates must be tested before the question is completely answered.

It is obvious, however, that a number of populations are contributing to the uptake of the label during experiments. Only by keeping the added substrate (A) as close to the S as possible can we obtain data on uptake in nature.

This last point deserves amplification for in natural samples it is possible to measure different uptake parameters for different levels of substrate added (A). For example, if uptake is measured at 2, 4, 6, and 8 μg glucose/liter, the uptake parameters (K, V_m) will be different from a measurement made at 20, 40, 60, and 80 μg/liter. Finally, there may be algal transport systems that can only be measured at the 500, 1000, 1500, and 2000 μg/liter levels. These algal systems, that may have a K of 1000 μg glucose/liter (Bennett and Hobbie 1972), at low substrate levels resemble diffusion or first-order kinetics. It is interesting that both V_m and K increase at higher levels of A.

From laboratory studies of cultures, it is known that some substrates, such as glucose, are taken up by a very specific transport system while other substrates, such as amino acids, may share a transport system with four or five other compounds. Several studies have shown that the same general patterns of transport systems can be found in natural populations (Wright 1973; Crawford et al. 1974) and that competitive inhibition of uptake is present. For example, Wright found that addition of 100 μg of lactate/liter changed the (K + S) of glycollate uptake by 100 μg but had no effect on the V_m. These two substrates evidently share the same transport system. Many believe that this inhibition introduces serious errors to the method and, indeed, it does mean that the value of K measured in nature (when S is determined separately) is not the true K of the natural bacteria. Instead, it is an effective K that is higher than the true K. In a similar manner the T measured

454

is not as low as the T that would occur if no competitive inhibition were occurring. However, we argue that the T measured is the true T for that substrate in nature as it is measured under natural conditions and with the natural inhibitors present (Crawford 1967; Hall et al. 1972). From an ecological viewpoint, this is the important measurement.

Work by Griffiths et al. (1974a) in the ocean and Wood and Chua (personal communication) in freshwater has elucidated still another problem, that of loss of labeled amino acids from bacterial cells after filtration onto the filter. According to the detailed work of Griffiths et al. (1974b) on a bacterial culture, two pools of amino acid are present. The first is an internal pool that stores amino acids resulting from transport into the cell and from de novo synthesis. The second pool appears to be loosely bound to the outside of the cell. Both pools are released by harsh treatments, such as fixing with formalin or acid or osmotically shocking the bacteria. The dramatic effect of an acid treatment is shown in the following table. Griffiths et al. (1974a) suggest that two series

Uptake Parameters for Two Marine Samples,
Each Incubated with Glutamic Acid[a]

	Sample 1		Sample 2	
	Acid	No Acid	Acid	No Acid
V_m (μg/liter/hr	0.52	1.21	0.24	0.39
T (hr)	3.4	0.8	51.2	26.0
(K + S) (μg/liter)	1.7	0.9	12.2	10.0

[a]Half the sample was treated with acid before filtration and half filtered without treatment (from Griffiths et al. 1974a).

of flasks be used, one for measurements of the retained
isotope (not fixed) and the other for studies of the respiration (fixed with acid). Wood and Chua greatly reduced
the problem by using a mixed fixative (trichloroacetic acid
plus $HgCl_2$).

The loss of label occurs only with amino acids. Wright
and Hobbie (1966) specifically tested for this type of
loss and did not find it for glucose or acetate. They
examined unfixed samples versus samples fixed with acid
Lugol's solution. Griffiths *et al.* (1974b) also found no
effect on glucose. Previously, Crawford (1967) looked for
this type of loss but did not find it for glutamic acid;
unfortunately, he used samples killed with Lugol's as his
control and found no effect of washing with TCA or dilute
HCl. Thus, we have indirect evidence that killing with
Lugol's will also cause loss of amino acids.

The solution of this methodological problem will depend upon the question being studied. If, for example,
respiration as a percentage of total uptake is being studied, then should the loosely bound amino acid on the outside of the cell be considered in the total uptake? This
problem was dealt with by Griffiths *et al.* (1974b), who
considered that the loosely bound pool was released by an
osmotic shock while the internal pool was TCA soluble.

Several authors (e.g., Arthur and Rigler 1967; Thompson and Hamilton 1973) found a filtration error when [14]C-
labeled plankton are filtered onto membrane filters. In
bacterial uptake studies, Thompson and Hamilton (1973)
suggested that this error resulted in underestimations of
(K + S) and V_m by a factor of two. This error is found
when different volumes (e.g., 0.1, 0.25, 0.5, 1.0, 5 ml)
of a water sample incubated with isotope are filtered
through membrane filters. When the CPM are all converted
to CPM/ml filtered, then the low volumes filtered always

456

have unexpectedly high counts. We believe that these results are artifacts of membrane filters and that there is
no filtration error. Our unpublished experiments (with
R. Daley) have shown that the CPM in particulate matter
decreases in direct proportion to the volume of water filtered. However, there are several processes that add a
constant amount of CPM to each filter independent of the
volume of water filtered. These include retention by the
filter of water containing ^{14}C (in the pores of the filter)
and adsorption of a small amount of ^{14}C onto the filter or
onto organic matter and then onto the filter. There is
evidence that the filter has a limited number of adsorption
sites. The end result is that the total CPM on each filter will contain a constant amount of CPM plus the particulate CPM. At very low volumes, this constant amount becomes an important fraction of the total CPM on the filter.
By correcting the CPM on each filter to CPM/ml, the volumes
less than 1 ml will be multiplied by factors that increase
as the volume filtered decreases. The factors will be 10,
4, 2, 1, and 0.2 for volumes of 0.1, 0.25, 0.5, 1, and 5
ml, respectively. Ward and Robinson (1974) also conclude
that the filtration error is an artifact as they found that
refiltration of filtrate produced the same results. Any
possible error caused by adsorbed isotope can be easily
corrected by adequate controls; perhaps the best control
would be one filtered immediately after the isotope is
added.

The final problem deals only with the measurements
of heterotrophic uptake in the sediments. As noted above,
the usual procedure is to mix the isotope into the sediments but it is obvious that this destroys the sediment
structure. Some results, for example turnover times for
glucose of fractions of a minute, are difficult to believe.
There is a possibility that many of the sediment bacteria

are not active for long periods but can be activated by disturbances, such as mixing. It is, of course, very difficult to obtain information on the bacterial activities in undisturbed sediments. The one attempt to compare an undisturbed with a disturbed core gave data that indicated an order of magnitude increase in activity as a result of disturbing the core (Hall *et al.* 1972).

RESULTS

Continued use of such complex and difficult methods can only be justified by the results. Our present judgment is that we have learned a great deal about the activities of heterotrophic bacteria despite these problems, and that we have no other tool sensitive enough to give any idea at all about the activity of these bacteria in situ.

A significant amount of the substrate taken up is respired; it is therefore important to measure $^{14}CO_2$ produced in all studies of heterotrophic activity. When the $^{14}CO_2$ is included in the total uptake, the V_m is increased, the T is decreased, while the (K + S) remains constant. This V_m increase is shown by the decrease in slope in Figure 14.2 [$V_m = (slope)^{-1}$] and amounts to a 20% change in this experiment. Although the studies of heterotrophy carried out in the 1960s did not include the $^{14}CO_2$ measurements, the data are often useful for comparison because the error is relatively small and approximately constant for a given system. For instance, Francisco (1970) found that glucose respiration averaged 30% with a range of 20-40% for 84 measurements over an entire year in the euphotic zone of a North Carolina reservoir.

The percentage of the substrate that is lost as CO_2 is different for different substrates but is relatively constant for each substrate in a wide variety of ecological

458

systems. For example, the percentage respired (of the
substrate taken up) varies from 3 to 60% for amino acids
(Crawford et al. 1974; Griffiths et al. 1974a), from 1 to
49% for sugars (Hobbie and Crawford 1969a; Wood 1970, 1973;
Thompson and Hamilton 1973; Williams 1970), and is around
70% for glycollate (Wright 1973). In all the amino acid
studies, glutamate respired was around 60%. The glutamate
measurements were made on acidified samples; if the glutamate pool that is taken up and not transported is considered, the percentage drops to 15-45% (Griffiths et al. 1974a).
About 20% of the total glucose taken up is respired except
in the case of estuarine sediments, where Wood (1970) found
1-12% respired and Hall et al. (1972) found 18-30% respired. The percentage respired of the amino acids appears
to depend on the length of the metabolic pathways necessary
for the compound to reach the Krebs cycle (Hobbie and
Crawford 1969a; Wright 1973).

Williams (1973) pointed out that the growth yields
(100 minus percentage respired) of natural populations are
greater than those for laboratory cultures. For example,
measurements in a variety of marine environments gave a
value for glucose of 71% and 77% for a mixture of amino
acids. He suggested that natural populations grow on a
mixture of substrates whereas cultures are usually grown
on a single substrate. Therefore, the cultured bacteria
must use energy to snythesize all the necessary organic
molecules. He observed that an estuarine population had
a growth yield of 64% at ambient glucose levels (10^{-7} M)
but when the glucose concentration was raised to 10^{-4} M
the yield fell to 51%.

The activity of the heterotrophic bacteria of the
plankton, measured as V_m, varies over four orders of magnitude in the complete spectrum of natural waters. The
highest values found are in polluted waters where Allen

(1969) found 20 μg glucose/liter/hr while the lowest are
in deep arctic lakes where Rodhe *et al.* (1966) measured
values of 0.0008 μg glucose/liter/hr. V_m reflects the
number of bacteria, as well as their activity, so such
extremely oligotrophic systems as the open ocean likely
have even lower V_m values than the one above. However,
the method does not work at these low levels and most of
the tests on ocean water do not give kinetic data that
follow the uptake model of equation [3]. The turnover
time of glucose also changes by four orders of magnitude;
the range is 15 min to 10,000 hr for the two studies mentioned above.

One use of the V_m has been to follow enrichment or
eutrophication of natural waters. The advantage of using
V_m is its great sensitivity and reproducibility. Hobbie
and Crawford (1969b) were able to show a marked increase
in the V_m of a slowly flowing tidal river caused by wastes
from a phosphate mine. In a similar fashion, Thompson and
Hamilton (1973) followed the fate of sucrose added to an
experimental lake at the rate of 5.4 g sucrose $C/m^2/yr$.
From laboratory experiments, it was found that 266 μg su-
crose/liter in the epilimnion (weekly addition) was likely
reduced to a few μg/liter within 1 day. These additions
caused marked fluctuations in microbial activity similar
to damped oscillations of a perturbed steady-state system.

It is perhaps easier to use tracer-level studies to
follow enrichment of bodies of water as a greater number
of stations can be sampled. In this way, Paerl and Goldman
(1972) used an A of 0.05 μg acetate/liter to follow movement of water masses in the extremely oligotrophic waters
of Lake Tahoe.

Uptake measurements have also been combined with actual measurements of S, the natural substrate concentration, to give rates of movement of organic compounds into

the bacteria (this is v_S). Crawford *et al.* (1973, 1974) did this with measures of V_m, T, and S for 15 amino acids in a shallow North Carolina estuary and estimated an annual flux of about 8 g amino acid $C/m^2/yr$. This was about 12% of planktonic primary production. Andrews and Williams (1971), using tracer levels of isotope and determinations of S in the English Channel, estimated a total annual flux of 8 g glucose C/m^2 and 60 g amino acid C/m^2 (about 30% of primary production). They also offered a rough estimate that perhaps 100 g C/m^2 were involved in the total annual flux of dissolved organic carbon.

The turnover times for individual organic compounds have provided insights into rates of recycling. Before the isotopic measurements were made, there were a number of studies that reported concentrations of individual sugars, amino acids, etc. The results were remarkably uniform for a wide variety of aquatic systems. Thus, Andrews and Williams (1971) pointed out that all the measurements of total amino acid concentrations in the sea spanned only a fivefold range. Their conclusion was that the bacteria are controlling the substrate concentrations in such a way that "only under unusual circumstances do high substrate levels occur." Glucose turnover rates in their study of the English Channel ranged from 60 to 1200 hr but polluted waters will have turnover times as low as 15 min. Thus, our present understanding is that there are small amounts of simple organic compounds that are readily taken up by the bacteria. They hold the concentration relatively constant by increasing their v_S as the concentration increases (see Fig. 14.1) or by rapidly growing and thus increasing the v_S of the total population (Thompson and Hamilton 1973).

Some information has come from a (K + S) determination about the maximum possible substrate concentration present.

In Lake Erken, for example, the value fell between 2 and 6 μg glucose/liter and 5 and 11 μg acetate/liter for the entire year. This may be correct for glucose as the transport system appears to be very specific, but for the amino acids the results are not very helpful since we know that competitive inhibition will increase the (K + S). The same caution should be applied to bioassay measurements using the kinetics of uptake as the indicator (Hobbie and Wright 1965).

It is difficult, at this stage in our understanding, to evaluate the sediment activity measurements. First, the sample is drastically disturbed during the experiment. Second, the V_m and T have to be extrapolated from very small samples. Third, the S available to the bacteria is usually unknown. Fourth, we have no way of knowing whether or not the bacteria are really active in undisturbed sediment or if they become active during the experiment. It is clear that extremely high V_m values are commonly measured in the sediment and that extremely short turnover times are found. Wood (1970) found times as short as 0.01 hr in estuarine sediment while Hall *et al.* (1972) found average values ranging from 0.05 to 0.4 hr in the sediment of Marion Lake, British Columbia. The V_m of uptake varied over the season mainly in response to temperature changes.

In general, the results of the sediment heterotrophic studies confirm what respiration measurements have already told us; namely, that in shallow, unpolluted systems most of the biological activity occurs in the sediments. From this viewpoint the sediment data are believable. The real test, however, would be tracer-level additions; the problem is to make these additions uniformly into the sediment without disturbing the structure of the sediments. It will be interesting to test the heterotrophic activity in the sediments of deeper aquatic systems. One would expect that a greater amount of mineralization would be carried out in the water column as the particulate material sinks. Thus, the deeper the system the smaller should be the heterotrophic activity of the sediments. Instead of the 1:75 ratio of planktonic heterotrophy/square meter: sediment heterotrophy/square meter, the ratio should be closer to 1:1.

In laboratory cultures, there is a direct correlation of V_m with numbers of bacteria. In nature, only rarely is this correlation seen. Cultures, however, are not good models of ecosystems and Postgate (1973) mentions that most bacteria seen in nature may be dead or inactive. Some counting methods, however, may count only larger active forms (e.g., erythrosin B stained preparations) while other methods (e.g., acridine orange with epifluorescence) also count tiny coccoid forms that may be inactive or dead. Viable counts (e.g., plate counts with dilute media) may count only the fastest growing forms. The only comparison of the V_m/cell for planktonic and sediment bacteria was made in a tundra pond by Rublee (1974), who used an acridine orange direct count. He found that the activity of a bacterial cell in the sediment was two orders of magnitude lower than that of a cell in the plankton. Evidently, most of the sediment bacteria measured with this direct count method were dead or inactive.

Another factor that would obscure any relationship between V_m and bacterial numbers is the dependence of V_m on temperature. Because the Q_{10} of V_m can be as high as 2.2 (Wright and Hobbie 1966), up to a fourfold increase in V_m can be expected because of seasonal temperatures alone. In an arctic lake, however, the temperature changed only a few degrees over the entire year and Morgan and Kalff (1972) found an excellent positive correlation between V_m and planktonic bacterial numbers (erythrosin stained). Wood (1970) also found a good correlation with erythrosin stained bacteria and V_m in the sediments (Fig. 14.3). In contrast, Francisco (1970) found no clear-cut correlation between these measurements in a reservoir. It should be kept in mind, however, that bacterial numbers will set the general level of V_m and that undoubtedly a good correlation between these measurements would emerge if we had adequate data to plot the log of the V_m against the log of the number of bacteria.

Not only is the V_m in most water bodies proportional to temperature, but it is also proportional to photosynthetic rates. These relationships are best seen in

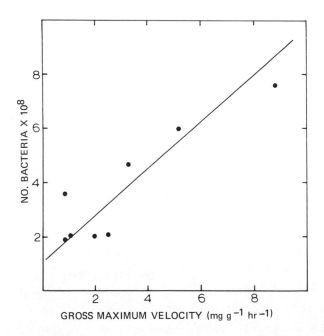

FIG. 14.3. The number of bacteria (in hundreds of millions) per g dry weight of estuarine sediments (from Wood 1970) versus the V_m of glucose. $Y = 1.15X + 1.21 \pm 1.27$.

situations where either temperature or daily photosynthesis
rates are constant. Thus, in a single tundra pond Miller
(personal communication) measured a diurnal change (Fig.
14.4) that gave a good correlation with temperature changes.
When the V_m in the plankton of 25 tundra ponds was measured
on the same day (temperature was constant), there was also
a significant correlation with phytoplankton primary productivity (Fig. 14.5).   Munro *et al.* (1973) also found
a good correlation between phytoplankton production in a
Scottish sea loch and the respiration of $[^{14}C]$-glucose.

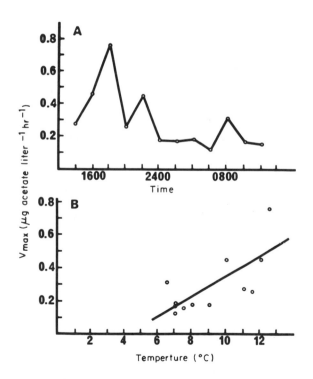

FIG. 14.4.   Diurnal cycle of V_m in a tundra pond (top) and the same
data plotted against temperature.   Data provided by M. C. Miller.

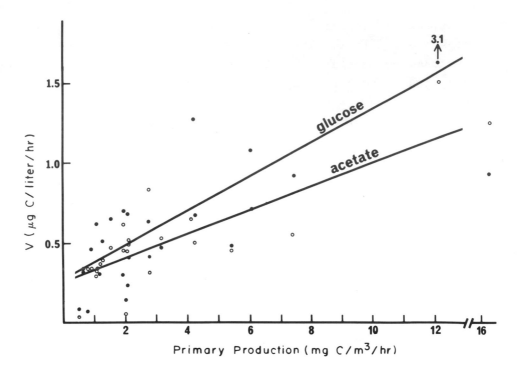

FIG. 14.5. Uptake of substrate at a single level of addition for a number of tundra ponds plotted against ^{14}C primary productivity. All measurements were made on August 9, 1973. Data provided by M. C. Miller.

The V_m of glucose uptake was closely coupled to two plankton blooms in a lake (Fig. 14.6) in another study. The lag in May was caused by the low temperatures (6 C) at that time that did not affect the photosynthesis but did drastically reduce the activity of the bacteria. The bacteria may have been responsible for the decrease in dissolved organic carbon (DOC) during May and a lack of bacterial activity responsible for the DOC peak in November. During October and November in this lake, the large beds of reeds (*Phragmites* sp.) release large quantities of DOC.

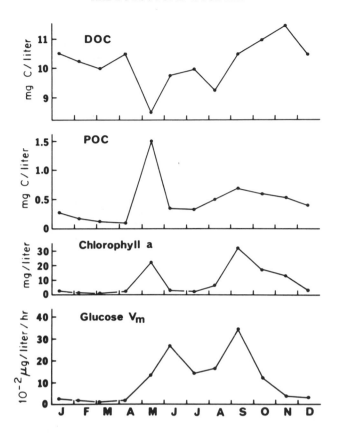

FIG. 14.6. Yearly cycle of bacteria activity, chlorophyll a, and dissolved and particulate organic carbon in Lake Erken (1964 to 1965).

The relationship between bacterial activity and primary productivity is striking when viewed over the entire range of both parameters (Fig. 14.7). In this figure, three of the water bodies (1, 2, and 15) are strongly influenced by sediment processes and two by pollution (8 and 15) and should be omitted from consideration. Also, it should be remembered that the primary production is presented here as an average. If the actual production were to be plotted against the actual V_m, then the relationship

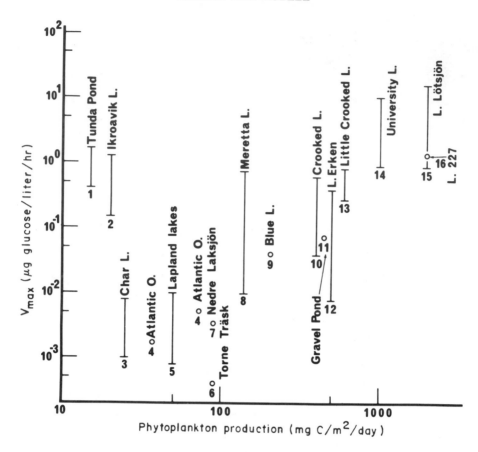

FIG. 14.7. Maximum velocity of uptake of glucose (seasonal range or daily value) versus planktonic primary production (average daily or single day's value) in various water bodies. (1) Rublee 1974; (2) Hobbie, unpublished; (3) Morgan and Kalff 1972; (4) Hobbie *et al.* 1972; (5) Hobbie and Wright 1968; Rodhe 1969; (6) and (7) Rodhe 1958; Rodhe *et al.* 1966; (8) Morgan and Kalff 1972; (9) Ward and Robinson (1974; (10) Wetzel 1966, 1968; (11) Wright 1973; (12) Hobbie and Wright 1968; Rodhe 1969; (13) Wetzel 1966, 1968; (14) Francisco 1970; (15) Allen 1969; Rodhe 1969; (16) Schindler and Fee 1973; Thompson and Hamilton 1973. Graph redrawn from Morgan and Kalff (1972) with many additions.

would be clearer. In spite of these problems, the relationship is evident and we conclude that the level of heterotrophy is indeed set by the level of algal productivity.
Several investigators are now attempting to find out how
closely the heterotrophic activity is tied to algal secretions or other losses of DOC.

Uptake of organic material may also be linked to populations of photosynthetic bacteria (Fig. 14.8). In the

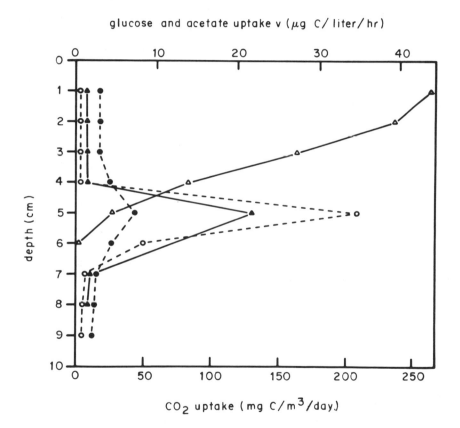

FIG. 14.8. Glucose (●) and acetate (○) uptake in the light, compared with light (△) and dark (▲) CO_2 fixation in a vertical pro-
file from Kolksee near Plön (from Overbeck 1974).

lake shown in this figure, the peak of acetate uptake occurred at the same place as a layer of purple sulfur bacteria.

CONCLUSIONS

The research of the past 13 years using isotopes of
organic compounds has resulted in a tremendous increase
in our understanding of the in situ activities of heterotrophic bacteria. Without these techniques we would know
little of the rates of turnover of dissolved organic compounds or of the importance of the small amounts of sugars
or amino acids. We now have some idea of the relationship
of bacterial numbers and temperature to bacterial activity
and a realization that the trophic state of the water body,
as measured by primary productivity, is of overriding importance in setting the level of bacterial activity. Perhaps the most important finding to come out of these studies is the quantitative demonstration of the importance
of bacteria in producing particulate matter from DOC.

As noted, many methodological problems have been discovered. Some are real, such as the loss of labeled amino
acids upon acidification. Some are experimental artifacts,
such as the filtration error. All problems of methods
must be judged against the needs of the particular experiment being run or hypothesis being tested. After all, a
20% or even 50% error in data in Figure 14.7 would make
no difference to the results. We think that the most valuable method of the future will be tracer-level additions
combined, if possible, with chemical measurements of S,
the natural substrate level.

The one ecological question that has barely been
touched upon concerns the production of bacteria. Because
of the great sensitivity to change of the bacterial populations, the method will likely be an isotopic one. Judging

from our experiences with the development of the heterotrophic uptake methods, the isotopic method will not be
an easy one analogous to ^{14}C measurements of algal productivity. Two methods, one using $^{14}CO_2$ and the other
using $^{35}SO_4$, have been proposed. The problem still to
be solved, however, is proof that the uptake values obtained by these methods are a measure of bacterial production.

REFERENCES

Allen, H. L. 1969. Chemo-organotrophic utilization of dissolved organic compounds by planktic algae and bacteria in a pond. *Int. Rev. Ges. Hydrobiol. 54*, 1-33.

Allen, H. L. 1972. Phytoplankton photosynthesis, micronutrient interactions, and inorganic carbon availability in a soft-water Vermont lake. In *Nutrients and Eutrophication*, G. E. Likens (Ed.). Special symposium, Vol. I. Am. Soc. Limnol. Oceanogr., pp. 63-80.

Andrews, P., and Williams, P. J. leB. 1971. Heterotrophic utilization of dissolved organic compounds in the sea. III. Measurement of the oxidation rates and concentrations of glucose and amino acids in sea water. *J. Mar. Biol. Assoc. U.K. 51*, 111-125.

Arthur, C. R., and Rigler, F. H. 1967. A possible source of error in the ^{14}C method of measuring primary productivity. *Limnol. Oceanog. 12*, 121-126.

Azam, F., and Holm-Hansen, O. 1973. Use of tritiated substrates in the study of heterotrophy in sea water. *Mar. Biol. 23*, 191-196.

Bennett, M. E., and Hobbie, J. E. 1972. The uptake of glucose by *Chlamydomonas* sp. *J. Phycol. 8*, 392-398.

Crawford, C. C. 1967. Heterotrophic uptake of dissolved amino acids by bacteria in natural waters. Unpublished M.S. Thesis, Dept. of Zoology, North Carolina State University, Chapel Hill, 39 pp.

Crawford, C. C., Hobbie, J. E., and Webb, K. L. 1973. Utilization of dissolved organic compounds by microorganisms in an estuary. In *Estuarine Microbial Ecology*, L. H. Stevenson and R. R. Colwell (Eds.). University of South Carolina Press, Columbia, S.C., pp. 169-180.

Crawford, C. C., Hobbie, J. E., and Webb, K. L. 1974. Utilization of dissolved free amino acids by estuarine microorganisms. *Ecology 55*, 551-563.

Francisco, D. A. 1970. Glucose and acetate utilization by the natural microbial community in a stratified reservoir. Unpublished Ph.D. Thesis, Dept. of Environmental Science and Engineering, University of North Carolina, Chapel Hill. 83 pp.

Griffiths, R. P., Hanus, F. J., and Morita, R. Y. 1974a. The effects of various water-sample treatments on the apparent uptake of glutamic acid by natural marine microbial populations. *Can. J. Microbiol 20*, 1261-1266.

Griffiths, R. P., Baross, J. A., Hanus, F. J., and Morita, R. Y. 1974b. Some physical and chemical parameters affecting formation and retention of glutamate pools in a marine psychrophilic bacterium. *Z. Allg. Mikrobiol. 14*, 359-369.

Hall, K. J., Kleiber, P. M., and Yesaki, I. 1972. Heterotrophic uptake of organic solutes by microorganisms in the sediment. In *Detritus and Its Role in Aquatic Ecosystems*, U. Melchiorri-Santolini and J. W. Hopton, (Eds.). Mem. Inst. Ital. Idrobiol. 29 Suppl., pp. 441-471.

Hamilton, R. T., and Austin, K. E. 1967. Assay of relative heterotrophic potential in the sea: the use of specifically labelled glucose. *Can. J. Microbiol. 13*, 1165-1173.

Harrison, M. J., Wright, R. T., and Morita, R. Y. 1971. Method for measuring mineralization in lake sediments. *Appl. Microbiol. 21*, 698-702.

Hobbie, J. E., and Crawford, C. C. 1969a. Respiration corrections for bacterial uptake of dissolved organic compounds in natural waters. *Limnol. Oceanog. 14*, 528-532.

Hobbie, J. E., and Crawford, C. C. 1969b. Bacterial uptake of organic substrate: new methods of study and application to eutrophication. *Ver. Int. Verein. Limnol. 17*, 725-730.

Hobbie, J. E., Holm-Hansen, O., Packard, T. T., Pomeroy, L. R., Sheldon, R. W., Thomas, J. P., and Wiebe, W. J. 1972. A study of the distribution and activity of microorganisms in ocean water. *Limnol. Oceanog. 17*, 544-555.

Hobbie, J. E., and Wright, R. T. 1965. Bioassay with bacterial uptake kinetics: Glucose in freshwater. *Limnol. Oceanog. 10*, 471-474.

Hobbie, J. E., and Wright, R. T. 1968. A new method for the study of bacteria in lakes: description and results. *Mitt. Int. Verein. Limnol. 14*, 64-71.

Kalff, J., Welch, H. E., and Holmgren, S. K. 1972. Pigment cycles in two high-arctic Canadian lakes. *Ver. Int. Verein. Limnol. 18*, 250-256.

Morgan, K. C., and Kalff, J. 1972. Bacterial dynamics in two high-arctic lakes. *Freshwater Biol. 2*, 217-228.

Munro, A. L. S., and Brock, T. D. 1968. Distinction between bacterial and algal utilization of soluble substances in the sea. *J. Gen. Microbiol. 51*, 35-42.

Munro, A. L. S., Williams, G. R., and Massie, L. C. 1973. Seasonal variation in microbial activity. *Bull. Ecol. Res. Comm. (Stockholm) 17*, 269-270.

Overbeck, J. 1974. Microbiology and biochemistry. *Mitt. Int. Verein. Limnol. 20*, 198-228.

Paerl, H. W., and Goldman, C. R. 1972. Heterotrophic assay in the detection of water masses at Lake Tahoe, California. *Limnol. Oceanog. 17*, 145-148.

Pardee, A. B. 1968. Membrane transport proteins. *Science 162*, 632-637.

Parsons, T. R., and Strickland, J. D. H. 1962. On the production of particulate organic matter by heterotrophic processes in sea water. *Deep-Sea. Res. 8*, 211-222.

Postgate, J. R. 1973. The viability of very slow-growing populations: a model for the natural ecosystem. *Bull. Ecol. Res. Comm. (Stockholm) 17*, 287-292.

Rodhe, W. 1958. Primarproduktion und seetypen. *Ver. Int. Verein. Limnol. 13*, 121-141.

Rodhe, W. 1969. Crystallization of eutrophication concepts in Northern Europe. In *Symposium on Eutrophication: Causes, Consequences, Correctives*. National Academy of Sciences, Washington, D. C., p. 50-64.

Rodhe, W., Hobbie, J. E., and Wright, R. T. 1966. Phototrophy and heterotrophy in high mountain lakes. *Mitt. Int. Verein. Limnol. 16*, 302-313.

Rublee, P. 1974. Production of bacteria in a pond. Unpublished M.S. Thesis, Dept. of Zoology, North Carolina State University, Raleigh, 39 pp.

Schindler, D. W., and Fee, E. J. 1973. Diurnal variation of dissolved inorganic carbon and its use in estimating primary production and CO_2 invasion in lake 227. *J. Fish. Res. Bd. Can. 30*, 1501-1510.

Thompson, B. M., and Hamilton, R. D. 1973. Heterotrophic utilization of sucrose in an artificially enriched lake. *J. Fish. Res. Bd. Can. 30*, 1547-1552.

Ward, F. J., and Robinson, G. G. C. 1974. A review of research on the limnology of West Blue Lake, Manitoba. *J. Fish. Res. Bd. Can. 31*, 977-1005.

Wetzel, R. G. 1966. Productivity and nutrient relationships in marl lakes of northern Indiana. *Ver. Int. Verein. Limnol. 16*, 321-332.

Wetzel, R. G. 1968. Dissolved organic matter and phytoplankton production in marl lakes. *Mitt. Int. Verein. Limnol. 14*, 261-270.

Williams, P. J. leB. 1970. Heterotrophic utilization of dissolved organic compounds in the sea. *J. Mar. Biol. Assoc. U.K. 50*, 859-870.

Williams, P. J. leB. 1973. On the question of growth yields of natural heterotrophic populations. *Bull. Ecol. Res. Comm. (Stockholm) 17*, 400-401.

Williams, P. J. leB., and Askew, C. 1968. The method of measuring the mineralization by micro-organisms of organic compounds in sea-water. *Deep-Sea. Res. 15*, 365-375.

Wood, L. W. 1970. The role of estuarine sediment microorganisms in the uptake of organic solutes under aerobic conditions. Unpublished Ph.D. Thesis, Dept. of Zoology, North Carolina State University, Raleigh, 75 pp.

Wood, L. W. 1973. Monosaccharide and disaccharide interactions on uptake and catabolism of carbohydrates by mixed microbial communities. In *Estuarine Microbial Ecology*, L. H. Stevenson and R. R. Colwell (Eds.). University of South Carolina Press, Columbia, pp. 181-197.

Wright, R. T. 1973. Some difficulties in using ^{14}C-organic solutes to measure heterotrophic bacterial activity. In *Estuarine Microbial Ecology*, L. H. Stevenson and R. R. Colwell (Eds.). University of South Carolina Press, Columbia, pp, 199-217.

Wright, R. T., and Hobbie, J. E. 1965. The uptake of organic solutes in lake water. *Limnol. Oceanog. 10*, 22-28.

Wright, R. T., and Hobbie, J. E. 1966. Use of glucose and acetate by bacteria and algae in aquatic ecosystems. *Ecology 47*, 447-464.

Chapter 15

EXPERIMENTAL STUDIES OF DISSOLVED ORGANIC
MATTER IN A SOFT-WATER LAKE

Harold L. Allen

Department of Biological Sciences
Dartmouth College
Hanover, New Hampshire

CONTENTS

INTRODUCTION

The functional significance of dissolved organic carbon compounds and bacterial interactions in freshwater ecosystems has been investigated experimentally over the past
decade (see, for example, Allen 1969, 1971a, 1973; Wetzel
and Allen 1972; Wetzel 1968; Saunders 1971; Hobbie 1967,
1971; Hobbie *et al.* 1968; Wright 1970, 1973a, 1973b; Wright
and Hobbie 1966; Robinson *et al.* 1973; Wood and Chua 1973;
among others). To some extent, although less successfully
than in more entrophic freshwater environments, dissolved
organic materials and bacterial interactions have been studied in marine ecosystems (see, for example, Hamilton and
Preslan 1970; Vaccaro and Jannasch 1966, 1967; Vaccaro *et.*
al. 1968; Williams 1970; Williams and Gray 1970; Andrews
and Williams 1971; among others). Much of this research
activity has resulted from the development and availability
of high-specific-activity organic radioisotopes and the
application of sophisticated analytical instrumentation to
field and laboratory ecological investigations.

In spite of the advancement of field and laboratory
procedures for assaying and studying dissolved organic compounds and the various direct and indirect roles they may
assume in aquatic ecosystems, there have been relatively
few data published from different types of ecosystems that
would indicate how bacterial metabolic activity may change
or alter the quantitative and qualitative spatial or temporal distribution of dissolved compounds present, or
otherwise influence some of their demonstrated indirect
roles (e.g., chelation-complexing capacity, provision of
essential metabolites). Interrelationships between dissolved organic matter and bacterial, or decomposer,

activity in general have for the most part only been superficially evaluated and studied. Yet, in order to understand aquatic ecosystem processes, including community
production, nutrient cycling, etc., to a point where meaningful and realistic mathematical models with predictive
capabilities can be developed, there will have to be more
intensified research focused on bacterial communities, especially as they relate to production, consumption, and
transformation of dissolved organic materials.

It is reasonably clear, at least in aquatic ecosystems
that undergo anaerobic conditions at the sediment-water
interface during summer thermal stratification, that bacteria metabolic activity is responsible for nutrient regeneration under reducing conditions and that these nutrients resupply phytoplankton-rich surface waters to promote
or enhance primary production. Identification of mechanisms by which the bacteria may metabolically alter or
change the dissolved organic carbon pool, and potential
role(s) of various constituents in it, is of major importance to understanding aquatic ecosystem processes.

The present experimental studies of dissolved organic
matter, based on field and laboratory experiments that were
conducted in a small soft-water New England lake, were designed to investigate (1) some quantitative properties of
the total dissolved organic carbon pool, in terms of characterizing the molecular weight size fractions present in
lakewater; (2) several properties of certain molecular
weight size fractions, including labile and refractory characteristics based on exposure to high intensity ultraviolet
(UV) light, and chelation-complexing capacity for inorganic
iron; and (3) bacterial chemoorganotrophic utilization of
a low molecular weight compound (glucose) and resulting
changes in dissolved organic carbon molecular weight size
fractions following bacterial release of extracellular

products. The latter experimental studies were also designed to demonstrate labile and refractory characteristics
of the bacterial extracellular products.

METHODS

All experimental studies described in this chapter
were conducted on lakewater samples collected from Star
Lake, a typical soft-water lake located in eastern Vermont
in close proximity to the Connecticut River basin. Nearly
all samples were collected from a depth of 0.5 m at the
center of the lake (station A in Fig. 15.1). General
physical, chemical, and certain biological characteristics
have been published elsewhere (Allen 1972). The same methods and apparatus as published previously were used for
physical and chemical limnological measurements in this
study.

Sample Collection and Treatment

Lakewater samples collected for analysis of total particulate organic (POC), dissolved organic (DOC), and total
inorganic carbon (IC), as well as dissolved organic carbon
molecular weight size fractionation, were collected from
0.5, 1.5, and 2.5 meters with an acid-cleaned, nonmetallic
Van Dorn-type sampler and returned immediately to the laboratory in 30-ml scintillation vials that were sealed with
polyvinyl cap inserts to exclude air bubbles. Small 2.5-
mm holes were made through the vial walls and these were
stoppered with tightly fitting serum injection ports.

A schematic diagram of laboratory sample preparation
for determination of POC, DOC, and IC is shown in Figure
15.2. The instrument configuration and flow diagram for
measuring actual concentrations of POC, DOC, and IC is
shown in Figure 15.3. Because both the laboratory treatment of samples and the methods of measurement have been

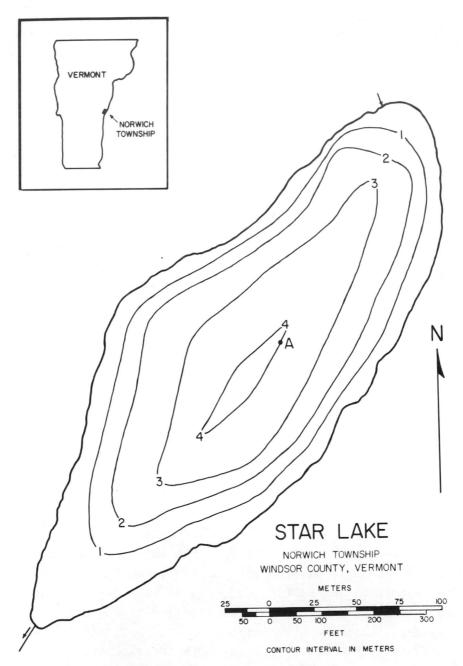

FIG. 15.1. Morphometric map of Star Lake, Vermont.

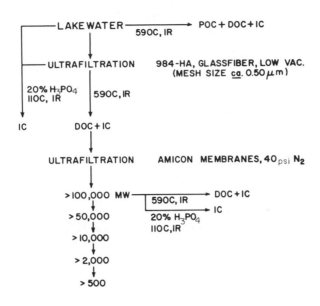

FIG. 15.2. Flow diagram for laboratory analysis of particulate organic (POC), dissolved organic (DOC), total inorganic carbon (IC), and dissolved organic carbon molecular weight size-fractions in lakewater samples, as used in this study. See text for explanation.

only recently developed and have not been described or published elsewhere, they are outlined in some detailed here.

Sample Preparation for POC, DOC, and IC Analysis

On return to the laboratory, aliquots of lakewater samples were filtered (under low vacuum; <0.5 atm) through precombusted (500 C for 1 hr) glassfiber ultrafilters (Reeve Angle 984HA; 2.4-mm diameter) that have a median effective retention size of approximately 0.50 μm (Sheldon 1972). Although there continues to be a discussion in the literature concerning filter retention characteristics and the arbitrary definition of "particulate" and "dissolved"

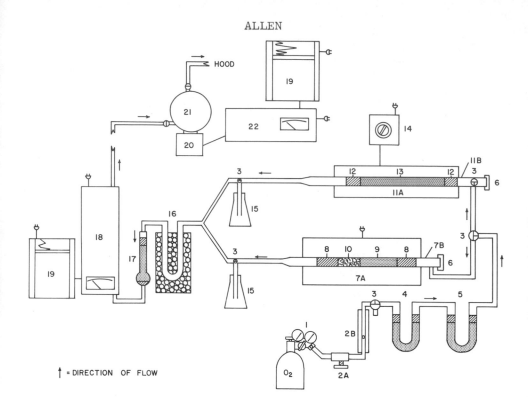

FIG. 15.3. Instrument configuration and flow diagram for measurement of particulate organic, dissolved organic, and total inorganic carbon in lakewater, and ^{14}C determination by gas-phase analysis, as used in this study. See text for explanation. Schematic code: (1) Three-stage regulator; (2A) needle valve; (2B) flowmeter (0-900 ml/min); (3) three-way valve; (4) magnesium perchlorate (Dehydrite) and lithium hydroxide water vapor trap (filtered with Pyrex glass wool); (5) Ascarite carbon dioxide trap (filtered with Pyrex glass wool); (6) serum injection port; (7A) Lindberg combustion tube furnace (0-1200C; Model 55035A); (7B) quartz combustion tube, with bent arm (3 mm ID); (8) quartz wool; (9) cobalt oxide catalyst (Coleman 29170.300); (10) silver wire; (11A) catalyst tube furnace (Leco 507-000); (11B) Pyrex catalyst tube (2.5 mm ID); (12) Pyrex wood plugs; (13) quartz beads (1-1.5 mm diameter); (14) voltage rheostat (110 V); (15) drain; (16) dry ice trap (20 cm U-tube in 665-ml Dewar flask); (17) $CaCO_3$ water vapor trap (filtered with Pyrex glass wool); (18) Beckman IR315-A infrared carbon dioxide analyzer, amplifier, and voltage regulator; instrument configuration 192415 (0-200 ppm CO_2), Beckman Instruments, Fullerton, California 92634; (19) Potentiometric single-channel recorder, 10 inches, Model 22741; Simpson Electric Company, Chicago, Illinois 60644; (20) Cary preamplifier/converter unit; Cary Instruments, Monrovia, California 91016; (21) Cary-Tolbert flow-ionization chamber, 275 ml (No. 3595600); (22) Cary Model 401 vibrating-reed electrometer with multiple input resistors (10^8, 10^{10}, 10^{12} ohms). Items (1)-(17) were supplied through Arthur H. Thomas Company, Philadelphia, Pennsylvania 19105. The pathway between flow components represents Tygon tubing.

organic carbon fractions on the basis of filtration procedures (cf. Sharp 1973), particulate organic and dissolved
organic carbon fractions are defined in this study as that
fraction retained and that fraction passing through a Reeve
Angle 984HA glassfiber ultrafilter, respectively. If data
presented by Sharp (1973) can be extrapolated to freshwaters, my definition probably overestimates the dissolved
organic carbon fraction because of the inclusion of colloidal organic carbon (see also discussion by Breger 1968).

Laboratory Measurement of POC, DOC, and IC

Aliquots (100 µl) from each control and glassfiber
ultrafiltered lakewater sample were injected through a
serum injection port into a high-temperature combustion
tube (cf. Fig. 15.3), following previous purging of the
tube with purified oxygen and zeroing of the infrared CO_2
analyzer, and combusted in an O_2 atmosphere at 590 C for
1 min to CO_2. Following thermal combustion, the CO_2 produced was passed as a "bullet" through dry ice and $CaCO_3$
water vapor traps and through an infrared CO_2 analyzer.
A flow rate of 900 ml/min was found to produce a sharp recorder deflection, indicating containment of the pulse of
CO_2 produced. Conversion of CO to CO_2 was catalyzed by
commercially available CoO. Although developed independently, this thermal combustion technique and system is
similar in principle to that used by Sharp (1973) and Van
Hall et al. (1963). The infrared CO_2 analyzer was calibrated at frequent intervals by injecting replicate microliter amounts of assayed CO_2 into the system. Overall reproducibility for replicate sample combustions was generally less than $\pm$ 5%.

Following instrument calibration, concentrations of
POC + DOC + IC and DOC + IC were calculated from peak
heights of evolved CO_2 passing through the infrared analyzer, representing 100-µl injections of unfiltered control

485

and ultrafiltered lakewater samples, respectively. Total
inorganic carbon (IC) was determined separately by injection of replicate 100-μl aliquots of control and ultrafiltered samples, after acidification of a 20-30 ml lakewater
sample with 1 μl of 20% H_3PO_4, into a small catalyst tube
filled with 1-1.5 mm quartz beads and maintained at 110 C
with a voltage rheostat. Evolution of total carbon dioxide
was quantitative after 1.0 min. Inorganic carbon concentrations were calculated after the infrared analyzer was
calibrated with replicate microliter injections of assayed
CO_2 gas. The low-temperature system was further calibrated
with known solutions of Na_2CO_3. By subtraction, concentrations of POC, DOC, and IC were obtained for each lakewater sample analyzed.

Combustion efficiency, reproducibility, and instrument
stability were found to be exceptionally good during development of the system. The only major problem encountered was interference of water vapor in the infrared analyzer; frequent draining (item 15 in Fig. 15.3) and cleaning
of the dry ice trap, and replacement of $CaCO_3$ absorbent
was necessitated in order to maintain a stable baseline.
A potential problem caused by inorganic salt interference
following high-temperature combustion of particulate and
dissolved organic carbon samples can be anticipated in
estuarine and marine samples (P. H. Rich, personal communication), or wherever high salinities are encountered. This
was not, however, found to be a significant problem in
Star Lake, undoubtedly because of the low electrolyte levels prevalent in the lake (specific conductance = 5-40
μmhos @ 20 C).

Dissolved Organic Carbon Molecular
Weight Size Fractionation

Lakewater samples were prepared initially for molecular weight size fractionation by ultrafiltration through

Reeve Angle 984HA glassfiber filters as described previously (cf. Fig. 15.2). Identical ultrafiltered aliquots (40
ml) were then filtered under 40 psi N_2 through a series of
membrane ultrafilters in an Amicon stirred cell pressure
system (Model 202). Characteristics of the five Diaflo
membrane filters that were used in this study, including
solute retention, pore sizes, and charges, are summarized
in Table 15.1. Stirring speeds, N_2 pressures, and methods
for filter storage between us followed the manufacturer's

Table 15.1

Characteristics of Amicon Diaflo

Membrane Ultrafilters Used in this Study[a]

A	B	C	D
XM 100A	100,000	60	Nonionic; negligible charge
XM 50	50,000	35	Nonionic; negligible charge
UM 10	10,000	14	Ionic sites available; net neutral charge
UM 2	2,000	12	Ionic sites available; net neutral charge
UM 05	500	10	Ionic sites available; net negative charge

[a]A = Amicon filter designation; B = approximate molec-
ular weight solute retention; C = poresize (radius in Å);
D = charge. Data and information from Amicon Corporation
(Lexington, Mass.), publication No. 403.

recommended procedures. The average filtration time per
sample depended on filter porosity and varied from 15 to
45 min. Replicate untreated control and acidified (20%
H_3PO_4) aliquots of these lakewater fractions were analyzed
for organic and inorganic carbon content by purging,

thermal combustion, and infrared procedures described previously. The organic carbon concentration of each molecular weight size fraction was calculated by subtracting from the concentration in the next larger size fraction.

Ultraviolt Photooxidation of Dissolved Organic Carbon Molecular Weight Size Fractions

Ultraviolet photooxidation or combustion of dissolved organic carbon molecular weight size fractions of lakewater was achieved in a water-cooled, all-quartz combustion vessel in which lakewater samples were separated from a 450 W mercury vapor lamp (Engelhard Hanovia 679A; Ace Glass, Ludlow, Massachusetts) by a thin quartz shell. The combustion apparatus was fabricated (Vitriforms, Inc., Brattleboro, Vermont) and similar in design to that described by Manny et al. (1971), with the exception of a single modification: a small drain tap was fabricated near the base of the outer shell, such that sample aliquots could be withdrawn during a time course of oxidation for determining DOC and IC with minimum potential contamination occurring.

The combustion process itself was also similar to that described by Manny et al. (1971), except there were no additions of H_2O_2, boric acid, or O_2, made to the lakewater fractions prior to combustion that would accelerate or facilitate the combustion process, or increase the efficiency of the oxidation. Efficiency of the combustion system was tested during development by adding ^{14}C-labeled organic compounds to distilled water and ultrafiltered lakewater and, following various periods of combustion, purging the [14]CO_2 evolved into a gas-phase radioassay system for determination of ^{14}C produced. Efficiencies determined in this manner were found to be comparable to those indicated in the above-mentioned paper.

Determination of ^{14}C in POC, DOC,
and IC by Gas-Phase Analysis

A vibrating-reed electrometer ionization chamber system was developed to measure ^{14}C in the gas phase (see Fig. 15.3). Although gas-phase procedures have been used on a limited basis in field and laboratory ^{14}C-incorporation ecological studies, there have been no such studies published that describe adaptations of the gas-phase technique in which ^{14}C is radioassayed at 100% counting efficiency in a flowthrough system. In most applications, ionization chamber systems have been used primarily for calibration of radioisotope batches or measurement of $^{14}CO_2$ in a static system following Van Slyke combustion of previously labeled organic materials (see Chase and Rabinowitz 1967, for theory, general principles, and applications). Such applications are tedious, frequently requiring 30-45 min per determination.

In the present experimental studies a 275-ml flowthrough ionization chamber, the smallest available from Cary Instruments, Inc., was used and connected in line with the infrared CO_2 analyzer, such that $^{12}CO_2$ and $^{14}CO_2$ passing through the infrared analyzer would also pass through the ionization chamber as a "bullet" of gas in a stream of purified O_2. The small volume of the flowthrough chamber, combined with the high flow rate, effectively contained the $^{14}CO_2$ within the chamber such that a sharp deflection could be recorded for ^{14}C activity. By using two identical potentiometric recorders, one connected to the infrared analyzer and the other to the vibrating-reed electrometer, two output signals were recorded nearly simultaneously for each sample purged through the entire system: one signal for the carbon content and the other for the ^{14}C activity, with the ^{14}C activity being recorded at 100% efficiency.

Both the infrared analyzer and the flowthrough ionization chamber system were calibrated with National Bureau of Standards (NBS) ^{14}C samples (Na_2CO_3, 1250 dps 1.000/g; with NBS accuracy estimated to be $\pm$1.5%). Diluted and undiluted standards were purged through the low-temperature catalyst tube and through the entire train, after acidification in a closed system; similarly, ^{14}C-labeled simple and complex organic compounds (including glucose and several straight-chain and ring-structured amino acids) were combusted in the high-temperature furnace for additional standardization. Recovery of the injected isotopes was determined by $Ba(OH)_2$ precipitation of the $^{14}CO_2$ existing from the flowthrough ionization chamber and compared against amounts injected initially. A comparison of these ratios generally approximated 99+% for inorganic carbonate and 98% for organic compounds. There was little indication, during development of the combined infrared and ionization chamber system, of adsorption of the isotope onto internal surfaces of the instruments or the interconnecting Tygon tubing.

Experimental Chelation-Complexing Capacity
Radioassay Developed for Dissolved Organic Carbon
Molecular Weight Size Fractions

An experimental radioassay was developed that would, theoretically, permit laboratory determination of whether or not different dissolved organic carbon molecular weight size fractions were capable of additional chelation or complexing of inorganic iron, thereby effectively "solubilizing" it. The theoretical basis upon which the radioassay procedure was developed was simply that ecologically relevant microgram concentrations of high specific activity and essentially carrier-free and soluble (in 0.1 M HCl) inorganic iron (^{59}Fe) could be added to lakewater

490

fractions, the fractions shaken, and subsequently ultrafiltered through a Millipore GS filter (porosity: 0.22 $\pm$
0.02 μm). Retention of radioactivity, following ultrafiltration, would indicate the formation of the insoluble
precipitate, $^{59}Fe(OH)_2$, or $^{59}FePO_4$. Radioactive iron added,
but not appearing on the filter (and not adsorbed onto
container or flow surfaces), would presumably have been
"solubilized" by finding an organic compound with an available "site" for actual chelation or complexation. In
effect, the assay demonstrates the competition that may
exist between potential chelation and complexation sites
and oxygen for the added ^{59}Fe.

Generally, in these radioassays, a 2-μl aliquot of
undiluted inorganic iron ($^{59}FeCl_3$ in 0.1 M HCl; 6.4 mCi/mg
Fe; Code ISS-1, Amersham/Searle, Arlington Heights, Illinois 60005; equivalent to a calculated final concentration
of 29.8 μg ^{59}Fe/liter based on additions of the labeled
iron to 10-ml lakewater fractions) was generally added to
10-ml aliquots of the different dissolved organic carbon
molecular weight size fractions of water from Star Lake.
Following shaking and ultrafiltration, 0.2 ml of each size
fraction was converted to a gel with Triton^2-X-100 (Packard
Instrument Co., Downers Grove, Illinois), and counted using
standard liquid scintillation techniques on a Packard Tri-
carbon scintillation counter. Since ^{59}Fe is both an β and
a γ emitter, it can be counted with conventional end-window
planchet counters, although at considerably reduced efficiency as compared to scintillation counting. (Counting
efficiency on this particular Tri-carb system was approximately 83%.) All ^{59}Fe data reported in this chapter have
been corrected for isotopic decay (half-life = 45 days)
and efficiency of counting before the μg Fe/liter chelated
or complexed per lakewater molecular weight fraction has
been calculated.

491

Potential pH changes in molecular weight size fractions following microadditions of $^{59}FeCl_3$ were monitored with Leeds and Northrup pH microelectrodes attached to a Cary pH switch and vibrating-reed electrometer. Actual pH shifts, accurate theoretically to 0.0005 pH units following calibration of the instrumentation with a National Bureau of Standards pH sample, were then calculated from the Nernst equation.

Bacterial Chemoorganotrophy and Release of Extracellular Metabolic Products

Bacterial chemoorganotrophic utilization of high-specific-activity $[^{14}C]$-glucose (uniformly labeled) in untreated lakewater samples from Star Lake was measured following field and laboratory techniques, including procedures for calculating various uptake parameters, which are described in detail by Allen (1969). In situ field incubation in the presence of added glucose (15 μg/liter) was completed over a 4-hr period. Samples were returned immediately without fixation to the laboratory and prepared for measurement of organic and inorganic carbon, including dissolved organic carbon molecular weight size fractionation, and ^{14}C content of all fractions using purging, thermal combustion, and gas-phase counting techniques described previously. Respiratory $^{14}CO_2$ production, and UV photooxidation of bacterial extracellular products formed during the field incubation on ^{14}C-labeled glucose were also determined by gas-phase analysis, circumventing the need for the time-consuming procedure of trapping the CO_2, followed by scintillation counting (see technique described by Hobbie and Crawford 1969).

RESULTS

Experimental studies of dissolved organic carbon were conducted during 1973 in Star Lake. These experimental studies consisted of (1) determining the molecular weight size fractions of dissolved organic carbon present, primarily in near surface water samples; (2) estimating labile or refractory characteristics of each of these molecular weight size fractions, based on exposure to high-intensity UV light; (3) developing a radioassay using labeled inorganic iron (^{59}Fe) that would, theoretically, indicate the potential or capacity for metal binding, or chelation-complexing capacity, of each of these dissolved organic carbon molecular weight size fractions; and (4) measuring bacterial chemoorganotrophic uptake of $[^{14}C]$-glucose and following the release of ^{14}C-labeled extracellular products into different molecular weight size fractions of dissolved organic carbon. The latter experimental studies also included determining labile and refractory characteristics of the ^{14}C-labeled extracellular products based on UV photooxidizability.

Dissolved Organic Carbon Molecular
Weight Size Fractionation

Lakewater samples collected from a depth of 0.5 m in Star Lake during September-December 1973 were size fractionated using Amicon Diaflo membrane ultrafilters in a stirred pressure chamber under 40 psi N_2, followed by thermal combustion and infrared CO_2 analysis. Concentrations of dissolved organic carbon in each of the six molecular weight size fractions, together with concentrations of total particulate organic carbon, dissolved organic carbon, inorganic carbon, and general limnological characteristics for eight different sampling periods are presented in Table 15.2.

Mean concentrations and standard deviations of dissolved organic carbon fractions from 0.5 m for all eight sampling dates were also calculated. In terms of absolute concentrations of dissolved organic carbon in each of the

Table 15.2

Molecular Weight Size Fractionation of Total Dissolved Organic Carbon (TDOC), Total Particulate Organic Carbon (TPOC), and Inorganic Carbon (TIC), and Selected Limnological Characteristics at a Depth of 0.5 m in Star Lake, Vermont, 1973[a]

Date	TDOC (mg C/ liter)	Molecular weight size-fractions (mg C/liter)						TPOC (mg C/ liter)	TIC (mg C/ liter)
		A	B	C	D	E	F		
24 September	9.06	3.08	0.90	1.36	0.23	0.65	2.84	3.10	8.84
27 September	10.41	3.22	1.47	1.94	0.15	0.21	3.42	2.89	5.46
28 September	9.64	0.54	2.42	0.99	0.18	1.42	4.09	5.10	9.65
16 October	7.65	2.68	1.83	1.35	0.40	0.25	1.14	1.94	5.83
22 October	11.09	3.76	2.32	0.95	0.02	1.16	2.88	2.14	6.18
27 November	9.75	2.81	2.25	1.18	0.26	0.10	3.15	1.16	7.64
8 December	7.65	0.82	1.47	2.13	0.55	1.22	1.46	0.92	7.60
13 December[b]	11.94	7.31	2.01	0.65	0.12	1.13	0.72	1.07	5.22
Mean ± SD	9.65 ± 1.52	3.03 ± 2.08	1.83 ± 0.52	1.32 ± 0.50	0.24 ± 0.27	0.77 ± 0.53	2.46 ± 1.20	2.29 ± 1.40	7.05 ± 1.64

Table 15.2 (Continued)

Date	Temp. (C)	Oxygen (mg/liter)	Alkalinity (mg CaCO$_3$/liter)^3	pH	Specific conductance (μmhos @ 25 C)	Light transmission (% surface intensity, I remaining at 0.5 m)^0
24 September	16.3	9.86	2.91	6.22	17.6	–
27 September	15.12	9.66	2.81	5.91	16.0	5.49
28 September	16.0	13.11	4.67	6.79	15.6	3.85
16 October	13.9	9.92	4.35	6.11	15.75	–
22 October	11.13	–	3.00	6.41	14.4	–
27 November	4.99	–	2.72	6.40	13.21	31.3
8 December	1.85	10.35	2.68	5.92	16.2	–
13 December^b	1.40	–	3.41	5.41	20.5	41.2
Mean ± SD			3.32 ± 0.77	6.15 ± 0.41	16.16 ± 2.18	

^{a}All samples collected between 0900 and 1030. Dissolved organic carbon molecular weight size fractions are as follows: A, Reeve Angle 984HA glassfiber ultrafilters (effective mesh size ca. 0.50 μm; Sheldon, 1972), 100,000; B, 100,000–50,000; C, 50,000–10,000; D, 10,000–2000; E, 2000–500; F, 500–0.

^{b}2 mm icecover over lake surface.

size fractions, the concentrations were highest for materials between 0.50 μm and an approximate molecular weight
of 100,000 (3.03 ± 2.08 mg C/liter), decreased as molecular weights decreased to a mean concentration of 0.24 ±
0.27 mg C/liter for the 10,000-2000 fraction, and then
increased significantly to 2.46 ± 1.20 mg C/liter for the
lower fraction of 500-0. Based on an analysis of standard
deviations for each of the fractions, there was considerably less variability in TDOC samples (sd = ±15.8% of the
mean), the 100,000-50,000 fraction (±28%), the 50,000-
10,000 fraction (±37.8%), and the 500-0 fraction (±48.8%),
than in all other fractions where standard deviations were
equivalent to at least ±50% of the mean. The 10,000-2000
fraction showed the greatest variability, with standard
deviations considerably greater (±112.5%) than the mean.
As a percentage of the mean total dissolved organic carbon
pool for all eight sampling dates, the molecular weight
fractions, from largest (A) to the smallest (F), accounted
for 31.4, 19.0, 13.7, 2.5, 8.0, and 25.5% of the total
pool. This suggests two important points: First, for these
samples the total dissolved organic carbon pool appears
to contain at least 25% of low molecular weight compounds;
and second, the largest fraction (A) is probably predominantly colloidal material (see Sharp 1973; Breger 1968;
Fox *et al.* 1963), since most colloidal organic material
can be expected to pass through the Reeve Angle 984HA
glassfiber with an effective retention of approximately
0.50 μm but to be trapped on the ultramembrane with a porosity of approximately 12 Å (radius). If a significant
fraction of the 0.50 μm-100,000 molecular weight fraction
were colloidal materials, there would be a tendency to
underestimate the size of the 500-0 fraction (a true "dissolved" fraction) and hence its quantitative importance
to the total pool.

DISSOLVED ORGANIC MATTER

The only other molecular weight size fractionation study based on the use of Amicon ultramembranes and conducted on lakewater samples overlying sediments in two extracted sediment cores (Schindler *et al.* 1972) demonstrated a disproportionately high (and equivalent to that reported in the present study) concentration of organic carbon in the largest (unfiltered-100,000) fraction, but it is not clear whether their samples were prefiltered with the equivalent to the Reeve Angle 984HA to remove the majority of true "particulate" materials present. If not, their largest fraction would include particulate, colloidal, and dissolved organic carbon and not be directly comparable to my data. Also, their use of the Menzel and Vaccaro (1964) persulfate oxidation technique for converting organic carbon to CO_2 may not have resulted in a 100% oxidation of refractory organic compounds and, hence, their data may represent underestimates of actual concentrations of certain size fractions and therefore not be comparable with data I am reporting.

Although metals and a variety of other ions have been analyzed in these molecular weight size fractions to determine metal-organic associations by the USGS, calculations of final concentrations have not yet been completed and the results are to be reported in a separate paper.

Several other observations are noteworthy in the data presented in Table 15.2. First, the ratios of POC, DOC, and TIC are typical for Star Lake. Frequently, DOC levels are two to eight times as high as POC and equivalent to TIC levels, in epilimnetic as well as hypolimnetic waters during summer thermal stratification. With the onset of cooler fall and early winter surface water temperatures, there is a parallel decrease in POC, indicative of decreasing biomass and production in the lake; alkalinity, however, remains essentially constant, with a standard

497

deviation of only +23.2% around the mean for the eight samples. Although seemingly constant, pH varies considerably on a diel basis and no particular significance should be attributed to these data, except that the pH in early to midmorning in Star Lake is frequently 5.5-6.5. The increase in percentage of surface light intensity reaching the depth of sample collection (0.5 m) parallels the decrease in POC and reflects the decreased biomass and particulate organic debris present in the surface waters with the onset of winter conditions.

Ultraviolet Photooxidation of Dissolved Organic Carbon Molecular Weight Size Fractions: An Indication of Labile and Refractory Characteristics

Lakewater samples were collected from 0.5, 1.5, and 2.5 m on June 5, 1973, and fractionated through the series of Amicon membrane ultrafilters. The lowest molecular weight size fraction (500-0) from each depth was subjected to high-intensity UV light for a period of 6 hr, with sample aliquots being removed at 10-min intervals for the first 0.5 hr, and at hourly intervals from 1 to 6 hr, for determination of organic carbon content. Data showing the amount of oxidation occurring after 0.5 and 4.0 hr, including the percent oxidized or combusted for this fraction at all three depths, are shown in Table 15.3.

First, this molecular weight size fraction, which would include such compounds as simple sugars, amino acids, represented 37.4, 32.2, and 54% of the TDOC pool existing at each of the respective depths, from 0.5 to 2.5 m. If colloidal organic matter concentrations were high, these percentages are likely underestimates of a fraction containing true "dissolved" organic carbon compounds. Carbon concentrations in this fraction, ranging from one-third to one-half of the total dissolved organic carbon pool,

Table 15.3

Total Particulate Organic (TPOC), Total Inorganic (TIC), and Total Dissolved Organic Carbon (TDOC) and Molecular Weight Size Fractionation and UV Photooxidation of Lakewater Samples from Star Lake, Vermont, June 5, 1973[a]

Depth (m)	TPOC (mg C/ liter)	TIC (mg C/ liter)	TDOC (mg C/ liter)	Molecular weight size fraction (F)	UV photooxidation of molecular weight size fraction (F)			
					0.5 hr	% oxidized	4 hr	% oxidized
0.5	2.56	5.35	13.49	5.04	2.64	46.8	2.14	57.5
1.5	4.28	4.13	14.26	4.59	1.42	69.1	0.89	80.6
2.5	2.44	3.98	13.20	7.12	1.85	74.2	1.56	78.1

[a]Dissolved organic carbon molecular weight size fraction F = 500-0, pH = 5.72, 5.85, and 5.46, at 0.5, 1.5, and 2.5 m, respectively; samples collected at 0930. See text for explanation.

were higher than anticipated but comparable with the concentration of carbon in the 500-0 fraction measured on numerous other occasions in Star Lake.

Ultraviolet photooxidation or combustion of this fraction shows that, minimally, 47% is oxidized after 0.5 hr,
suggesting approximately half of this fraction is labile
and half is refractory. The fact that the percent oxidized
does not increase appreciably after 4-hr exposure to UV
light suggests that materials not quickly oxidized are
highly refractory to decomposition by this means. Although
there have been other UV photooxidation studies of dissolved organic matter in lakewater samples (see, for example, Manny *et al*. 1971), the TDOC pools were not fractionated to show labile and refractory characteristics of
specific molecular weight ranges. In spite of the fact
that too few data were obtained to estimate any significance
that might be attributed to it, low molecular weight size
fractions from 1.5 and 2.5 m appeared to be composed of
materials that were more labile, since approximately 80%
was oxidized after 4-hr exposure in both samples in comparison to 50% for the surface sample. This may suggest
that there are important qualitative differences in the
compounds within this fraction with depth, and not just
quantitative differences.

Data from another UV photooxidation time-course experiment, in which a single sample from 0.5 m was size
fractionated through six Amicon membrane ultrafilters before being treated to high-intensity UV light, are shown
in Table 15.4.

Concentrations of dissolved organic carbon in each of
the size fractions before exposure to UV light accounted
for 26.0, 19.2, 1.7, 3.3, 16.3, and 33.2% of the total dissolved organic carbon pool present. These percentages are
similar to those reported in Table 15.2, with a dispropor-
tionate concentration of organic carbon in the largest (A)

Table 15.4

Time Course of UV Photooxidation and Decomposition
of Molecular Weight Size Fractions of Dissolved Organic
Carbon from a Depth of 0.5 m in Star Lake, Vermont,
December 14, 1973[a]

| Molecular weight size fractions | UV photooxidation time (hr) | | | | Percent oxidized per fraction after 4 hr |
| | 0 (Control mg C/liter) | 0.5 | 1.0 | 4.0 | |
		Residual DOC (mg C/liter)			
A	1.90	1.32	1.25	1.11	41.6
B	1.40	0.81	0.66	0.60	57.0
C	0.13	0.10	0.14	0.14	0.0[b]
D	0.24	0.25	0.17	0.20	17.0
E	1.19	1.00	0.84	0.82	31.0
F	2.42	1.12	0.87	0.89	63.3
Total dissolved organic carbon	7.28	4.60	3.93	3.76	
Percent oxidation of total dissolved organic carbon		36.8	46.1	48.4	

[a]Sample collected at 1000. Total particulate organic
carbon = 1.26 mg C/liter; total inorganic carbon = 6.12
mg C/liter; pH = 6.09. Dissolved organic carbon molecular
weight size fractions same as in Table 15.2.

[b]8% increase in DOC over control.

and smallest (F) size fractions. (Colloids present in the
largest fraction, as pointed out previously, may cause a
significant underestimation of the percentage composition
of the total pool in the other fractions.) As in most of
the UV photooxidation studies completed on Star Lake samples, oxidation appears to proceed rapidly for the first

0.5 hr and then slows down considerably. After 4 hr of exposure to UV light, approximately half of the two largest fractions, A and B (down to a molecular weight range of 50,000), was oxidized. The fraction of lowest quantitative significance (molecular weight range = 50,000-10,000) showed an increase which undoubtedly was caused by sample contamination. Regardless of the explanation, there was no apparent oxidation of organic compounds in this fraction. Percentage oxidation increased after 4 hr in successively lower molecular weight size fractions until 63.3% was oxidized in the 500-0 fraction. The pattern of percentage UV photooxidation or combustion seen in this experiment after 4 hr suggests the percent of UV refractory organic compounds in each fraction increases as molecular weight ranges decrease until the 50,000-10,000, essentially refractory, fraction is reached and then decreases as the lower molecular weight ranges (E and F) are approached. These differential oxidation data also suggest there are significant differences in the qualitative nature of each of these fractions. In terms of the size of the total pool of dissolved organic carbon present, approximately half was UV oxidized after a 4-hr period of exposure.

Chelation-Complexing Capacity Radioassay for Inorganic Iron as Demonstrated by Safe Additions to Different Dissolved Organic Carbon Molecular Weight Size Fractions

Ecologically relevant concentrations of labeled inorganic iron (^{59}Fe) were added to dissolved organic carbon molecular weight size fractions of lakewater collected from 0.5 m in Star Lake on September 26, 1973. As the theoretical basis for the development of an experimental radioassay to estimate chelation or complexing "capacity" or "potential," it was reasoned that addition to carrier-free, soluble inorganic iron would (1) either immediately precipitate as an insoluble $^{59}Fe(OH)_2$, or $^{59}FePO_4$; or (2) become chelated or complexed onto organic compounds and, therefore, become "solubilized," when added to lakewater sample fractions. (Adsorption onto the surfaces of mono-carbonates, a mechanism that could also account for pre-cipitated loss in a hard-water lake, was not considered possible because of the low prevalent pH of Star Lake wa-ter and the lack of $CaCO_3$ or $MgCO_3$ in any quantitatively significant concentration.) This hypothesis was testable by ultrafiltering each of the molecular weight size frac-tions following ^{59}Fe removed by the filtration process (through a Millipore GS filter; porosity: 0.22 ± 0.02 μm) or remaining in the filtered lakewater fraction. Re-sults of a single such experiment are shown in Table 15.5.

Molecular weight size fractionation of the dissolved organic carbon pool showed that 19.8, 30.3, 13.5, 4.2, 2.1, and 30.2% of the total pool occurs in each of the molecular weight size fractions, from the largest (A) to the smallest (F), respectively. This percentage quanti-tative distribution is similar to other fractionations re-ported in this chapter. With 29.8 μg of ^{59}Fe/liter added (calculated final concentrations based on additions of 10 ml of each fraction), the fractions were vigorously shaken for 1.0 min on a Vortex tube mixer and immediately filtered under low-vacuum pressure onto a Millipore GS filter. From resultant ^{59}Fe activity data, corrected for an 83% scin-tillation counting efficiency and isotopic decay (^{59}Fe half-life = 45 days), concentrations of Fe (μg/liter) "solubilized," or passing through the membrane filter were calculated for each fraction. Although isotopic dilution may have occurred, with natural iron that may have been chelated or complexed in control fractions, this was not taken into account in the calculations. Based on specific

Table 15.5

Chelation-Complexing Capacity for Inorganic Iron by Different Dissolved Organic Carbon Molecular Weight Size Fractions, as Demonstrated by the Addition of 29.8 µg ^{59}Fe/liter to a Size Fractionated Lakewater Sample from a Depth of 0.5 m in Star Lake, Vermont, September 26, 1973[a]

Molecular weight size fractions	mg C/liter	Addition of 29.8 µg ^{59}Fe/liter to each fraction	
		cpm of ^{59}Fe solubilized in 10-ml fraction[b]	Calculated µg Fe l^{-1} solubilized per fraction
A	1.73	4760	4
B	2.65	8330	7
C	1.18	2380	2
D	0.37	35	Trace
E	0.18	98	Trace
F	2.64	1190	1
Total(s) =	8.75		14+

[a] Dissolved organic carbon molecular weight size fractions same as in Table 15.2. See text for explanation.

[b] Net cpm above background, based on duplicate samples; actual counting efficiency for ^{59}Fe = 83%; data are corrected for isotopic decay and efficiency.

Table 15.5 (Continued)

Molecular weight size fractions	UV photooxidation 4 hr		Addition of 29.8 μg ^{59}Fe/liter to each fraction	
	Residual DOC (mg C/liter)	Percent Oxidized	cpm of ^{59}Fe solubilized in 10-ml fraction[b]	Calculated μg Fe/liter solubilized per fraction
A	1.08	37.6	7,140	6
B	0.72	72.8	9,520	8
C	0.94	20.0	10,710	9
D	0.25	32.0	5,950	5
E	0.17	6.0	43	Trace
F	0.49	81.0	3,570	3
Total(s) =	3.16			31+

Percent oxidized of total = 63.9

activity of the ^{59}Fe at the time of the experiment, concentrations of inorganic iron from a trace (<1 µg/liter)
to 7 µg/liter were chelated or complexed by the dissolved
organic carbon present in each fraction, to give a total
of 14+ µg Fe/liter for the total dissolved organic carbon
in the sample. Also, the higher molecular weight materials appear to have a considerable number of "sites" available for iron chelation or complexation that were not saturated. It is likely that the largest fraction (A) contains considerable colloidal materials, and that chelation
or complexing has not been caused by true "dissolved" materials present. The absolute Fe concentration in the
total dissolved organic carbon fraction in the control
sample was 2.3 µg Fe/liter.

It should be emphasized that no significant pH changes
were observed following microadditions of ^{59}FeCl$_3$ to fractionated lakewater samples. The natural chelation strength
of individual molecular weight fractions therefore was not
affected by the acidified inorganic iron additions.

As a part of the same experiment, a portion of each
control size fraction was exposed to UV light for a period
of 4 hr. There were significant changes in the organic
carbon concentrations in most individual fractions following exposure to UV light, with as much as 81% of the low
molecular weight organic compounds being oxidized. The
same equivalent concentration of labeled iron (^{59}Fe) was
added to 10-ml aliquots of these fractions; they were
shaken vigorously, ultrafiltered, and radioassayed. The
levels of radioactivity remaining in the filtered lakewater fractions and the calculated levels of iron chelated
or complexed increased markedly and nearly in all fractions, in spite of the 63.9% oxidation of the total dissolved organic carbon pool. This demonstrates that UV
oxidation may change the quantity of "sites" available for

chelation or complexation, and it undoubtedly does this through both qualitative and quantitative changes in the organic compounds in some of the fractions, since significant changes occur, in comparison to nonirridiated control fractions, in the C (50,000-10,000), D (10,000-2000), and F (500-0) fractions. The total iron chelated or complexed increased 139% over that solubilized in the non-UV-treated total organic carbon pool, although, again, 63.9% of the total dissolved organic carbon pool was oxidized following 4 hr of UV exposure.

Bacterial Chemoorganotrophy and Extracellular Release: Effects on Dissolved Organic Carbon Molecular Weight Size Fractions

The measurement of bacterial uptake of 15 µg/liter of $[^{14}C]$-glucose (uniformly labeled) by chemoorganotrophy in total darkness in a lakewater sample from 0.5 m in Star Lake was used to determine changes occurring in the different dissolved organic carbon molecular weight size fractions resulting from bacterial metabolic activity. Data resulting from this experiment are presented in Table 15.6.

A control lakewater sample was size fractionated with the Amicon ultramembranes. Concentrations of dissolved organic carbon in each fraction accounted for 33.6, 16.6, 5.5, 2.7, 11.2, and 30.4% of the total dissolved organic carbon pool present, from the largest (A) to the smallest (B) fractions, respectively. Following a 0.5-hr in situ incubation in the presence of ^{14}C-labeled glucose, the dissolved organic carbon pool was fractionated to determine carbon concentration and ^{14}C activity in each of the molecular weight size fractionations, using both infrared CO_2 analysis and gas-phase counting techniques.

Table 15.6

Bacterial Chemoorganotrophic Utilization of High-Specific-Activity [^{14}C]-glucose (Uniformly Labeled), and Subsequent Molecular Weight Size Fractionation of the Total Dissolved Organic Carbon Pool in a Lakewater Sample from 0.5 meter in Star Lake, Vermont December 7, 1973[a]

Molecular weight size fraction	Control (mg C/liter)	Following 30 min bacterial incubation with [^{14}C]-glucose		Following 4 hr bacterial incubation with [^{14}C]-glucose	
		mg DOC/liter	^{14}C-DOC per 10-ml fraction[b]	mg DOC/liter	^{14}C-DOC per 10-ml fraction[b]
A	2.87	2.97	1364	2.47	3942
B	1.42	1.64	2196	2.14	6143
C	0.47	0.35	317	0.31	1029
D	0.23	0.17	196	0.24	465
E	0.96	1.11	1043	1.26	4129
F	2.60	2.43	3.8×10^5	2.64	2.1×10^5
Total(s) =	8.55	8.67		9.06	

[a]TPOC = 1.64 mg C/liter; TIC = 7.63 mg C/liter; pH = 6.11. Dissolved organic carbon molecular weight size fractions same as in Table 15.2. See text for explanation.

[b]Based on 100% gas-phase counting efficiency.

Table 15.6 (Continued)

Molecular weight size fraction	mg DOC/ liter	Percent DOC oxidized	UV photooxidation for 4 hr following 4 hr bacterial incubation on [^{14}C]-glucose	
			^{14}C-DOC per 10-ml fraction	Percent ^{14}C-DOC oxidized
A	1.37	44.5	2947	25.2
B	1.04	51.4	4931	23.1
C	0.33	0.0^c	924	10.2
D	0.16	33.3	326	29.9
E	0.98	22.2	1864	55.0
F	0.89	66.3	0.9×10^4	97.6
Total(s) =	4.77	47.4		

^{c}6% increase in DOC over control.

509

Dissolved organic carbon concentrations changed in the fractionated lakewater sample incubated for 0.5 hr in the presence of labeled glucose, in comparison to initial controls [increases or decreases of +3.4, +14.1, -25.0, -26.1, +15.6, and -6.5%, in comparison to carbon concentrations in initial control molecular weight size fractions, from largest (a) to smallest (F), respectively].

The total dissolved organic carbon pool size increased from 8.55 to 8.67 mg C/liter, an increase of 1.4% over the initial control. The ^{14}C label, introduced into lakewater into fraction F (500-0), appeared in all other molecular weight fractions—indicative of bacterial production of ^{14}C-labeled byproducts or extracellular products into the medium. These ^{14}C products were predominantly found in fractions A (>100,000), B (100,000-50,000), and E (2000-500), after 0.5 hr labeling.

Following a 4-hr incubation in the presence of [^{14}C]-glucose, concentrations of dissolved organic carbon in each of the fractions accounted for 27.3, 23.6, 3.4, 2.6, 13.9, and 29.1% of the total pool, from the largest fraction (A) through the smallest (F), respectively. The total pool increased to 9.06 mg C/liter, an increase of 4.5% over the 0.5 hr incubated sample and 5.9% over the initial untreated control. The ^{14}C label, aside from the F (500-0) fraction into which the glucose was diluted initially, became concentrated again in the A, B, and E molecular weight size fractions. There is some indication, from the A and B molecular weight ranges, that production of ^{14}C-labeled extracellular mucopolysaccharides might account for the high isotope levels occurring in these two fractions, although this is speculative. This portion of the experiment demonstrated that (1) bacteria using a low molecular weight compound can produce metabolic products that have higher molecular weights, and (2) bacterial extracellular production may be responsible for rapid (<4 hr) changes in the total dissolved organic carbon pool. The remainder of this experiment was designed to determine whether these bacterial byproducts or extracellular products were labile or refractory through exposure to UV light.

Duplicate dissolved organic carbon molecular weight size fractions, following the 4-hr incubation in the presence of [^{14}C]-glucose, were exposed for 4 hr to UV light. Each molecular weight size fraction was then analyzed for carbon and ^{14}C content. Using the 4-hr ^{14}C-labeled fractions as controls, the carbon content of each fraction changed -44.5, -51.4, +6.1, -33.3, -22.2, and -62.3%, in comparison to non-UV-treated controls for each fraction from the largest (A) to the smallest (F) fractions, respectively. The total dissolved organic carbon pool decreased from 9.06 to 4.77 mg C/liter, a decrease of 47.4% in 4 hr.

The importance of the results of this part of the experiment are seen in a comparison of changes in the ^{14}C content of each fraction to the change occurring in the carbon content of that fraction following the 4-hr exposure to UV light. Oxidation of ^{14}C-labeled dissolved organic compounds in the higher molecular weight ranges (>50,000) resulted in a ^{14}C loss of approximately 25% of the fraction, while nearly half of the carbon was oxidized. The fractions with generally the lowest carbon concentrations, C (50,000-10,000) and D (10,000-2000), which are frequently the most refractory to UV light, released lower levels of ^{14}C activity. As molecular weights are decreased to the E (2000-500) and F (500-0) fractions, the ^{14}C activity lost because of oxidation or combustion increased. These data suggest (1) that there are significant qualitative differences in the organic compounds in these different

511

molecular weight fractions, and (2) that there may be a
pattern to the labile and refractory characteristics
throughout the molecular weight spectrum of TDOC that reflects these qualitative differences, since UV absorption
and combustion are directly related to specific bonding
and structural properties of the organic compounds themselves.

DISCUSSION

Dissolved Organic Carbon Molecular
Weight Size Fractionation

Molecular weight size fractionation of the total dissolved organic carbon pool from a depth of 0.5 m in Star
Lake over a 4-month period in 1973 demonstrated a distinct
pattern in the distribution of carbon concentrations over
the different molecular weight ranges. Based on the eight
samples fractionated, there were consistently higher carbon concentrations in the largest fractions (A-C; equivalent to >100,000-10,000) and in the lowest fraction (F;
500-0), than in the intermediate size fractions (D, E;
10,000-500). Although there was no attempt made in this
study to verify the solute retention characteristics for
each of the membrane ultrafilter designations that are
given by the Amicon Corporation, the means and standard
deviations of carbon concentrations per size fraction were
low enough to suggest the filters have reasonably well-
defined and discrete cutoff points. The amount of error
that might have been introduced from adsorption onto the
surfaces of certain filters that have slight charges (cf.
Table 15.1) would have been obscured by the process of subtracting solute concentrations to obtain levels for each
of the molecular weight ranges. This error was probably
minimal, since the sample was continually stirred during
filtration under pressure.

There is a strong possibility that the largest fraction (A) was contaminated with colloids and colloidal-forming substances, such as clays. This potential error was not evaluated, but if it were significant (e.g., 1.5-2.0 mg C/liter) the net effect would be to cause an underestimation of the percentage of the total dissolved organic carbon pool present in each of the other size fractions. Instead of the low molecular weight materials comprising 20-30% of the total dissolved organic carbon pool, their contribution might be 40%, or greater. This is not inconsequential, since bacterial metabolic activity of the smallest fraction probably proceeds at the highest rates in comparison to potential use of the larger and more complex organic compounds.

It is especially noteworthy that the TDOC pood does not change quantitatively in a manner that parallels particulate organic carbon changes, or in a way that reflects the seasonal changes in water temperature, or the increased light penetration with decreased surface primary production and the onset of winter conditions.

There are virtually no published data available on specific molecular weight size fractions of dissolved organic carbon from other freshwater (or even marine) ecosystems, with which my data can be compared. Data provided by Schindler et al. (1972) are considered to be incomparable since their oxidation process (Menzel and Vaccaro 1964) relied on the use of potassium persulfate, which may have resulted in less than 100% oxidation of complex compounds. Also, their largest fraction (A) was possibly contaminated with particulate organic carbon in addition to colloidal and dissolved organic materials. Carbon concentrations in fractions smaller than the largest (cf. Schindler et al. 1972) appear to be comparable with those I have obtained, although their C fraction frequently

contained higher carbon concentrations than those observed
in B, D, or E fractions. Four out of five of their samples
showed comparatively higher concentrations than were present in fraction F. Other similar data are not presently
available for comparison.

No hypothesis has been formulated that can explain
the pattern observed in the carbon concentrations of molecular weight size fractions analyzed in Star Lake. The development of a hypothesis is made difficult if one considers the rate at which the carbon contents of individual
size fractions may be produced, consumed, or transformed
by the bacteria. Measurement of bacterial chemoorganotrophic uptake of organic compounds of known molecular
weight, and following qualitative as well as quantitative
changes resulting from bacterial release of extracellular
products, such as was employed in one of the experiments
to be discussed, offers the potential for calculating turnover and conversion rates for some of these compounds and/
or fractions. A similar approach has been described by
Fleischer (1974) that shows considerable promise.

Labile and Refractory Characteristics of Dissolved
Organic Carbon Molecular Weight Size Fractions

Three lakewater samples (0.5, 1.5, and 2.5 m) were
ultrafiltered to provide the lowest molecular weight size
fraction of dissolved organic compounds (F; 500-0) and
exposed to UV light over a 6-hr period to determine labile
and refractory properties of this fraction with depth in
the water column. In these particular samples levels of
TDOC were unusually high and were probably derived in part
from the large phytoplankton standing crop present through
release of extracellular products and decomposition,
through bacterial activity, and by lateral diffusion of
organic materials from the littoral water column because

of higher aquatic plant metabolism (cf. Allen 1972). Carbon concentrations of this molecular weight size fraction (500-0) were frequently about half this amount in lakewater samples from Star Lake.

Exposure to high-intensity UV light showed that 57.5-80.6% of this fraction from all depths was oxidized during a 4-hr period. Data for comparison are not presently available in the published literature. It has been demonstrated (Manny et al. 1971), however, that low molecular weight ^{14}C-labeled compounds, such as glycine, when added to lakewater samples can be UV combusted and recovered partially, with the natural TDOC level playing an undefined role in the efficiency of the combustion process itself. The studies by Manny et al. (1971) are not directly comparable with my data since their oxidation studies have been conducted on unfractionated TDOC lakewater samples. Preliminary data, from this experiment alone, suggest the low molecular weight size fraction is composed of approximately 20-35% refractory materials, with the remainder being labile. Since UV absorption and cleaving of intramolecular bonds is probably structure specific, there may be some slight qualitative differences between the 0.5-m sample and samples collected from 1.5 and 2.5 m, based on the percent oxidation occurring.

A 0.5-m sample of lakewater collected from Star Lake in mid-December under 3 mm of ice cover was size fractionated and exposed to UV light over a 4-hr period. To my knowledge, these are the first data to be published on labile and refractory characteristics of individual dissolved organic carbon molecular weight size fractions, as determined through UV combustion. These data demonstrated that (1) the lowest (F) and the highest (A) molecular weight size fractions contain the largest percentages of UV-labile organic compounds, and (2) molecular weight size fractions C and D (50,000-2000) are highly resistant to decomposition by this method. These differential oxidation characteristics further suggest significant qualitative differences in each of the size fractions. Last, the greatest percentage of oxidation, for those fractions that were highly labile (>40% oxidized after 4 hr exposure) occurred in the first 0.5 hr of exposure. It is further important to note that in spite of the fact that the TDOC pool lost 48.4% of the organic carbon to CO_2, this did not reflect a uniform percentage of oxidation of all of the size fractions.

Chelation-Complexing Capacity of Inorganic Iron by Different Dissolved Organic Carbon Molecular Weight Size Fractions

Chelation or complexation of inorganic iron is one of the most critical roles assumed by dissolved organic compounds in both fresh (see, for example, Allen 1971b; Wetzel 1968; Shapiro 1966, 1967a, 1967b; Wetzel and Allen 1972; Schelske 1962; among others) and marine waters (Menzel and Ryther 1961; Menzel et al. 1963; Barber and Ryther 1969; Barber 1973; among others). Since inorganic iron is characteristically insoluble at the redox and pH values commonly encountered in most natural waters (cf. Hutchinson 1957), the potential phytoplankton primary productivity is extremely dependent on and indeed controlled or regulated by, the availability of iron that is maintained in a "physiologically available" state through organic chelation or complexation.

In order to demonstrate chelation-complexing capacity for inorganic iron by dissolved organic compounds, a lake-water sample was collected from 0.5 m, size fractionated, and treated experimentally by adding ecologically relevant concentrations (29.8 µg Fe/liter of high specific

activity, essentially carrier-free iron (^{59}Fe) to 10-ml
aliquots of each of the size fractions. Following vigorous
shaking and ultrafiltration, the radioactivity remaining
in solution was converted via calculation to the absolute
concentration of iron "solubilized" in each fraction by
chelation or complexing onto organic compounds. My experimental data, in size fractions otherwise untreated, indicated that (1) high molecular weight organic compounds
(>100,000-10,000) possessed the largest capacity for iron
binding, and (2) that there were considerable "sites"
available that had not been previously saturated. In effect, this radioassay demonstrated the competition existing between sites available for chelation or complexation,
and oxygen.

Shapiro (1967b) showed in a rather elegant experimental study of four lakewaters of differing limnological
characteristics that the greatest iron-holding capacity
was associated with constituents in the highest molecular
weight size fraction, but his fractionation of the TDOC
pool was based on ion-exchange, concentration, and gel-
filtration techniques, which makes a direct comparison
with my size fractions impossible. However, it is important to note that Shapiro (1967b) was able to provide evidence that the high molecular compounds were ten times
more effective than the lower molecular weight compounds
in holding iron in solution.

A second part of this experiment was designed to determine whether there were changes in iron-binding capacity following exposure of the molecular weight size
fractions to UV light. A 4-hr exposure of the same molecular weight size fractions, as used previously, to UV
light resulted in a 63.9% decrease in the TDOC pool, although the largest size fraction-specific losses were in
the B (100,000-50,000) and F (500-0) fractions. Addition

of labeled iron to 10-ml aliquots of the UV-treated dissolved organic carbon size fractions showed that (1) there
was considerably more iron chelated or complexed onto the
individual molecular weight size-fractionated organic compounds, and (2) that the total iron "solubilized" increased
139% over controls (to 31+ μg Fe/liter) in the TDOC pool
of that lakewater sample. These data, although preliminary, demonstrate that exposure to UV light may enhance
the binding capacity by modification of charge or structural characteristics of compounds within the individual
fractions. Such data have not been previously reported.
As a final point, most of Shapiro's data, and that of
others, suggest that inorganic iron is naturally associated with high molecular weight organic compounds. Data
obtained in this experiment suggest a type of iron binding
that is associated with low molecular weight compounds
and may suggest a weak complexing with such compounds as
amino acids (see also Wetzel 1972). One additional study
by Schindler *et al*. (1972), in which concentrations of inorganic iron associated with dissolved organic carbon molecular weight size fractions were determined in sediment-
water interface lakewater samples, suggests iron was associated primarily with size fractions of >100,000-10,000,
with iron appearing in fractions as low as 2000. My Fe
radioassay data and the data of Schindler *et al*. (1972)
are comparable.

In Star Lake, the chelation or complexing effects of
added iron, as shown in this experiment, parallel responses
observed in radiobioassay studies of nutritional factors
controlling or regulating phytoplankton primary productivity rates (see Allen 1972). A comparison of the ^{59}Fe
radioassay data with nutritional bioassays suggests the
low levels of inorganic iron available to the phytoplankton
are primarily governed by the availability of suitable

organic chelators or complexing compounds, since inorganic iron additions to natural lakewater samples generally result in a strongly stimulated primary productivity response.

Bacterial Chemoorganotrophy and Extracellular Release
Effects on Dissolved Organic Carbon Molecular Weight
Size Fractions

A simple experiment, in which the natural heterogeneous bacteria in a 0.5-m lakewater sample was incubated in the presence of ^{14}C-labeled glucose (uniformly labeled), was designed to determine how bacterial metabolic activity might change the composition of the TDOC pool and individual dissolved organic carbon molecular weight size fractions through release of extracellular products.

Results of a 0.5-hr incubation on 15 µg of high-specific-activity [^{14}C]-glucose under conditions of total darkness showed that the natural bacteria (and heterotrophic phytoplankton) could (1) increase the size of the TDOC pool (+4.5%), and (2) produce extracellular ^{14}C products of considerably higher molecular weights than the initial glucose. A 4-hr incubation in the presence of the isotope showed the same pattern of ^{14}C release into higher molecular weight ranges, although the absolute concentration of ^{14}C extracellular products had increased markedly with the extended incubation—with the label concentrated predominantly in the A–C (>100,000–10,000) and E (2000–500) fractions.

The remaining part of this experiment was designed to determine labile and refractory characteristics of bacterially released, ^{14}C-labeled extracellular products, following a 4-hr period of exposure to UV light. This UV treatment resulted in a reduction of the TDOC pool by 47.7%, with maximum organic carbon oxidation occurring in the A, B (>100,000-10,000), and F (500-0) fractions. The ^{14}C extracellular products oxidized did not parallel the pattern of organic carbon oxidation, except that oxidation of ^{14}C-labeled materials was more effective for the lower molecular weights. Although it is speculative, it is possible that the refractory characteristics observed in the high molecular weight size fractions may represent complex mucopolysaccharides that may be resistant to UV photooxidation. In any event, the difference in carbon and ^{14}C content oxidation observed are probably indicative of qualitative differences existing between the bacterial extracellular products and the natural organic compounds in these molecular weight size fractions.

Although based on a different experimental approach, studies by Fleischer (1974) on sugar turnover permit the determination of bacterially released ^{14}C extracellular products appearing in lakewater—also following incubation on ^{14}C-labeled simple sugars.

Future Research

Based on the development of ultrafiltration procedures, thermal combustion of organic compounds followed by infrared CO_2 measurement, and the ^{14}C gas-phase counting techniques as used in this study, it should be possible to design field and laboratory experimental investigations that more directly document functional interactions existing between bacterial metabolic activities under natural conditions and changes occurring in dissolved organic matter pools. It should further be possible to evaluate to a greater extent how the bacteria influence important indirect roles of dissolved organic compounds, including changes in chelation-complexation capacity for metals, and provision of extracellular metabolites, through various applications of these sensitive

techniques. Since bacterial activity ultimately regulates
organic and inorganic nutrient regeneration rates and,
hence, through time, the rate at which aquatic ecosystems
age or eutrophicate, development of these types of experimental studies becomes even more important.

Since it is virtually impossible to study the individual bacterium, the community metabolic approach still holds
the most potential for ultimately providing an understanding of aquatic microbial processes in general. In spite
of the fact that the Michaelis-Menten kinetic approach
has been demonstrated ineffective in dilute, often highly
oligotrophic, marine aquatic ecosystems (see, for example,
Vaccaro and Jannasch 1966; Hamilton and Preslan 1970), experimental evidence has continued to accumulate that suggests the kinetic procedure provides valid bacterial utilization parameters, including turnover times for solutes
assayed, in the more eutrophic freshwater ecosystems
(Robinson et al. 1973; Allen 1969, 1971a, 1973; see also
pertinent discussions in Williams 1973; Wright 1973b).
Other more sensitive approaches may ultimately be developed, based on radioisotope applications, that can measure
bacterial activity in dilute marine environments.

Although a considerable amount of aquatic microbiological research emphasis should be placed on identifying
and documenting important metabolic processes that are
operative under natural field conditions in a variety of
different aquatic ecosystem types, there should be some
emphasis placed on describing the nature of the dissolved
organic matter pool, both in terms of spatial and temporal
changes and in terms of qualitative as well as quantitative
characteristics.

It should be emphasized that in spite of the progress
that has been made on earlier defined research needs in
aquatic microbial ecology (cf. Heukelekian and Dondero

1964), there still remain many unanswered questions regarding interactions between bacteria and dissolved organic matter pools.

ACKNOWLEDGMENTS

The financial assistance of the National Science Foundation (Grant GB-35374) is gratefully acknowledged. Certain of the field and laboratory equipment used in this study was provided through financial assistance from the Cramer Fund, Department of Biological Sciences, Dartmouth College, and the Richard K. Mellon Foundation.

Ms. Signe K. Anderson assisted the author in the field on numerous occasions and independently performed the majority of the dissolved organic carbon molecular weight size fractionations. Her thoroughness, attention to experimental detail, the dedication to the work is greatly appreciated.

The author would like to thank Drs. Luigi Provasoli, Joseph Shapiro, and Edward Meyer for critically reading an abbreviated version of this chapter.

REFERENCES

Allen, H. L. 1969. Chemoorganotrophic utilization of dissolved organic compounds by planktonic bacteria and algae in a pond. *Int. Rev. Ges. Hydrobiol. 54* (1), 1-33.

Allen, H. L. 1971a. Dissolved organic carbon utilization in size-fractionated algal and bacterial communities. *Int. Rev. Ges. Hydrobiol. 56*(5), 731-749.

Allen, H. L. 1971b. Primary productivity, chemoorganotrophy, and nutritional interactions of epiphytic algae and bacteria on macrophytes in the littoral of a lake. *Ecol. Monogr. 41*, 97-127.

Allen, H. L. 1972. Phytoplankton photosynthesis, micronutrient interactions, and inorganic carbon availability in a soft-water Vermont lake. *Am. Soc. Limnol. Oceanog. Spec. Symp. Vol. Nutrients and Eutrophication 1*, 63-80.

Allen, H. L. 1973. Dissolved organic carbon: patterns of utilization and turnover in two small lakes. *Int. Rev. Ges. Hydrobiol. 58*(5), 617-624.

Andrews, P., and Williams, P. J. LeB. 1971. Heterotrophic utilization of dissolved organic compounds in the sea. III. Measurement of the oxidation rates and concentrations of glucose and amino acids in sea water. *J. Mar. Biol. Assoc. U.K. 51*, 111-125.

Barber, R. T. 1973. Organic ligands and phytoplankton growth in nutrient-rich seawater. In *Trace Metals and Metal-Organic Interactions in Natural Waters*, P. C. Singer (Ed.), Ann Arbor Sci. Publ., Ann Arbor, Michigan, pp. 321-338.

Barber, R. T., and Ryther, J. H. 1969. Organic chelators: factors affecting primary production in the Cromwell Current upwelling. *J. Exp. Mar. Biol. Ecol. 3*, 191-199.

Breger, I. A. 1968. What you don't know can hurt you: organic colloids and natural waters. In *Organic Matter in Natural Waters*, D. W. Hood (Ed.). Inst. Mar. Science Occasional Publ. *1*, Univ. Alaska, College, Alaska, pp. 563-574.

Chase, G. D., and Rabinowitz, J. L. 1967. Radiation detection based on ion collection. In *Principles of Radioisotope Methodology*, (3rd ed.), G. D. Chase and J. L. Radinowitz (Eds.). Burgess, Minneapolis, Minn., pp. 244-282.

Fleischer, S. 1974. Sugar turnover in lakewater and sediment. Int. Assoc. Theor. and Appl. Limnol., 19th Congress, Winnipeg, Canada, August 22-29, 1974. Publ. Abst. of Proc., p. 55.

Fox, D. L., Oppenheimer, C. H., and Kittredge, J. S. 1963. Microfiltration in oceanographic research. *J. Mar. Res. 12*, 233-243.

Hamilton, R., and Preslan, J. E. 1970. Observations on heterotrophic activity in the eastern tropical Pacific. *Limnol. Oceanog. 15*, 395-401.

Heukelekian, H., and Dondero, N. C. 1964. Research in aquatic microbiology: trends and needs. In *Principals and Applications in Aquatic Microbiology*, H. Heukelekian and N. C. Dondero (Eds.). Wiley, New York, pp. 441-452.

Hobbie, J. E. 1967. Glucose and acetate in freshwater: concentrations and turnover rates. In *Chemical Environment in the Aquatic Habitat*, H. L. Golterman and R. S. Clymo (Eds.). Proceedings of an IBP Symp. held in Amsterdam and Nieuwerslius, October 10-16, 1966. N. V. Noord-Hollandsche Uitgevers Mattschappij, Amsterdam, pp. 245-251.

Hobbie, J. E. 1971. Heterotrophic bacteria in aquatic ecosystems: some results of studies with organic radioisotopes. In *The Structure and Function of Freshwater Microbial Communities*, J. Cairns, Jr. (Ed.). Virginia Polytechnic Institute and State University Press, Blacksburg, Va., pp. 181-194.

Hobbie, J. E., and Crawford, C. C. 1969. Respiration corrections for bacterial uptake of dissolved organic compounds in natural waters. *Limnol. Oceanog. 14*, 528-532.

Hobbie, J. E., Crawford, C. C., and Webb, K. L. 1968. Amino acid flux in an estuary. *Science 159*, 1463-1464.

Hutchinson, G. E. 1957. *A Treatise on Limnology. Vol. 1. Geography, Physics and Chemistry*. Wiley, New York, 1015 p.

Manny, B. A., Miller, M. C., and Wetzel, R. G. 1971. Ultraviolet combustion of dissolved organic nitrogen compounds in lake waters. *Limnol. Oceanog. 16*(1), 71-85.

Menzel, D. W., and Ryther, J. H. 1961. Nutrients limiting the production of phytoplankton in the Sargasso Sea, with special reference to iron. *Deep-Sea Res.* 7, 276-281.

Menzel, D. W., Hulburt, E. M., and Ryther, J. H. 1963. The effects of enriching Sargasso Sea water on the production and species composition of the phytoplankton. *Deep-Sea Res.* 10, 209-219.

Menzel, D. W., and Vaccaro, R. F. 1964. The measurement of dissolved organic and particulate carbon in sea water. *Limnol. Oceanog.* 9, 138-142.

Robinson, G. C. C., Hendzel, L. L., and Gillespie, D. C. 1973. A relationship between heterotrophic utilization of organic acids and bacterial populations in West Blue Lake, Manitoba. *Limnol. Oceanog.* 18(2), 264-269.

Saunders, G. W. 1971. Carbon flow in the aquatic system. In *The Structure and Function of Freshwater Microbial Communities*, J. Cairns, Jr. (Ed.). Virginia Polytechnic Institute and State University Press, Blacksburg, Va., pp. 31-45.

Schelske, C. L. 1962. Iron, organic matter, and other factors limiting primary productivity in a marl lake. *Science 136*, 45-46.

Schindler, J. E., Alberts, J. J., and Honick, K. R. 1972. A preliminary investigation of organic-inorganic associations in a stagnating system. *Limnol. Oceanog.* 17(6), 952-957.

Shapiro, J. 1966. The relation of humic color to iron in natural waters. *Ver. Int. Verein. Limnol. 16*, 477-484.

Shapiro, J. 1967a. Iron available to algae. In *Chemical Environment in the Aquatic Habitat*, H. L. Golterman and R. S. Clymo (Eds.). Proceedings of an IBP Symp. held in Amsterdam and Nieuwerslius, October 10-16, 1966. N. V. Noord-Hollandsche Uitgevers Mattschappij, Amsterdam, pp. 219-228.

Shapiro, J. 1967b. Yellow organic acids of lakewater: Differences in their composition and behavior. In *Chemical Environment in the Aquatic Habitat*, H. L. Golterman and R. S. Clymo (Eds.). Proceedings of an IBP Symp. held in Amsterdam and Nieuwerslius, October 10-16, 1966. N. V. Noord-Hollandsche Uitgevers Mattschappij, Amsterdam, pp. 202-216.

Sharp, J. H. 1973. Size classes of organic carbon in seawater. *Limnol. Oceanog. 18*(3), 441-447.

Sheldon, R. W. 1972. Size-separation of marine seston by membrane and glass-fiber filters. *Limnol. Oceanog. 17*, 494-498.

Vaccaro, R. F., and Jannasch, H. W. 1966. Studies on heterotrophic activity in seawater based on glucose assimilation. *Limnol. Oceanog. 11*, 596-607.

Vaccaro, R. F., and Jannasch, H. W. 1967. Variations in uptake kinetics of glucose by natural populations in seawater. *Limnol. Oceanog. 12*, 540-542.

Vaccaro, R. F., Hicks, S. E., Jannasch, H. W., and Carey, F. G. 1968. The occurrence and role of glucose in seawater. *Limnol. Oceanog. 13*, 356-360.

Van Hall, C. E., Satvanko, J., and Stenger, V. A. 1963. Rapid combustion method for determination of organic substances in aqueous solutions. *Anal. Chem. 35*, 315-319.

Wetzel, R. G. 1968. Dissolved organic matter and phytoplanktonic productivity in marl lakes. *Mitt. Int. Verein. Theor. Angew. Limnol. 14*, 261-270.

Wetzel, R. G. 1972. The role of carbon in hard-water marl lakes. *Am. Soc. Limnol. Oceanogr*. Spec. Symp. Vol. Nutrients and Eutrophication. *1*, 84-97.

Wetzel, R. G., and Allen, H. L. 1972. Functions and interactions of dissolved organic matter and the littoral zone in lake metabolism and eutrophication. In *Proc. IBP-UNESCO Symp. on Productivity Problems of Freshwaters*, Z. Kajak and A. Hillbricht-Ilkowska (Eds.). Warszawa-Krakow, Poland, pp. 333-347.

Williams, P. J. LeB. 1970. Heterotrophic utilization of dissolved organic compounds in the sea. I. Size distribution of population and relationship between respiration and incorporation of growth substrates. *J. Mar. Biol. Assoc. U.K. 50*, 859-870.

Williams, P. J. LeB. 1973. The validity of the application of simple kinetic analysis to heterogeneous microbial populations. *Limnol. Oceanog. 18*, 159-165.

Williams, P. J. LeB., and Gray, R. W. 1970. Heterotrophic utilization of dissolved organic compounds in the sea. II. Observations on the responses of heterotrophic marine populations to abrupt increases in amino acid concentration. *J. Mar. Biol. Assoc. U.K. 50*, 871-881.

Wood, L. W., and Chua, K. E. 1973. Glucose flux at the sediment-water interface of Toronto Harbour, Lake Ontario, with reference to pollution stress. *Can. J. Microbiol. 19*, 413-420.

Wright, R. T. 1970. Glycollic acid uptake by planktonic bacteria. In *Organic Matter in Natural Waters*, D. W. Hood (Ed.). Inst. Mar. Sci. Occasional Publ. No. 1, University of Alaska Press, College, Alaska, pp. 521-536.

Wright, R. T. 1973a. Mineralization of organic solutes by heterotrophic bacteria. In *Second U.S.-Japan Seminar on Marine Microbiology*, R. R. Colwell and R. Y. Morita (Eds.).

Wright, R. T. 1973b. Some difficulties in using ^{14}C-organic solutes to measure heterotrophic bacterial activity. In *Estuarine Microbial Ecology*, H. L. Stevenson and R. R. Colwell (Eds.). The Belle W. Baruch Symp. in Mar. Sci., No. 1, University of South Carolina Press, Columbia, pp. 199-217.

Wright, R. T., and Hobbie, J. E. 1966. The use of glucose and acetate by bacteria and algae in aquatic ecosystems. *Ecology 47*, 447-464.

Chapter 16

THE SIGNIFICANCE OF BACTIVOROUS PROTOZOA
IN THE MICROBIAL COMMUNITY OF DETRITAL PARTICLES

Tom Fenchel

Laboratory of Ecology
Institute of Zoology
University of Aarhus
Dk-8000 Arhus C, Denmark

CONTENTS

ABSTRACT

The microbial community on the surface of detrital particles originating from the tissue of higher plants is described quantitatively. Experiments with homogenously ^{14}C-labeled hay particles inoculated with bacteria and various protozoan grazers of bacteria show that (1) protozoa control bacterial numbers on the particles; (2) the decomposition rate of detrital material is strongly enhanced in the presence of protozoa; and (3) the breakdown rate often is limited by phosphorus. Grazed systems have higher rates of phosphorus cycling, which is mainly a result of a higher rate of P uptake and release in grazed than in nongrazed bacteria, whereas, the P excretion of the protozoa is not significant. Some mechanisms that might explain the found relationship between bacterial metabolism and protozoan grazing are discussed.

INTRODUCTION

Rates of protozoan consumption of bacteria measured in monoxenic cultures (e.g., Barsdate *et al*. 1974; Curds and Cockburn 1968) and the large quantitative importance of protozoa, especially ciliates and zooflagellates, in many microbial communities (e.g., Bick 1958; Fenchel 1969) indicate the significance of protozoa as grazers of natural bacterial populations. Thus, Fenchel (1969) in a quantitative study of the ciliate communities of various types of marine sediments found between 5×10^6 and 3×10^7 ciliates/m^2 (corresponding to 0.3-1.5 g wet weight). They are responsible for a very important and sometimes the largest part of the fraction of community respiration contributed by animals. In coarser sands and in noncapillary sediments (mud, clay, detritus) they play a smaller but still important role. Small zooflagellates seem to play a large role in detrital sediments. Also, the studies of Bick (1964) on the microbial successions following addition of organics to lakewater demonstrated the importance of protozoa in controlling bacterial numbers. Fenchel (1975) attempted to quantify the protozoan grazing on bacteria and the microflora in the sediment of an arctic tundra pond. By measuring the rate of digestion of bacterial cells in different kinds of protozoa, counting the number of bacterial cells in the vacuoles of animals collected in the field, and quantifying the protozoa in the sediments, it was estimated that in the summer (at about 12 C) the protozoa are responsible for a 5-10% turnover of the bacterial population per day.

In addition to reducing population sizes of bacteria, protozoan grazers may affect the species composition and physiology of the bacterial populations and thus a number of community properties, such as nutrient cycling and community metabolism. Johannes (1965) suggested that protozoan grazing of bacteria is important for the regeneration of phosphorus in aquatic systems. Studies on the function of biological wastewater treatment have revealed that protozoa enhance not only flocculation and a reduction of bacteria but also increase the efficiency of mineralization (e.g., McKinney 1971). Other authors (Fenchel 1970, 1972; Hargrave 1970) have demonstrated that metazoan "detritus feeders" may increase the metabolic activity of the microorganisms on which they feed. This effect was attributed to nutrient regeneration and to various mechanical effects.

Such "feedback" effects of grazers may be easy to demonstrate but are difficult to analyze with regards to natural, microbial communities or laboratory microcosms that resemble field communities. First of all natural, microbial communities are often very complex. Thus many protozoa are very specialized in their choice of food, eating only certain species or size groups of bacteria, while other species feed exclusively on certain kinds of microalgae or attack other protozoa or small metazoans (Fenchel 1968 and references cited therein). Fenchel (1969) found that samples of fine sand representing 1 cm^2 of shallow water sediment may contain about 50 species of ciliates alone, each having specialized food requirements or preferring special microhabitats. Together with other types of protozoa, micrometazoa, microalgae, and bacteria communities are formed that resemble coral reefs and rain forests in complexity.

In addition it is technically difficult to manipulate microbial communities in certain ways such as can be done when studying populations of metazoa; i.e., removal of one or a few species from the system or keeping the populations under known, constant sizes.

This chapter describes some experiments regarding the significance of protozoa living on detrital particles and subsequent effects on bacterial population sizes, decomposition of the substrate, and nutrient cycling.

Detrital particles derived from macrophyte tissue with their microbial communities are of great significance in shallow water environments, where they often make up a large part of the sediment. The energy flow channeled through dead macrophyte tissue is important in many ecosystems and the rate of decomposition is paramount for the overall function of many aquatic systems. The microbial communities of the detrital particles constitute the food of detrital feeding animals. Furthermore, the microbial communities on such detrital particles under aerobic conditions seem to be relatively simple and they are more easily reproduced and studied in laboratory systems than are most other natural, microbial communities (Fenchel 1970, 1972; Odum and De la Cruz 1967).

THE MICROBIAL COMMUNITY OF DETRITAL PARTICLES AND THE COMMUNITY SUCCESSION OF PREVIOUSLY STERILIZED AND REINOCULATED PARTICLES

The quantitative and, at least with regard to genera or types of microorganisms, qualitative composition of the microbial communities of detrital particles derived from plant material is very constant, irrespective of the exact nature of the particles and the source of aquatic environment. Thus debris of leaves of mangroves, *Thalassia, Zostera,* and macroalgae collected in seawater; debris of *Carex* from freshwater; as well as previously sterilized particles of the mentioned plants or particles derived from hay or pure cellulose in sea- or freshwater inoculated with a natural source of detritus and left for some days, all reveal bacterial populations of between 5 and 15 cells/100 μm^2 surface of the particles (Fenchel 1970, 1972 and unpublished observations). The number of bacteria is thus dependent on surface area and therefore on particle size. For particles between 62 and 1000 μm it will generally mean that between 10^{10} and 10^9 bacteria are found per gram dry material. Another constituent of the microbial community is small, bactivorous zooflagellates, e.g., such genera as *Monas, Oikomonas, Bodo, Rhynchomonas*, and choanoflagellates. Usually between three and 10 cells are found per 10^4 μm^2 corresponding to numbers between 5 x 10^7 and 5 x 10^8 cells/g detritus. Ciliates are restricted to relatively few genera, especially such hypotrichs as *Euplotes, Holosticha,* and *Aspidisca* and also some holotrichs; usually there are between 10^4 and 10^5 cells/g detritus. Less

important than the above-mentioned groups are filamentous
bacteria, fungi (mainly chytrids), small amoebas, rotifers,
gastrotrichs, and nematodes. Detritus exposed to light
will usually contain diatoms and blue-green algae; up to
10^5 algal cells/g may be found.

If previously sterilized detrital particles of water-
extracted, dried plant material is placed in sea- or freshwater and inoculated with some natural source of detritus,
a characteristic succession may be followed (Fig. 16.1).
According to temperature, bacteria are found on all particles within 5-10 hr and their numbers increase to 20-30
cells/100 μm^2 of surface area after 100-150 hr. The numbers then decline again to about half of this value. This
decline coincides with a rapid increase in zooflagellates,
which usually show a maximum value after 200-300 hr. Ciliates appear later and also show a maximum value to decrease
in numbers again. Most of the other groups of organisms
appear somewhat later but in general the numbers and types
of organisms remain relatively stable after about the first
300 hr and until 80-90% of the detrital material has been
decomposed.

The numbers of organisms and the rate of decomposition
of the substrate is dependent upon oxygen availability and
the nutrient level (N and P) in the water. All experiments
were bubbled to maintain aerobic conditions and to create
turbulence, and the O_2 factor has not been studied in detail. Addition of N and P (in these experiments always
as $NaNO_3$ and NaH_2PO_4) has a stimulating effect on the systems (Figs. 16.1 and 16.5). Initial concentrations up to
10 mg $P-PO_4^{-3}$ and 16 mg $N-NO_3^{-}$/liter will still stimulate
the systems with regard to microbial biomass and breakdown
rates. In systems started with low values of phosphate
(e.g., about 50 μg P/liter) the concentration decreases
to 2-3 μg/liter within 100-200 hr.

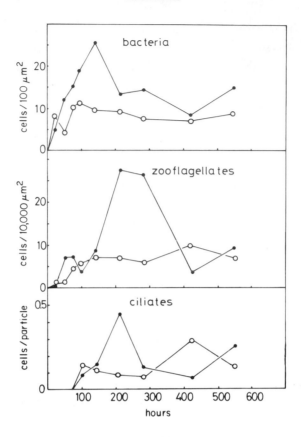

FIG. 16.1. The succession of bacteria, zooflagellates, and ciliates
on hay particles in sea water inoculated with natural detritus. Filled
circles; enriched with N and P. Open circles; only enriched with N.
Each count of bacteria and flagellates is based on counts of 50 fields
with an area of 100 μm^2 under the epifluorescence microscope after
vital staining with acridine orange. Numbers of ciliates were estimated from the number of individuals per 100 particles.

BACTERIAL POPULATION SIZES, NUTRIENT CYCLING, AND THE RATE OF DECOMPOSITION IN GRAZED AND IN UNGRAZED SYSTEMS

In order to measure the rate of decomposition, ground, water-extracted, homogenously ^{14}C labeled barley hay is used in an experimental setup as shown in Figure 16.2. The hay has a specific activity of 35 µCi/g. The particle size used in most experiments is 200-500 µm; these particles contain 6.8 mg Kjeldahl N/g, 380 µg total/g and 648 mg carbohydrate/g (as glucose units measured with the anthrone reagent method). With regard to the microbial populations evolving on these particles, they do not differ in any way from what has been found on similar particles derived from other plants. The "system" consists of a 500-ml erlenmeyer flask with 200 ml seawater, which can be enriched with P and N to a desired level, and 100 mg of the radioactive hay particles. The system can be autoclaved prior to in-oculation with microorganisms. During the experiment moist air passed through the system and the activity of the collected CO_2 was taken as a measure of the rate of decomposition. At intervals samples of particles could be removed,

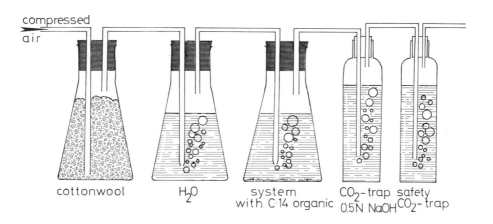

compressed
air

cottonwool H_2O system CO_2-trap   safety
                                 with C-14 organic   0.5N NaOH CO_2-trap

FIG. 16.2. Experimental setup for measuring the breakdown rate of ^{14}C-labeled organic material.

employing a sterile technique, and the microbial populations quantified.

A large number of experiments with the hay particles
as well as with particles of xylan and cellulose prepared
from the hay have been carried out. The results of some
of these experiments will be reported elsewhere. The general behavior of the system is the following: there is an
initial lag time of about 50 hr in which the decomposition
rate is very low, corresponding to the time it takes for
the particles to become colonized. Thereafter and usually
until 80-90% of the carbon of the particles has been recovered as CO_2, the decomposition process can be approximated as a first-order reaction when grazers are present
(Fig. 16.4). This was also found by Otsuki and Hanya
(1972) in a study of the decomposition of killed *Scenedes-*
mus cells. The actual rate is dependent on nutrient level
and type of inoculum.

In a number of experiments the bacterial populations
and the decomposition rates were compared to systems with
only bacteria and to systems that contained either a natural assemblage of protozoa or single, selected species.
In order to obtain protozoa-free systems containing a natural community of bacteria, an inoculum was made by passing water in which detritus had been suspended twice
through a 3-μm Millipore filter (in some cases small zooflagellates turned up later during the experiments, indicating that a few cells may sometimes pass the filters;
such contamined experiments were discarded).

Figures 16.3 and 16.4 show an experiment in which two
systems with a mixed protozoan fauna are compared with two
pure bacteria systems. It can be seen that protozoan grazing keeps the bacterial biomass on a level that is approximately half of that found in ungrazed systems. While the
grazed system had a consistently high rate of decomposition,

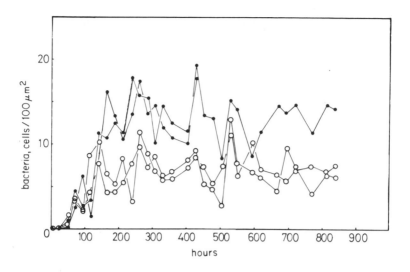

FIG. 16.3. The numbers of bacteria on hay particles in two systems grazed by protozoa (open circles) and in two systems without protozoa (filled circles). The systems were 100 mg hay particles in 200 ml sea water enriched with 5 mg NaH_2PO_4 and 20 mg $NaNO_3$.

the ungrazed system showed a lower, decreasing rate. After 1000 hr 80% of the detritus in the grazed system had been recovered as CO_2, whereas the ungrazed system had decomposed only slightly more than 20% of the substrate. Similar experiments in which single species of protozoan grazers (*Euplotes vannus* and an unidentified choanoflagellate) were used instead of a mixed inoculate of protozoa gave the same result: a reduction in the number of bacteria and an increase in the decomposition rate (Fig. 16.5). However, the effect was somewhat less pronounced than for a mixed fauna of grazers.

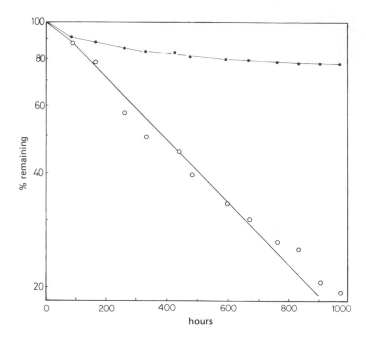

FIG. 16.4. The decomposition of the hay particles of two of the systems shown on Figure 16.3. Protozoa included (open circles); protozoa not included (filled circles).

The results of studies on the phosphorus cycling in
such systems are reported in detail elsewhere (Barsdate
et al. 1974) and only the main results are given here.
Barsdate *et al.* (1974) studied the phosphorus cycling in
microcosms based on dried, ground *Carex* leaves, sterile
pond water, and an inoculum of mixed bacteria with or without a bacterial grazer, the ciliate *Tetrahymena pyriformis*.
Using $^{32}PO_4^{-3}$ as a tracer and quantifying the pools of
phosphorus in the system, it was found that such systems
have a rapid turnover of phosphorus. In systems with a
low concentration of dissolved PO_4^{-3}, fractional turnover
rates as low as 2 min were found. The overall cycling of

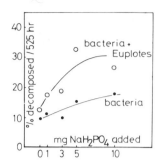

FIG. 16.5. The percentage of hay decomposed in 525 hr by five systems with bacteria and five systems with bacteria and the bactivorous ciliate *Euplotes vannus* and with different degrees of enrichment with P. All systems were enriched with 20 mg $NaNO_3$.

P in grazed systems was somewhat higher than in ungrazed systems. Grazed bacteria had a phosphate uptake per cell on the average four times higher than that of ungrazed bacteria. However, it was found that the ciliate grazers were responsible for only a small part (about 4%) of the phosphorus regeneration of the system. This is intelligible since the ciliates were responsible for a turnover of the bacterial population approximately once every 24 hr, whereas the bacteria had a turnover of phosphorus about once an hour. Thus it was not possible to confirm the results of Johannes (1965) that protozoan grazing keep the bacteria active by regenerating phosphorus that would otherwise be accumulated in bacterial biomass. Also, ungrazed bacteria have a high rate of phosphorus excretion. Thus it seems that grazing of the bacteria change their physiology. Finally it was also shown that mineralization preceded about four times faster in grazed than in ungrazed systems. Rahn and Fenchel (unpublished results) have been able to support the above-mentioned results in the systems consisting of

^{14}C hay and seawater as discussed previously. It was found that the grazed systems deplete the water of dissolved PO_4^{-3} to values as low as or even lower than ungrazed systems. Figure 16.5 shows the effect of different levels of P enrichment to five grazed and five ungrazed systems. It can be seen that both systems respond to the enrichment but that the response is greater for the grazed systems. If the bacterial activity was stimulated because of P excretion of the grazers, a larger difference between grazed and ungrazed systems could be expected at the lowest PO_4^{-3} concentrations. This experiment suggests that grazed systems utilize the enrichment more efficiently than ungrazed systems do.

DISCUSSION AND CONCLUSIONS

The present results demonstrate that protozoan grazing of bacteria has a significant effect, not only controlling bacterial size but also in controlling the metabolic activity as measured by the breakdown of structural carbohydrates of plant tissue and in the cycling of mineral nutrients. Although the conclusions are drawn from experimental laboratory microcosms, there is no reason to assume that similar mechanisms should not be as important in the field, considering the identity of the microbial communities of the studied detrital particles with those found on similar particles in nature. It therefore seems that bactivorous protozoa play a significant, if indirect, role in the rate at which structural carbohydrates are broken down in nature.

It has not yet been possible to make clear the exact mechanism responsible for the enhancement of bacterial activity by protozoan grazers. Much evidence indicates that the regeneration of nutrients by the grazers is not a significant factor, although it cannot be totally ruled out that

the protozoans excrete some growth-promoting substance.
Other possibilities are that the mechanical activity of
the protozoans leads to "microturbulence" and thus increase
the availability of nutrients or oxygen in the immediate
vicinity of the bacteria, or that grazing selects for
quickly growing forms among the mixed assemblage of bacteria. In order to understand the system fully it will
be necessary to understand which factors limit the bacteria
growing on particles consisting of structural carbohydrates.
Even in systems that are enriched by nutrients and in the
absence of grazers only 20% or so of the detrital particles
are covered by bacteria. When grazed, whether in natural
samples or in experimental systems only about 10% of the
particle surface is covered by bacteria that then, however,
appear to be much more active. Further experimentation
will elucidate these questions.

For the moment the most important conclusions of the
described experiment are, next to demonstrating the significance of protozoa in aquatic ecosystems, that ecosystem
components may have important functions that are not described by their metabolic rates and that nutrient cycling
does not necessarily follow the carbon cycle. Thus it is
exemplified that the currently popular ecosystem models of
carbon cycling or energy flow may well be valuable from
a descriptive point of view but give only a limited insight
in the function of the systems and may well be of little
value as predictive models.

ACKNOWLEDGMENTS

My gratitude is expressed to Mr. Preben G. Sørensen
for technical help in the investigation. The work was
supported by grants from "Det naturvidenskabelige forskn-
ingsråd."

REFERENCES

Barsdate, R. J., Fenchel, T., and Prentiki, R. T. 1974. *Oikos 25*, 239-251.

Bick, H. 1958. *Arch. Hydrobiol. 54*, 506-542.

Bick, H. 1964. *Min. ELF des Landes Nordrhein/Westfalen, Düsseldorf*, p. 139

Curds, C. R., and Cockburn, A. 1968. *J. Gen. Microbiol. 54*, 343-358.

Fenchel, T. 1968. *Ophelia 5*, 73-121.

Fenchel, T. 1969. *Ophelia 6*, 1-182.

Fenchel, T. 1970. *Limnol. Oceanog. 15*, 14-20.

Fenchel, T. 1972. *Ver. Deutsch. Zool. Ges. 65 Jahresversamml. 14*, 14-22.

Fenchel, T. 1975. *Hydrobiologia 46*, 445-464.

Hargrave, B. T. 1970. *Limnol. Oceanog. 15*, 21-30.

Johannes, R. E. 1965. *Limnol. Oceanog. 10*, 434-442.

McKinney, R. E. 1971. In *The Structure and Function of Freshwater Microbial Communities*, J. Cairns, Jr. (Ed.). Research Division Monograph 3, Virginia Polytechnic Institute and State University, Blacksburg, Va., pp. 165-179.

Odum, E. P., and De la Cruz, A. A. 1967. In *Estuaries*, A. H. Lauff (Ed.). Publ. Am. Assoc. Adv. Sci. 83, pp. 333-388.

Otsuki, A., and Hanya, T. 1972. *Limnol. Oceanog. 17*, 248-257.

Chapter 17

OXYGEN UPTAKE OF MICROBIAL COMMUNITIES ON SOLID SURFACES

Barry T. Hargrave

and

Georgina A. Phillips

Department of the Environment
Fisheries and Marine Service
Marine Ecology Laboratory
Bedford Institute of Oceanography
Dartmouth, Nova Scotia B2Y 4A2
Canada

CONTENTS

ABSTRACT

A review of the literature provides evidence that
colonization of solid surfaces by aquatic microorganisms
is related to both the physical characteristics of the
attachment surface and the nutrient conditions in the surrounding medium. The integrated effect of these factors,
which may affect both numbers and biomass of colonizing
microorganisms, may be represented by measures of oxygen
consumption by intact communities on solid substrates.

Rates of oxygen uptake by algal and bacterial populations attached to sand, pebble, rock, and detrital surfaces are directly related to surface area under constant
conditions of temperature, pH, and oxygen concentration.
Microbial communities consume between 0.1 and 10 x 10^{-4}
mg O_2/cm^2/hr when expressed on an areal basis. Changes
in community respiration on rock surfaces from a marine
bay are positively correlated with changes in temperature
over an annual cycle but the relationship is nonlinear for
experimental changes in temperature. Also, when community
oxygen uptake is measured at 10 C, there is a significant
depression in areal respiration during summer. There is
no simple linear relation between community respiration
and organic content per unit area, but oxygen uptake by
communities attached to substrates from different locations
can be separated when consumption is expressed on an organic weight basis. An inverse relation between respiration per unit organic weight and the organic content of
the attached community is then apparent. Community respiration on a total weight basis is directly related to
the percentage of carbon and nitrogen in the community and

oxygen uptake per unit weight of carbon is linearly related
to uptake per unit weight of nitrogen over a wide range of
metabolic rates. Temporal changes in rates of oxygen uptake by microbial communities on recolonized rock and glass
surfaces expressed on an areal and organic weight basis
span the same range and reach asymptotic levels over 14
months equivalent to those measured with natural substrates.

The broad similarities in measures of community metabolism on natural and artificial substrates support the
idea that respiratory energy flow through algal and bacterial populations on solid surfaces is a conservative property of the intact community and is largely controlled by
the availability of space. Given time and constant nutrient conditions succession will produce communities with
stable maximum biomass and metabolism per unit area even
through respiration per unit organic matter in the community decreases. These parameters provide a useful measure
of heterotrophic activity for communities colonizing solid
surfaces in any aquatic system.

INTRODUCTION

Colonization of Submerged Surfaces

The propensity for aquatic microorganisms to become
associated with surfaces has been known since the work of
Krüger, which showed that the addition of powdered substances to water reduced the number of bacterial cells in
suspension (Rubentschik *et al*. 1936). The economic interest in marine fouling during the late 1930s stimulated
studies of bacterial attachment to inert substrates. Waksman and Carey (1935) and ZoBell and Anderson (1936) observed in the laboratory that bacterial numbers increased
proportionately as the size of the container decreased.
Several studies (Lloyd 1937; Stark *et al*. 1938; Heukelekian
and Heller 1940; Harvey 1941) indicated that solid surfaces

concentrated nutrients from dilute solutions and thus stimulated bacterial attachment and growth.

ZoBell (1943) was foremost in studying how surfaces
may influence bacterial activity. He observed that most
aquatic bacteria appear to grow on solid surfaces by producing a mucilaginous exudate that may serve to accumulate
nutrients and prevent diffusion of exoenzymes away from
the cell surface. Extracellular carbohydrate materials
serve in adhesion and the formation of bacterial films
(Corpe 1974), and it has been demonstrated by electron
microscopy that many microorganisms are imbedded in a con-
tinous gelatinous matrix of polysaccharide and water with
the cell volume forming only a small part of the total
layer (Jones *et al*. 1969).

Exogenous factors, such as nutrient concentration and
substrate surface properties, also facilitate bacterial
attachment to surfaces (ZoBell 1943; Jannasch and Pritchard
1972; Characklis 1973). Increased surface area has no
effect on cell attachment or respiration when solutions
contain more than 5 mg dissolved organic matter per liter
of easily oxidized substrate. Most metabolic activity in
these solutions results from cells in suspension. The nature of the organic matter and its availability for oxidation is of considerable importance in stimulating attachment. Caseinate, lignoprotein, and emulsified chitin enhance attachment and oxygen uptake while equal concentrations of either glucose, glycerol, or lactate have no influence (ZoBell 1943). Reduced attachment of cells in the
presence of colloidal (gelatin) material, in contrast to
substrates in true solution, probably reflects reduced nutrient availability (Meadows 1971).

Surface charges may also play an important role in
the attachment of microorganisms on substrates. Floodgate
(1972) reported clumping of bacteria on acid-washed glass

549

coverslips, with cells clumped along lines of "stress" in the glass, but uniform colonization occurred on mica flakes, which have a uniform uncharged crystalline structure. Bacteria, algae, and detrital particles in seawater generally have an electronegative surface charge (Neihof and Loeb 1972) and therefore may be repulsed by glass and other artificial substrates that also carry negative surface potentials. There is some evidence to suggest that organic polymers adsorb to electronegative surfaces and this would offset repulsive forces apparent with acid-cleaned substrates (Characklis 1973).

Colonization of submerged surfaces generally occurs in a defined sequence (Marshall *et al*. 1971; Floodgate 1972). The first "reversible" phase lasts only a few hours and since physicochemical factors predominate detachment may occur. In the second "irreversible" stage bacteria secrete polysaccharide cementing compounds that anchor them to the surface. The third phase is characterized by cell growth and subsequent colonization by diatoms, other algae, and fungi and finally a resultant metozoan animal community.

O'Neill and Wilcox (1971) classify each successional stage by its predominant organisms and show qualitative and quantitative differences in the biotic succession dependent on season, location, and substrate type. Butcher (1932), Patrick *et al*. (1954), and Peters (1959) observed that glass and plastic substrates were not selective and supported the same algal flora as natural substrates. Stockner and Armstrong (1971) noted a well-defined succession of diatoms on both natural rocks and glass slides with only minor differences in the relative abundance of some species on glass. Tippett (1970), in contrast, observed twice as many species of diatoms on natural surfaces as occurred on glass slides with differences in the relative abundance of various species. Sieburth *et al.* (1974) has documented the microcolonization of *Zostera, Spartina,* various seaweed species, driftwood, and polypropylene rope using scanning electron microscopy. These authors describe distinct substrate-specific colonization patterns that were notably different on living and dead plants, with surfaces of living tissue having a reduced diversity in comparison to communities developing on glass slides. The effect is opposite to that observed by Tippett (1970) and is perhaps caused by toxic substances being secreted by living macrophytes that prohibit "fouling." Thus, while colonization patterns may appear regular for any given substrate, substantial differences in species composition can occur that may be related to the nature of the solid surface.

Limitations in Size of Attached Communities

The above discussion does not provide any suggestion of regularities in community structure that might be expected to result from boundary conditions imposed by stratification on solid surfaces. Numerous studies have reported the numbers of organisms attached to substrates on a surface area basis, but it was Fenchel (1970) who first observed that the numbers of bacteria, zooflagellates, and diatoms on detritus were proportional to surface area. *Thalassia* detritus was covered by 3×10^6 bacterial cells/ cm^2 (33 μm^2/cell) after 6 days of aging, which was similar to densities observed by direct counts on intact dead leaves and various sized particles. Other estimates of areal coverage by bacteria on different substrates show a wide range (Table 17.1). The most accurate estimates are those derived from direct counts and even these show considerable variation depending on the length of exposure for colonization. These data indicate considerable spacing

Table 17.1

Comparison of Areal Coverage by Bacterial Populations on Different
Substrates in Various Habitats

Substrate	Habitat	μm^2 available per cell	Comments	Reference
Sand and silt	Marine intertidal sandflat	130-411	Natural substrates[a]	Dale (1974)
Pebbles	Marine intertidal	300	Natural substrates[a]	Batoosingh and Anthony (1971)
Clay-silt	Lakes	70-311	Natural substrates[a]	Tsernoglou and Anthony (1971)
Salmon eggs	Streams	$4-300 \times 10^2$	Natural substrates[b]	Bell et al. (1971)
Polyethylene spheres	Streams	$2-40 \times 10^2$	Natural substrates[b]	Bell et al. (1971)
Etched glass spheres	Streams	400×10^2	Artificial substrates exposed in streams[b]	Bell et al. (1971)
Glass slides	Marine	200	4-days colonization[a]	Marshall et al. (1971)
Glass slides	Marine	30-60	4-8 days colonization[a]	O'Neill and Wilcox (1971)

Table 17.1 (Continued)

Substrate	Habitat	μm^2 available per cell	Comments	Reference
Plexiglass	Marine	20-40	4-8 days colonization[a]	O'Neill and Wilcox (1971)
Zostera detritus	Marine	4-50	With and without inorganic salt additions[a]	Fenchel (1972)
Thalassia detritus	Marine	10-100	Natural and artificial detritus[a]	Fenchel (1970)
Sand grains	Marine intertidal	4-40	Natural substrates[a]	Anderson and Meadows (1968)
Glass slides	Marine	$1-36 \times 10^2$	1-7 days exposure[a]	ZoBell and Allen (1935)

[a]Indicates direct counts.

[b]Estimates derived from plate cultures.

between cells or adjacent microcolonies on well-colonized surfaces, a view supported by direct observation of sand grains (Meadows and Anderson 1968), suspended particles from different marine areas (Wiebe and Pomeroy 1972), and lake detritus (Paerl 1973).

Far greater densities per unit surface can be achieved on agar and nutrient broth cultures and at the air-water interface of stagnant water than on submerged solid surfaces (Fenchel 1972; DiSalvo 1973). Surface films contain greater concentrations of organic matter in comparison to natural waters. Inorganic nutrient additions to water can also cause dense colonization. Fenchel (1972) observed that densities of bacteria on decaying *Zostera* were increased from 2 to 25 cells $100/\mu m^2$ by the addition of phosphate and nitrate salts. This stimulatory effect could explain the large epiphyte communities observed on living *Zostera* by Sieburth *et al.* (1974). Organic photosynthetic products released by macrophytes (Allen 1971) and inorganic nutrient release (Reimold 1972) could provide continuous sources of enrichment to microorganisms and allow dense colonization.

Data in the literature also suggest that the total biomass of attached communities varies within fixed limits (Table 17.2). Highest standing stocks of organic matter are achieved under laboratory conditions where light and turbulence can be manipulated to maximize production (McIntyre 1968). Communities in nature have an order of magnitude lower standing stock, presumably because of suboptimal conditions for growth.

The areal content of chlorophyll a in benthic algal communities also has an upper boundary (Moss 1968). Chlorophyll a ranges from 200 to 2000 mg/m^2, with a more restricted range for populations developing under similar conditions of substrate stability and water turbulence.

Table 17.2

Organic Weight Present in Attached Climax Communities
Formed on Various Artificial Substrates

Substrate	Maximum mg organic matter/cm2	Comments	Reference
Glass cylinder	0.8	Unilluminated flowing seawater tank	Present study
Glass slides	6.6	Laboratory stream	Kehde and Wilhm (1972)
Glass slides	1.2	Precambrian Shield lakes- natural exposure	Stockner and Armstrong (1971)
Wooden trough	0.8	Diverted hot spring effluent	Stockner (1968)
Stream substrate	5.0	Artificial stream communities	McIntyre (1968)
Acrylic plastic plates	3-17	Laboratory microcosm, maximum- minimum light exposure	McIntyre and Wulff (1969)
Plexiglass plates	1.2[a]	Laboratory streams	Kevern et al. (1966)
Plexiglass plates	0.5	Natural streams	King and Ball (1966)
Various plastics	2.2[a]	Marine, excluding large metazoans	Pomerat and Weiss (1946)

[a]Total dry weight uncorrected for ash content.

These observations imply an upper limit to the pigment concentration and community biomass per unit area. It is perhaps significant that standing crops of phytoplankton chlorophyll a integrated per square meter for the euphotic zone of lakes and oceans have a similar range to those observed for attached communities on solid substrates (Moss 1968).

The Rate-Limiting Role of Surfaces in Energy Flow Through Organisms and Communities

Production and respiration of organisms are usually calculated per unit biomass. Such formulation has led to the realization that allometric equations, where a power function describes the constancy of increase of one variable relative to that of another, empirically describe many weight-metabolism relationships in organisms. Various explanations involving diffusion processes have been proposed for the well-known decrease in weight-specific metabolic rate with increasing size, but the absolute value of the power function seldom corresponds to that theoretically expected if metabolic rate per organism increases in proportion to surface area (Hemmingsen 1960; Bertalanffy 1964).

Rubner's classical law states that metabolism in organisms remains approximately constant per unit surface area (Bertalanffy 1964). This has seldom been measured and considerable controversy has arisen over the past 90 years through the assumption that body size (weight or nitrogen content) to the two-thirds power is proportional to surface area. A direct test of the law for organisms necessitates measurement of surface area across which gaseous exchange occurs.

Just as metabolism in organisms may be related to surface area, the magnitude of energy flow through aquatic and terrestrial ecosystems may be governed by their exposed surface. The conversion of solar radiation and the exchange of gases and nutrients are examples. Surface-dependent metabolism might also be expected in communities of microorganisms attached to substrates. Nutrient adsorption, the action of exoenzymes, and oxygen diffusion are three surface-related processes that must ultimately limit growth rates of attached communities. Such limitation was explicitly stated by Sanders (1967), who theorized that the growth rate of attached microorganisms becomes limited when surfaces become covered with a single layer of cells. Thereafter, only a linear increase in community biomass can occur and when a thickness corresponding to the critical level for oxygen diffusion is reached, the organisms enter a steady state of growth. To date there has been no experimental evidence to clearly demonstrate conservative properties in metabolism of attached communities that would result from these surface-dependent constraints in growth.

Numerous methods have been employed to quantify metabolism in natural microbial communities (Brock 1966). Few measures, however, allow a holistic view of total community performance. Measurements of substrate uptake kinetics, for example, only document the response of some members of the community to the added substrate. Oxygen consumption, in contrast, reflects total aerobic oxidative metabolism of all organisms in a community and it is related, although not in a constant way, to growth and nutrient uptake for a variety of organisms. Measurements of community oxygen consumption might thus be expected to show conservative features that result from surface-dependent limitations in community size.

Several studies with natural and artificial aquatic ecosystems have shown that community oxygen consumption is regulated and may be conservative in nature (Copeland 1965; Cooke 1971). Intact bottom sediments and stirred suspensions of detritus and sediment also consume oxygen at rates that suggest limitations caused by surface area. Oxygen supply to benthic communities is dependent in part on dissolved oxygen concentration in overlying water and thus oxidative processes may be diffusion limited (Hargrave 1969). Also, oxygen uptake on a total and ash-free weight basis by communities of microorganisms associated with detritus and sediment particles decreases with increasing particle size, which parallels decreases in surface area (Hargrave 1972). The implication of both sets of observations is that surfaces available for oxygen consumption may regulate the ability of communities to utilize a given concentration of oxidizable sediment organic matter.

Submerged surfaces provide solid interfaces for which community oxygen consumption can be directly related to measured surface area. Thus, the aim of the present study was to examine differences in community oxygen uptake per unit area and per unit organic matter for a variety of substrates colonized by natural communities of bacteria, algae, and protozoa and to demonstrate a uniformity in rates that would indicate conservatism in aerobic metabolism. No attempt was made to identify the species components of microbial communities developing on different substrates. Instead, measurements of oxygen uptake were intended to provide a standardized measure of heterotrophic activity on the basis of surface available for colonization that would characterize the aerobic metabolic activity of different attached communities.

MATERIALS AND METHODS

Rocks, pebbles, sand, and plant detritus were collected from various lakes, marine bays, and salt marshes around

Dartmouth, Nova Scotia, on the Canadian Atlantic Coast.
The lakes, located in and around Dartmouth, are oligotrophic except for localized enrichment from storm sewage.
They generally overlie granitic rock, and some are highly
colored with humic compounds leached from surrounding coniferous forests. The chemical characteristics of water
and sediments typical of these lakes have been described
by Hayes and Anthony (1958). Sand, mud, and detritus were
also collected from West Lawrencetown salt marsh (Duff and
Teal 1965) and from intertidal and exposed sand beaches
at Eastern Passage along the approaches to Halifax Harbour.
Intertidal rocks and pebbles were taken from the shores
of Bedford Basin, a eutrophic marine bay (17 km^2 surface
area) surrounded by urban development of Halifax and Dartmouth, separated from the sea by Halifax Harbour. All
substrates to be incubated for measurements of oxygen consumption were kept submerged in water from the collection
site at approximately comparable temperatures and experiments were performed within 1 hr of collection. Experiments to consider the effects of holding or drying provided exceptions to this routine and procedures were altered accordingly.

Sand, pebbles, and pieces of detritus cut into disks
by a 1-cm diameter cork borer were placed in 30-ml glass-
stoppered erlenmeyer flasks. Large pebbles and rocks were
incubated in glass Mason jars (450 ml). Rubber stoppers
containing a tuberculin vial septum and a magnetic stirring
bar suspended on a pivot from a stainless steel rod were
used to cap the jars. Control (blank) and experimental
flasks and jars were held at incubation temperature and
filled by siphoning membrane (Millipore-HA) filtered water
taken from the collection area also equilibrated to the
same temperature. This procedure insured a homogeneous
distribution of dissolved oxygen throughout all samples
and avoided the enclosure of air bubbles.

Incubations were carried out in darkness in a refrigerator with temperature adjusted to seasonal levels or to 10 C, a standard used throughout the year. Incubations lasted 2-3 hr and flasks were rotated on their long axis (10 rpm). Mason jars were placed adjacent to a large magnet mounted on the drive shaft of a variable speed motor. The rotating magnet turned the stirring bars inside incubation jars by induction at 20 rpm. Dissolved oxygen concentrations were followed with time by drawing 1- to 2-ml aliquots in syringes inserted through the rubber septa in the stoppers. Replacement of water equal in volume to the sample withdrawn was provided from a water-filled syringe inserted through the septum throughout the experiment. The small volume of this water relative to the total in the jar would not introduce a significant error. Glass-stoppered flasks could not be sampled during the incubation period and dissolved oxygen determinations were only made before and after an experiment.

A Blood Gas Analyzer (Radiometer, Copenhagen) with a thermostated electrode injection chamber was used to determine dissolved oxygen in syringe samples immediately after they were taken. Measurements were often made in duplicate but single determinations provided reproducibility of $\pm$ 0.1 mg/liter. Sample size was varied so as to achieve a 10-20% reduction in total dissolved oxygen present and since experiments always began with air-equilibrated water, oxygen concentrations never fell below 70% of saturation.

Many experiments with pebbles and rocks from various locations were conducted with water containing 1% buffered (pH 6.8) formalin to insure that all oxygen consumption was caused by biological respiration. Oxygen consumption by all samples totally ceased in the presence of formalin except for those that had been partly buried in reduced

sediments. Continued oxygen uptake by these samples, although at reduced rates, indicated chemical (or abiotic)
oxidative processes and they were not used in estimates
of community respiration reported here.

Absolute changes in dissolved oxygen in flasks or
jars, after correction for volume displacement and experiment duration, were divided by the surface area of the
substrate present and the organic content of the periphytic
community adsorbed to the solid surface. No differentia-
tion between organic matter as attached epiphytes and as
substrate was possible for detritus and oxygen uptake was
related to total organic content in these samples. Disks
(1 cm diameter) cut from decaying vegetation, seaweed,
and marsh grass were assumed to provide a nominal surface
area of 1.56 cm^2. The planar surface of irregular pieces
of detritus was measured wet on graph paper or under a
binocular microscope. Total organic content of detritus
and attached microbial communities was assumed equivalent
to loss in dry weight on ignition (550 C for 3 hr), although some loss of inorganic material may have occurred
on combustion. Carbon and nitrogen were determined by
combustion (900 C) of pulverized dry samples in a Perkin
Elmer 240 elemental analyzer.

Adsorbed organisms and organic surface films on rocks,
pebbles, and glass surfaces were removed by sonication
in a beaker containing membrane (Millipore-HA) filtered
water placed in an ultrasonic cleaner. Over 90% of the
organic material dislodged from stone and glass surfaces
with cleaning for 3 hr was removed during the first 60
min of sonication and this was used as a standard duration
of cleaning. After washing with distilled water and centrifugation, material collected was quantitatively transferred to dry, preweighed aluminum weighing pans and dried
(80 C) to constant weight overnight for total weight

determinations. Subsamples were used for determination
of weight loss on ignition, and carbon and nitrogen as
described above.

Estimates of surface area of rocks and pebbles, after
sonication and thorough rinsing in distilled water, were
provided by a method suggested by Zeidman (1948) for determination of the surface area of small irregular objects.
The technique was adapted to quantify the surface area of
pebbles by Batoosingh (1964) and consists of determining
the amount of sodium chloride deposited after immersion
of a wettable surface in a 0.9% salt solution. The amount
of sodium chloride deposited is subsequently measured by
dipping the object in 0.15% silver nitrate with a volume
dependent on the surface area present (25 ml for up to
150 cm^2). A silver chloride precipitate forms decreasing
the amount of silver nitrate by the quantity of chloride
introduced adsorbed to the immersed surface. The amount
of silver nitrate remaining in solution is then determined
after addition of 5 ml concentrated nitric acid and 0.6 g
of ferric ammonium sulfate and titrating with 0.15% ammonium thiocyanate. Estimates of surface area determined by
measuring pieces of aluminum foil that had been tightly
pressed around rocks or pebbles, gave variable results but
of the same order as those determined by adsorption of
sodium chloride.

Pyrex glass tubing (3.4 cm inside diameter) cut into
5-cm lengths (113 cm^2 combined area of inside and outside
surfaces) was used as an artificial substrate to follow
temporal changes in community respiration and organic content during colonization. Glass cylinders were held on
end in a flowing seawater aquarium for up to 14 months
under ambient laboratory light conditions. Temperature
fluctuations corresponded to those occurring at 30 m in
Bedford Basin (0-10 C over an annual period), the site

from which seawater was drawn before passing through sand filter beds into the laboratory. Fourteen cylinders were placed in the seawater at the start of the colonization period. On each sampling occasion two cylinders were removed for measurements of oxygen uptake as described above except that jars of 200-ml capacity were used. Total organic, carbon, and nitrogen content present on both inner and outer surfaces of cylinders was determined after removal by sonication.

RESULTS

Community Metabolism and Substrate Surface Area

The depletion of dissolved oxygen from sealed jars and flasks containing rocks, pebbles, and detritus of varying surface area was linear over the first 2-3 hr and related to surface area (Fig. 17.1). The rate of consumption, however, lessened with longer incubation periods and this time-dependent decrease was greatest in samples with the most surface area. Maximum rates of uptake for all substrates were obtained during the first 2-3 hr. The linear decrease in dissolved oxygen over this period permitted comparison with estimates of uptake by pieces of detritus incubated in 30-ml flasks. Absolute rates of oxygen consumed by attached communities on different surfaces were linearly related to substrate surface area despite differences in substrate types and location of collection (Fig. 17.2).

These observations suggest that measures of respiration of communities on all substrates should fall within a narrow range when expressed on an areal basis. Previous calculations (Hargrave 1972) place limits on community metabolism on this basis between 0.1 and 10×10^{-4} mg O_2/cm/hr for sand, pebbles, and detritus. Substitution of

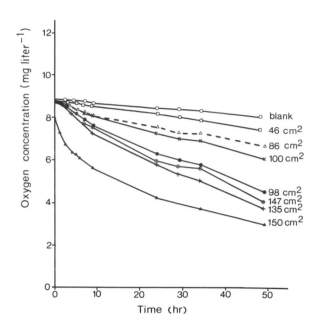

FIG. 17.1. Temporal changes in dissolved oxygen in 450-ml glass Mason jars containing rocks and pebbles and detritus of differing surface area collected from various intertidal marine beaches near Dartmouth, Nova Scotia. Incubation at 10 C.

different surface area values in the equation derived in Figure 17.2 yields an average rate of 5×10^{-4} $O_2/cm^2/hr$.

Factors Affecting Measures of Community Metabolism on Solid Surfaces

The predictability of values of community metabolism suggested above implies some limit to energy flowthrough communities associated with solid surfaces. Consideration should be given to methodology, however, to insure that experimental artifacts do not impart some degree of regularity to the results.

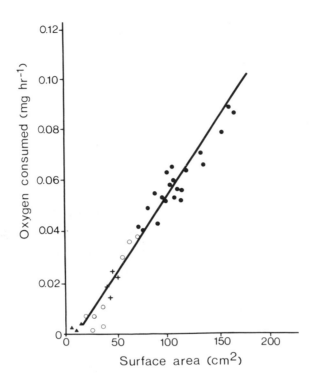

FIG. 17.2. Comparison of absolute amounts of oxygen consumed by various types of solid substrates ($\bigcirc$, detritus; +, pebbles; $\blacktriangle$, sand; $\bullet$, rocks) with different surface area. Incubations of 4-hr duration at 10 C in 450-ml jars or 30-ml flasks, depending on total surface available per sample. $y = -0.005 + 0.0006x$; $r = 0.98$.

Effects of Experimental Conditions

It was apparent in initial experiments that the holding of samples under laboratory conditions resulted in a decrease in oxygen consumption per unit area. Respiration by communities on rocks of various sizes from Eastern Passage and Bedford Basin decreased continuously over 15 days when held in a flowing seawater aquarium. Community metabolism also decreased during 48-hr experiments of

Days	mg O_2 x 10^{-4}/cm^2/hr					
	1	2	4	6	10	15
Bedford Basin	11.5	10.8	7.6	6.5	5.0	4.8
Eastern Passage	5.3	4.9	4.0	3.5	2.5	2.1

continuous incubation (Fig. 17.1) although, as mentioned above, part of this effect may have resulted from lowered oxygen tensions because of consumption by samples with large surface area. No attempt was made to quantify the effect of oxygen concentration on rates of uptake.

Submerged rocks, pebbles, and detritus air dried before use always consumed less oxygen per unit area than duplicate non-air-dried samples. This reduction was not always predictable and it could not be related to the length of time of exposure. Metabolism of communities on rock and pebble surfaces from intertidal areas appeared to be less affected by drying than samples collected from permanently submerged areas.

Oxygen depletion was not linear when samples were left unstirred during incubation. Even in experiments where agitation was provided prior to withdrawing samples for oxygen determination, there was often no measurable oxygen consumption. Subsequent experiments with the same samples mixed during incubation provided measurable rates of uptake. The stimulatory effect of agitation on rates of oxygen uptake and nutrient exchange by suspended sediment and detritus has been described previously (Lee 1970; Hargrave 1972). A boundary layer effect around solid surfaces in stagnant water would explain the results if depletion of oxygen or nutrients immediately adjacent to the surface limits further uptake. The factors affecting oxygen supply to communities on solid substrates were not studied, thus air-saturated water and mixing techniques were used to maximize community metabolism.

Even with the above standardized design, it was apparent that considerable variation existed between replicate measurements with samples treated in a similar fashion. Ten rocks collected at one time from the littoral zone of Lake Charles and incubated separately had a mean areal community respiration rate of $1.2 \pm 0.5 \times 10^{-4}$ mg $O_2/cm^2/hr$ (coefficient of variation, $\sigma/\bar{x} \cdot 100$, of 42.4%). Similar determinations with rocks from Beford Basin on one sampling occasion yielded a mean value of 5.9 ± 1.4 (CV of 23.6%). These represented the maximum range for similar estimates with rock and pebble substrates from various areas.

Seasonal Effects

Rates of oxygen consumption by microbial communities on rocks from midintertidal levels of a beach in Bedford Basin incubated at ambient temperatures immediately after collection showed more than an order of magnitude variation over an annual period (Fig. 17.3). The range of values was within that observed for other substrates and there was a significant linear relation between community respiration and temperature ($y = 0.96 + 0.33x$, $r = 0.75$). These measurements, however, were made over an annual cycle and they combine the effects of changes in community size, composition, and physiological state with the direct effects of temperature change on community metabolic rate. It was hoped that the effect of temperature alone might be more clearly shown by a series of measurements of oxygen uptake by samples collected at one time and incubated at various temperatures (Fig. 17.4). No acclimation period was used and thus rates of community respiration reflect

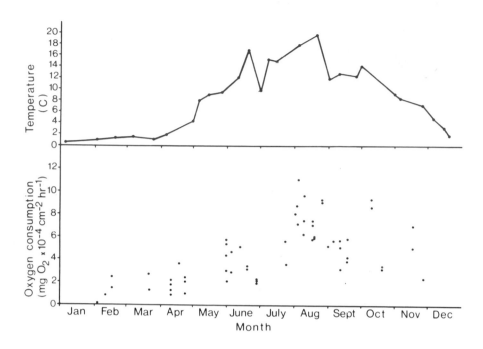

FIG. 17.3. Changes in surface water temperature and areal respiration rates of communities attached to intertidal rocks from Bedford Basin held at these temperatures throughout 1973. Each point represents a single determination of oxygen consumption by communities associated with a single rock.

an acute response to altered temperature during the 3-hr holding period and 3-hr incubation. Obviously, no simple linear correlation between community respiration and temperature existed in the experiment.

The significant nonlinear effect of temperature on community respiration precludes calculations of Q_{10} values for community metabolism from these measurements and necessitates standardization of incubation temperature prior to comparison on any other basis. The use of a standard

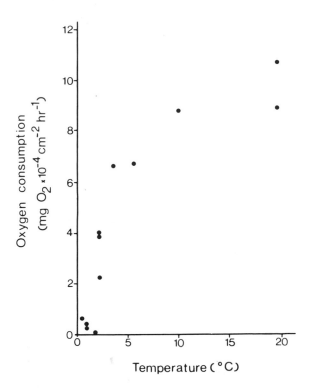

FIG. 17.4. Effect of altered incubation temperature on areal respiration rates of communities attached to rocks collected from an intertidal beach in Bedford Basin, Nova Scotia, during August 1973. Each
.point represents a single determination with one rock exposed to the
incubation temperature for 3 hr prior to measurement.

10 C incubation temperature in the present study was an
attempt to minimize the direct effect of temperature on
metabolic rate. However, this involved large increases
and decreases from ambient conditions in winter and summer,
respectively, even though changes occurred slowly over
2-3 hr immediately before incubation. An unknown temperature effect would result from these abrupt changes in

temperature. Despite this possible experimental artifact, it is of value to consider areal rates of community respiration measured at 10 C over an annual cycle (Fig. 17.5). A clear depression in oxygen consumption occurs during summer. Since this happens at a time of temperature maxima (Fig. 17.3), community metabolism per unit area must be determined by parameters in addition to temperature. Change in community composition and biomass is the most probable cause.

The expression of community respiration in sediments and detritus on an organic (ash-free) basis allows comparison of oxygen uptake by a variety of substrates with

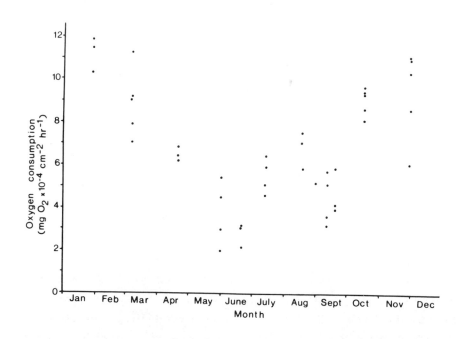

FIG. 17.5. Areal respiration of communities attached to rocks collected from an intertidal beach in Bedford Basin, Nova Scotia, throughout 1973. Each point represents a single determination with one rock incubated at 10 C.

different organic content (Hargrave 1972). Measured areal
respiration rates in the present study were thus similarly
expressed by using the weight of organic matter removed
per unit area. Seasonal and site-specific rates of areal
community oxygen uptake were unrelated to the amount of
organic matter adsorbed per cm on different substrates
and no single regression existed that fit all of the data
significantly (Fig. 17.6). Also no clear seasonal trend
in organic matter per unit area occurred that might have
been correlated with seasonal changes in community oxygen
consumption.

Despite the lack of a relationship to describe community respiration as a function of organic content per
unit area, a separation of measures of oxygen uptake by
substrates from various areas existed (Fig. 17.6). Community metabolism per unit organic weight in the community
on substrates from the three lakes was generally lower
than those rates measured with substrates from marine areas.
The separation is more clearly shown when community respiration is recalculated per gram of organic matter and
compared with organic content in the community (Fig. 17.7).
Respiration per unit organic matter decreases as the organic content in the community increases. Thus, although
microbial communities attached to freshwater rocks and
pebbles were higher in organic content per unit area than
those from marine areas, the metabolism per gram of organic matter was lower.

A more complete description of the dependence of oxygen consumption on organic content in attached communities
may be provided by consideration of the carbon and nitrogen
content and not only total organic matter present. Since
the percent of carbon and nitrogen in the community is
positively correlated with the organic content, a more
independent comparison is made by expressing community

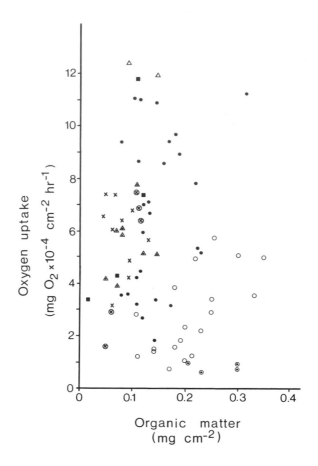

FIG. 17.6. Comparison of areal community respiration rate and organic content per unit area of rocks, pebbles, and plant detritus from various marine and freshwater locations. Samples were collected during summer months and incubated at 10 C. Open circles, Lakes (Banook, MicMac, and Charles); solid dots, Bedford Basin intertidal and subtidal; circled x, Lawrencetown beach and salt marsh; triangles and triangles with dots, Eastern Passage intertidal beaches; x, intertidal beach Northwest Arm, Halifax; solid squares, recolonized glass cylinders exposed in laboratory tanks. All samples incubated immediately after collection except those from Lake Banook held aerated in the laboratory for 2-6 days (circles with dots).

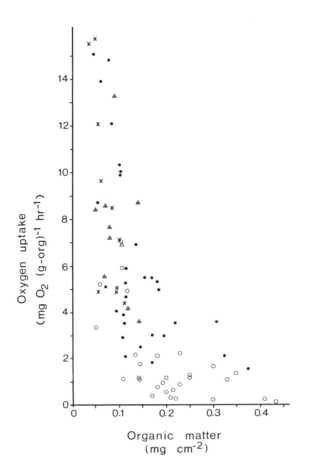

FIG. 17.7. Comparison of community respiration per gram organic (loss on ignition) weight and organic content per unit area on various solid substrates from different marine and freshwater locations. Incubated at 10 C. Symbols as described in Figure 17.6.

oxygen uptake on a total weight basis (Fig. 17.8). Although fewer data exist for comparison, significant positive cor- relations exist between community respiration per unit weight and both carbon and nitrogen content. Correlation

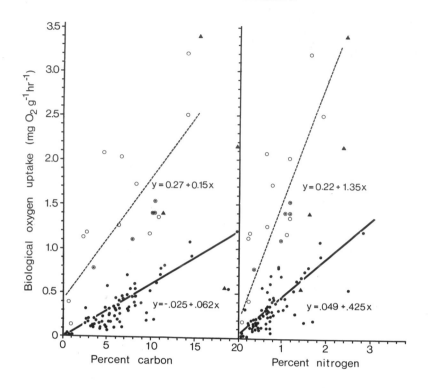

FIG. 17.8. Rates of community respiration per unit weight on various substrates expressed as a function of percentage carbon and nitrogen present in the community and adsorbed to different substrates. Incubations at 10 C. Solid points, fine suspended sediments from Bedford Basin; circles with dots, rocks from intertidal areas of Bedford Basin and Eastern Passage; open circles, rocks from various Dartmouth lakes; triangles, recolonized glass cylinders held in flowing seawater in the laboratory for various lengths of time. Solid regression lines are calculated using only data from suspended sediment collected in traps in Bedford Basin; dotted lines from all other data excluding recolonized glass cylinders.

coefficients for carbon (0.80) and nitrogen (0.84) are approximately equal, although the small sample number (n = 20) forms a poor basis for comparison. No significant regression existed between oxygen uptake per unit weight and carbon:nitrogen ratio in different attached communities.

Oxygen consumption per gram dry weight in communities attached to various substrates is much higher for a given carbon or nitrogen content than those rates observed for samples of fine suspended silt caught in sediment traps in Bedford Basin (Fig. 17.8, lower regression lines). Although lowered rates of community respiration could result from decomposition of material in sediment traps during the 3- to 4-week collection period, changes in carbon and nitrogen content should also occur. These sediment samples were collected throughout the year at various depths with considerable changes in percentages of carbon and nitrogen. Yet, for comparable carbon and nitrogen levels, oxygen consumption per unit weight by communities of organisms associated with sedimented silt and detritus was significantly lower than measures of oxygen uptake by communities stratified on solid rock and pebble surfaces expressed on a similar basis.

Measurement of carbon and nitrogen in various communities associated with different solid substrates permits comparison of metabolism on the basis of carbon and nitrogen content (Fig. 17.9). The linear relation between measures of community respiration calculated on this basis is not surprising when one considers that a close correlation between carbon and nitrogen content usually exists. The slope value (b = 9.5) in fact represents the mean carbon:nitrogen ratio of the various samples. What is noteworthy, however, is the linear nature of the relationship over a wide range of values of community oxygen consumption and the similarity on this basis of measures of metabolism in communities associated with a variety of types of substrate.

Colonization of Artificial Substrates

Temporal changes in community biomass, organic composition and oxygen uptake were followed during

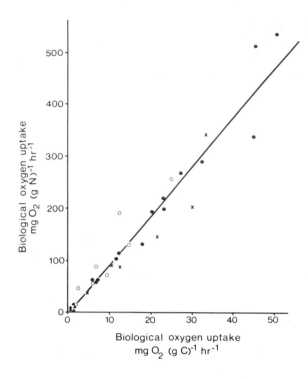

FIG. 17.9.   Respiration of epipelic communities attached to various solid substrates (▲, sand; ○, detritus; ●, rock surfaces; x, re-colonized glass surfaces) under natural and laboratory conditions expressed as oxygen consumption per unit carbon and nitrogen present in the community.

colonization of clean glass and rock surfaces held in flow-ing seawater in the laboratory for up to 14 months. While organic content and community oxygen uptake per unit area of glass and rock surface increased with time, metabolism per gram of organic matter decreased (Fig. 17.10).   The greatest changes in addition of organic matter to surfaces, through both direct adsorption and the establishment of attached communities, occurred during the first 3-5 weeks of colonization.   While areal respiration increased,

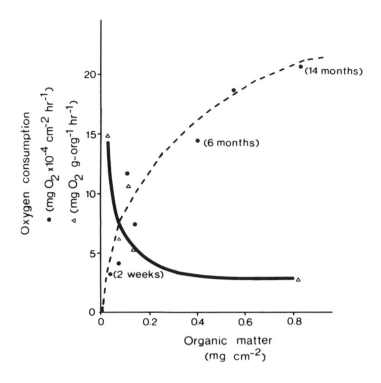

FIG. 17.10. Community respiration on an areal and organic weight
basis as a function of organic content of communities attached to
glass cylinders (113 cm^2) held in flowing seawater in the laboratory
for various lengths of time. Temperature varied between 1 and 10 C
during 14 months of holding. Experiments were begun in May 1972 with
temperatures about 5 C, increasing throughout the summer and falling
during the winter (see Fig. 17.3). Each point represents the mean
of duplicate determinations made with different glass cylinders.
Similar experiments performed with ultrasonically cleaned rocks produced comparable results, although maximum organic contents of
attached communities reached 1.1 mg/cm^2.

metabolism per gram of organic weight decreased rapidly
during this period. Subsequent changes in these measures
over the following 10 months were asymptotic and small in
comparison to those changes that occurred initially. While
organic matter continued to be slowly added to the attached

communities and total areal respiration increased, oxygen uptake on an organic weight basis decreased slightly.

The pattern of changes in community organic weight and metabolism during colonization were similar for rock and glass substrates and the range in these measures was comparable to those observed with freshly collected substrates (Fig. 17.10 compared with Figs. 17.6 and 17.7). No linear correlation existed, however, between oxygen uptake per unit weight in these communities and their carbon and nitrogen content although maximum respiration on a weight basis did coincide with the highest percentage of nitrogen in the community (Fig. 17.8). Also, despite the artificial conditions under which these communities developed, oxygen uptake per unit carbon was linearly related to consumption per unit nitrogen as found for natural substrates (Fig. 17.9).

DISCUSSION

Values of community oxygen uptake per unit area and per gram organic matter observed in the present study lie within narrow limits and demonstrate that space availability and community size constrain respiratory energy loss by attached microbial communities. The variability associated with seasonal effects (Figs. 17.3 and 17.5), with differences in carbon and nitrogen content (Fig. 17.8) and with temporal changes during succession and colonization (Fig. 17.10) overlies a linear relationship between community oxygen consumption and substrate surface area that is common for all substrates (Fig. 17.2). Thus, while species composition may differ with environmental conditions, the level of oxidative metabolism of the community is unchanged.

Consideration of experimental factors that alter areal rates of oxygen uptake may indicate mechanisms that

control community metabolism within narrow limits. The
time-dependent decrease in respiration observed in the
present study was not caused by oxygen or nutrient depletion or accumulation of metabolites, since rocks were
held in flowing aerated seawater between measurements.
Changes in species composition and relative abundance is
response to stable holding conditions could be expected,
however. Rocks for these experiments came from intertidal
areas and community species composition was probably substantially altered by continuous submergence. Organic
weight/cm^2 did not change during the 15 days the rocks were
kept in the laboratory, however, and thus reductions (up
to twofold) in areal metabolism, if they were caused by
changes in species abundance, may give some indication of
variation in respiration possibly because of typological
effects.

The stimulatory effect of agitation on metabolism of
attached microbial communities (Hargrave 1972; McIntyre
1968) clearly demonstrates the importance of diffusion of
oxygen for the maintenance of a specific level of respiration. Once the thickness of an attached community exceeds
the limiting thickness of oxygen diffusion, approximately
21 µm in bacterial films, reduced oxygen supply may limit
aerobic respiration even in the presence of adequate nutrient supply (Sanders 1967). The existence of anaerobic
regions in colonies of *Pseudomonas stutzeri*, which permit
denitrification and nitrogen release in the presence of
oxygen (Jannasch and Pritchard 1972), demonstrate small-
scale oxygen gradients within clumps of aggregated cells.
Cell size, shape, and orientation and the nature of the
organic matrix surrounding cells, however, can alter the
actual depth of penetration from that theoretically calculated only on the basis of diffusion. The presence of
oxygen to depths of 200 µm in some bacterial films (Sanders

et al. 1971) is presumably caused by such enhanced penetration.

These considerations permit a reevaluation of the
stimulatory effect that inert and organically available
substrates provide for microbial attachment and growth.
Viewed with respect to the surface area-dependence of community size and metabolism, the availability of solid surfaces appears to regulate and limit community metabolic
processes. The initial sparse microbial populations on
the glass cylinders (Fig. 17.10) have a level of metabolism
per gram organic matter as high as has ever been previously
observed for natural communities developed on solid surfaces and, interestingly, equivalent to those rates measured on recolonized mud, sand, and pebble surfaces (Hargrave 1972). During succession, as community biomass increases, metabolism per unit organic weight decreases
rapidly to a lower level that may indicate physical-chemical limitations of diffusion. Thus, while the presence
of surfaces stimulate the formation of the attached community, the long-term effect is one of regulation and
stabilization of both community biomass and metabolism.

Seasonal changes in community respiration per unit
area on solid substrates from any one area reflect the
composite effects of species composition, temperature
change, and nutrient supply (Fig. 17.3). Seasonal changes
in community metabolism that are independent of temperature
(Fig. 17.5), but not entirely dependent on organic content,
or the percentage of carbon and nitrogen in the community
(Figs. 17.7 and 17.8), may indicate the importance of nutrient flux in maintaining high levels of community respiration. High rates of oxygen uptake by communities on
rocks from intertidal areas of eutrophic Bedford Basin as
compared with those measured for communities with a similar organic content on rocks from oligotrophic lakes may

thus reflect, among other things, differences in nutrient
supply.

Seasonal and site-specific differences in community
metabolism, corrected for differences in organic content,
could also result from differences in species composition
during colonization or different successional states. Mature stable communities, under constant nutrient regimes,
achieve a maximum biomass with decreased oxygen uptake
per unit organic matter (Fig. 17.10). Environments with
changing physical-chemical conditions maintain communities
in an early stage of succession and metabolism per unit
weight is increased. This would account for the reduced
respiration of microbial communities associated with sediment particles in Bedford Basin when compared, on the
basis of carbon and nitrogen content, with the metabolism
of communities on substrates from littoral and intertidal
areas (Fig. 17.8).

Energy Conservatism in Ecological Systems

Kerr (1974) argues for the existence of constraints
present in all ecological systems that control community
structure through interactions of microsystem and macro-
system properties. He emphasizes that past reductionist
views of ecological systems, while providing empirical
data on the behavior of components, have not permitted
recognition of properties that are inherently coupled and
observable only through holistic approaches. Uniformities
in fish yield for a given level of primary production and
the existence of roughly equal concentrations of material
at all particle sizes in marine pelagic communities are
cited as examples that suggest that both "concentration
and yield of energy are conserved within any particular
size fraction of an unstressed community." Restrictions
in size, numbers, and growth of component populations must

occur if such conservative features exist in natural ecological systems.

The tendency for attached communities to develop stable biomass and metabolic levels supports Kerr's contention
for constraints that limit community energy content and
flow. Limits to growth and metabolism in attached communities seem obvious, but the degree of uniformity imposed
and the similarity for different substrates is surprising.
It is also significant that constraints in metabolism in
communities associated with substrates observed in the
present study, and previously (Hargrave 1969), are observable on the basis of respiratory energy loss. Similar
conservatism may exist in production, particularly since
natural ecological systems appear to evolve toward equal
rates of production and respiration (Cooke 1971). The
lack of standardized measures of production, however, may
prevent recognition of similar uniformities. Thus, while
production is usually used as a basic index for comparison
of biological activity in natural ecosystems, oxygen uptake provides an equally valuable and easily measured para-
meter. Expression of these measurements on the basis of
carbon and nitrogen forms a standardized index of heterotrophic activity that may lead to identifiable properties
of energy conservatism in ecological systems.

ACKNOWLEDGMENTS

Dr. G. Harding offered valuable criticism and comments
on the chapter.

REFERENCES

Allen, H. L. 1971. Primary productivity, chemo-organo-
trophy and nutritional interactions of epiphytic algae
and bacteria on macrophytes in the littoral of a lake.
Ecol. Monogr. 41, 97-127.

Anderson, J. G., and Meadows, P. S. 1968. Bacteria on
intertidal sand grains. *Hydrobiologia 33*, 33-46.

Batoosingh, E. 1964. The bacteriology of marine pebbles.
M.S. Thesis, Dalhousie University, New Brunswick, 67
pp.

Batoosingh, E., and Anthony, E. H. 1971. Direct and indirect observations of bacteria on marine pebbles.
Can. J. Microbiol. 17, 655-664.

Bell, G. R., Hoskins, G. E., and Hodgkiss, W. 1971. Aspects of the characterization, identification, and
ecology of the bacterial flora associated with the
surface of stream-incubating Pacific salmon (*Oncorhyn-
chus*) eggs. *J. Fish. Res. Bd. Can. 28*, 1511-1525.

Bertalanffy, L. von. 1964. Basic concepts in quantitative
biology of metabolism. *Helgoläander Wiss. Meeresunters
9*, 5-37.

Brock, T. D. 1966. *Principals of Microbial Ecology*.
Prentice-Hall, Englewood Cliffs, N.J.

Butcher, R. W. 1932. An apparatus for studying growth
of epiphytic algae with special reference to the River
Tees. *Trans. nth. Nat. Un. 1*, 1-15.

Characklis, W. G. 1973. Attached microbial growths — I.
Attachment and growth. *Water Res. 7*, 1113-1127.

Cooke, G. D. 1971. Aquatic laboratory microsystems and
communities. In *The Structure and Function of Freshwater Microbial Communities*, J. Cairns, Jr. (Ed.).
Monograph 3, Res. Div., Virginia Polytechnic Institute
and State University, Blacksburg, Va., pp. 47-85.

Copeland, B. J. 1965. Evidence for regulation of community metabolism in a marine ecosystem. *Ecology 46*,
563-564.

Corpe, W. A. 1974. Periphytic marine bacteria and the
formation of microbial films on solid surfaces. In
*Effect of the Ocean Environment on Microbial Activi-
ties*, R. R. Colwell and R. Y. Morita (Eds.). University Park Press, Baltimore, pp. 397-417.

Dale, N. G. 1974. Bacteria in intertidal sediments: Factors related to their distribution. *Limnol. Oceanog.* *19*, 509-518.

DiSalvo, L. H. 1973. Contamination of surfaces by bacterial neuston. *Limnol. Oceanog.* *18*, 165-168.

Duff, S., and Teal, J. M. 1965. Temperature change and gas exchange in Nova Scotia and Georgia salt-marsh muds. *Limnol. Oceanog.* *10*, 67-73.

Fenchel, T. 1970. Studies on the decomposition of organic detritus derived from the turtle grass *Thalassia testudinum*. *Limnol. Oceanog.* *15*, 14-20.

Fenchel, T. 1972. Aspects of decomposer food chains in marine benthos. *Ver. Dtsch. Zool. Ges. 65*, 14-22.

Floodgate, G. D. 1972. The mechanism of bacterial attachment to detritus in aquatic systems. *Mem. Inst. Ital. Idrobiol. 29*, 309-323.

Hargrave, B. T. 1969. Similarity of oxygen uptake by benthic communities. *Limnol. Oceanog. 14*, 801-805.

Hargrave, B. T. 1972. Aerobic decomposition of sediment and detritus as a function of particle surface area and organic content. *Limnol. Oceanog. 17*, 583-596.

Harvey, H. W. 1941. On changes taking place in sea water during storage. *J. Mar. Biol. Assoc. U.K. 25*, 225-233.

Hayes, F. R., and Anthony, E. H. 1958. Lake water and sediment. I. Characteristics and water chemistry of some Canadian east coast lakes. *Limnol. Oceanog. 3*, 299-307.

Hemmingsen, A. M. 1960. Energy metabolism as related to body size and respiratory surfaces, and its evolution. *Rep. Steno. Hosp., Copenhagen 9*, 1-110.

Heukelekian, H., and Heller, A. 1940. Relation between food concentration and surface for bacterial growth. *J. Bacteriol. 40*, 547-558.

Jannasch, H. W., and Pritchard, P. H. 1972. The role of inert particulate matter in the activity of aquatic microorganisms. *Mem. Inst. Ital. Idrobiol. 29*, 289-308.

Jones, C. H., Roth, I. L., and Sanders, W. M. 1969. Electron microscopic study of a slime layer. *J. Bacteriol. 99*, 316-325.

Kehde, P. M., and Wilhm, J. L. 1972. The effects of grazing on community structure of periphyton in laboratory streams. *Am. Midl. Nat. 87*, 8-24.

Kerr, S. R. 1974. Structural analysis of aquatic communities. In *First Int. Symp. on Ecology*, The Hague, pp. 69-74.

Kevern, N. R., Wilhm, J. L., and VanDyne, G. M. 1966. Use of artificial substrata to estimate the productivity of periphyton. *Limnol. Oceanog. 11*, 499-502.

King, D. L., and Ball, R. C. 1966. A qualitative and quantitative measure of *aufwuchs* production. *Trans. Am. Micros. Soc. 85*, 232-240.

Lee, G. F. 1970. *Factors Affecting the Transfer of Materials Between Water and Sediments*. Publ. Water Res. Center, Eutrophication Inform. Prog., University of Wisconsin, Madison.

Lloyd, B. 1937. Bacteria in stored water. *J. Roy. Tech. Coll. (Glasgow) 4*, 173-177.

Marshall, K. C., Stout, R., and Mitchell, R. 1971. Mechanism of the initial events in the sorption of marine bacteria to surfaces. *J. Gen. Microbiol. 68*, 337-348.

Meadows, P. S. 1971. The attachment of bacteria to solid surfaces. *Arc. Microbiol. 75*, 374-381.

Meadows, P. S., and Anderson, J. G. 1968. Microorganisms attached to marine sand grains. *J. Mar. Biol. Assoc. U.K. 48*, 161-175.

McIntyre, C. D. 1968. Structural characteristics of benthic algal communities in laboratory streams. *Ecol. Monogr. 49*, 520-537.

McIntyre, C. D., and Wulff, B. L. 1969. A laboratory method for study of marine benthic diatoms. *Limnol. Oceanog. 14*, 667-678.

Moss, B. 1968. The chlorophyll a content of some benthic algal communities. *Arch. Hydrobiol. 65*, 51-62.

Neihof., R. A., and Loeb, G. I. 1972. The surface charge of particulate matter in seawater. *Limnol. Oceanog. 17*, 7-16.

O'Neill, T. B., and Wilcox, G. L. 1971. The formation of a "primary film" on materials submerged in the sea at Port Hueneme, California. *Pac. Sci. 25*, 1-12.

Paerl, H. W. 1973. Detritus in Lake Tahoe: structural modification by attached microflora. *Science 180*, 496-498.

Patrick, R., Hohn, M. H., and Wallace, J. H. 1954. A new method for determining the pattern of the diatom flora. *Not. Nat. (Acad. Nat. Sci. Phila.) 259*, 1-12.

Peters, J. C. 1959. An evaluation of the use of artificial substrates for determining primary production in flowing waters. M.S. Thesis, Michigan State University, East Lansing. 106 pp.

Pomerat, C. M., and Weiss, C. M. 1946. The influence of texture and composition of surface on the attachment of sedentary marine organisms. *Biol. Bull. 91*, 57-65.

Reimold, R. J. 1972. The movement of phosphorus through the salt marsh cold grass, *Spartina alterniflora* Loisel. *Limnol. Oceanog. 17*, 606-611.

Rubentschik, L., Roisin, M. B., and Bieljansky, F. M. 1936. Adsorption of bacteria in salt lakes. *J. Bacteriol. 32*, 11-31.

Sanders, W. M. III. 1967. The growth and development of attached stream bacteria, Part I. Theoretical growth kinetics of attached stream bacteria. *Water Resour. Res. 3*, 81-87.

Sanders, W. M. III, Bungay, H. R. III, and Whalen, W. J. 1971. Oxygen microprobe studies on microbial slime films. *Chem. Eng. Prog. Symp. Ser. 67*, 69-74.

Sieburth, J. McN., Brooks, R. D., Gessner, R. V., Thomas, C. D., and Tootle, J. L. 1974. Microbial colonization of marine plant surfaces as observed by scanning electron microscopy. In *Effects of the Ocean Environment on Microbial Activities*, R. R. Colwell and R. Y. Morita (Eds.). University Park Press, Baltimore, pp. 418-432.

Stark, W. H., Stadler, J., and McCoy, E. 1938. Some factors affecting the bacterial population of freshwater lakes. *J. Bacteriol. 36*, 653-654.

Stockner, J. G. 1968. Algal growth and primary productivity in a thermal stream. *J. Fish. Res. Bd. Can.* *25*, 2037-2058.

Stockner, J. G., and Armstrong, F. A. J. 1971. Periphyton of the Experimental Lakes Area, Northwestern Ontario. *J. Fish. Res. Bd. Can.* *28*, 215-229.

Tippett, R. 1970. Artificial surfaces as a method of studying populations of benthic micro-algae in freshwater. *Brit. Physiol. J.* *5*, 187-199.

Tsernoglou, D., and Anthony, E. H. 1971. Particle size, water-stable aggregates, and bacterial populations in lake sediments. *Can. J. Microbiol.* *17*, 217-227.

Waksman, S. A., and Carey, C. L. 1935. Decomposition of organic matter in sea water by bacteria. I. Bacterial multiplication in stored sea water. *J. Bacteriol.* *29*, 531-543.

Wiebe, W. J., and Pomeroy, L. R. 1972. Microorganisms and their association with aggregates and detritus in the sea: A microscopic study. *Mem. Inst. Ital. Idrobiol.* *29*, 325-352.

Zeidman, I. 1948. A simple method of measuring the surface area of small objects of irregular shape. *Science* *27*, 214-215.

ZoBell, C. E. 1943. The effect of solid surfaces upon bacterial activity. *J. Bacteriol.* *46*, 39-56.

ZoBell, C. E., and Allen, E. C. 1935. The significance of marine bacteria in the fouling of submerged surfaces. *J. Bacteriol.* *29*, 239-251.

ZoBell, C. E., and Anderson, D. Q. 1936. Observations on the multiplication of bacteria in different volumes of stored sea water and the influence of oxygen tension and solid surfaces. *Biol. Bull.* *71*, 324-342.

Chapter 18

EXTRACELLULAR RELEASE IN FRESHWATER ALGAE AND
BACTERIA: EXTRACELLULAR PRODUCTS OF ALGAE AS
A SOURCE OF CARBON FOR HETEROTROPHS

C. Nalewajko

Department of Botany
Scarborough College
University of Toronto
West Hill, Ontario
Canada

CONTENTS

ABSTRACT

Extracellular release (defined as the liberation of
soluble organic compounds by living organisms) occurs commonly in microorganisms; however, far more information is
available on qualitative and quantitative aspects of the
process in phytoplankton than in bacterioplankton. On the
basis of data for clinically and commercially important bacteria, excretion of antibiotics, enzymes, proteins, and
other macromolecules, and in particular of polysaccharides,
appears to be widespread. As in algae, extracellular polysaccharides are derived from capsules and mucilage sheaths.
Some low molecular weight organic acids and alcohols are
also excreted. Certain freshwater and marine bacteria excrete vitamin B_{12} and thiamin. Mixed bacterioplankton
growing on glycollate respired most of the substrate but
after 48 hr 0.5% of the total carbon utilized (13.8% of
the carbon retained by the bacteria) was excreted as unidentified large molecules. *Pseudomonas fluorescens* growing on glucose excreted 1.3% of the carbon utilized (3.7%
of the carbon retained) after 1 hr. Much lower extracellular release rates were observed with fructose and sucrose
as substrates. Polysaccharides appear to be the predominant extracellular products when sugars are used as substrates.

Culture filtrates containing algal extracellular products selectively support the growth of a variety of bacteria. In mixed algal-bacterial culture, some algal extracellular products are rapidly and selectively consumed by
bacteria, resulting in significantly lower apparent excretion rates than in axenic algal cultures. The composition of extracellular products is also modified toward

a predominance of high molecular weight, including colloidal, substances. A similar utilization by bacteria of algal extracellular products is likely to occur in natural waters and may be responsible for the general predominance of large over small molecules. During ^{14}C algal primary production experiments, ^{14}C algal extracellular products respired by bacteria would be lost as $^{14}CO_2$, while ^{14}C-DOC retained by bacteria would be estimated in the particulate fraction. Regardless of the proportions respired and retained, true or gross extracellular release rates of algae will be underestimated in the presence of heterotrophs capable of utilizing algal excretion products. Measured rates of extracellular release are probably net values, detecting only the substances not immediately utilized by bacteria. It is suggested that the sum of net extracellular release rates and simultaneously measured DOC utilization (of substances known to be excreted by the phytoplankton population) by heterotrophs, may give the true or gross extracellular release values. Extracellular products of phytoplankton are involved in two processes:

1. An unknown proportion consist predominantly of low molecular weight substances readily used by heterotrophs. Here, uptake is rapid and practically simultaneous with excretion, as exemplified by glucose and acetate, which at low concentrations show very short turnover times.

2. In addition, binding of low molecular weight compounds to colloids may occur, by processes that may involve a phosphorus-mediated polycondensation, resulting in abiotic removal of these substances from solution. High molecular weight extracellular products also may be bound to each other, to colloids, or to detritus and eventually be removed from lakewater by sedimentation, in addition to their inclusion into food chains.

EXTRACELLULAR RELEASE

EXTRACELLULAR PRODUCTS OF ALGAE

The terms "extracellular release" or "excretion" are used to describe the loss of soluble organic compounds by healthy cells of microorganisms. This definitiion attempts to separate "true" extracellular substances from organic compounds released by moribund organisms or liberated during cell lysis, although in practice no clear-cut distinction may exist.

Extracellular release appears to be a widespread phenomenon in microorganisms. Interest in extracellular products of algae arose in the mid-1950s and much information has accumulated since then. Several recent reviews give excellent coverage of extracellular release in both freshwater (Fogg 1971) and marine algae (Fogg 1966; Hellebust 1974); hence the review here is very brief.

Much of the information on the chemical composition of algal extracellular products has been obtained using unialgal or axenic cultures. Substances excreted by algae include organic acids, carbohydrates, and various nitrogenous compounds, such as amino acids and peptides. In addition, excretion of vitamins, various growth-modifying substances, enzymes, phenolic substances, various volatile substances, sex factors, and toxins has been reported (Fogg 1971; Hellebust 1974). Although the above represents a great variety of compounds, in general one species may excrete only one or several compounds. The dissolved organic matter (DOM) excreted by algae may consist of fewer compounds than DOM from a decaying or lysed population. Many compounds known to be present in algal cells are never excreted. This selectivity confirms that release of extracellular products is under some metabolic control and is not the result of gross cell disorganization and lysis.

Fogg (1966) suggested a convenient classification of extracellular products into type I, low molecular weight

(MW) metabolic intermediates, glycollate, and other acids, including amino acids, and type II (high MW substances, which may be considered end products of metabolic pathways, polysaccharides, polypeptides). Both types appear to be liberated by passive diffusion from regions of high concentration (inside the cell) to lower concentrations (in the medium), but a release process involving fusion of vesicles containing macromolecules with cell membrane has also been suggested for type II compounds (Whaley *et al.* 1972). In addition, release of macromolecules may occur during certain specialized processes, such as gliding movements (Walsby 1968) and during reproduction (Hellebust 1974).

The physiology of extracellular release is best known for glycollate, clearly a type I extracellular product, formed early in photosynthesis (Pritchard *et al.* 1962). Rates of glycollate excretion are proportional to the concentration inside the cell (Fogg 1971); among other factors, the concentration is increased by high light intensity, high oxygen concentrations, and low CO_2 concentrations, as well as by inhibitors of the further metabolism of glycollate. Exponential phase cultures of *Chlorella* excrete more glycollate than lag phase or senescent cultures. Glycollate is by far the major substance excreted by *Chlorella pyrenoidosa* Chick (Nalewajko and Lean 1972) and is excreted by a great variety of algae, freshwater and marine (Hellebust 1974), but it is seldom the major extracellular product. Although only a small proportion of freshwater phytoplankton species has been studied in culture, Watt (1966), on the basis of analyses of mixed populations, has concluded that polysaccharides are the commonest extracellular products. The polysaccharides appear to be derived from solution of capsular or sheath material and accumulate in filtrates as cultures age (Lewin 1955, 1956; Moore and Tischer 1964; Watt 1969).

EXTRACELLULAR PRODUCTS OF PHYTOPLANKTON PHOTOSYNTHESIS

The reports that labeled glycollate is excreted by
Chlorella during fairly short photosynthesis experiments
with $^{14}CO_2$ (Tolbert and Zill 1956) raised the possibility
that phytoplankton also excrete photosynthate and hence
that the ^{14}C method underestimates primary production to
a certain extent (Fogg 1958). Numerous investigators have
since confirmed that the loss of ^{14}C-labeled photosynthate
by phytoplankton is a universal phenomenon in many types
of aquatic habitats. Expressed as a percentage of the
total carbon fixed during photosynthesis, extracellular
release values between 0.3% and 35% are common (Fogg *et al.*
1965; Watt 1966; Wright 1968; Fogg and Horne 1970; Samuel
et al. 1971; Saunders 1972a; Nalewajko and Schindler 1976).
Following the demonstration by Watt (1966) that the percentage extracellular release (PER) was inversely proportional to phytoplankton biomass and, on the basis of reported PER values in a variety of habitats (Anderson and
Zeutschel 1970; Thomas 1971; Saunders 1972a), it has been
suggested that, in general, oligotrophic waters are characterized by higher PER values than eutrophic ones. It
is important to note, however, that the amounts of carbon
excreted and the rates of excretion are far more useful
parameters than PER. For example, an increase in PER values may be caused either by an increase in excretion rate
or by a decrease in fixation by the particulate fraction,
excretion remaining unchanged. In this context, amounts
of carbon excreted in eutrophic lakes are much greater
than in oligotrophic lakes. An approximate calculation,
based on annual primary production estimates of 75-250 g
C/m^2/year (Rodhe 1969), and maximum excretion of 8%, gives
6-20 g C/m^2/year of extracellular carbon production in
eutrophic lakes. In oligotrophic lakes with primary production of 7-25 g C/m^2/year and 35% maximum PER, extracellular release amounts to 2.5-8.8 g C/m^2/year. These are,

of course, approximations that do not take into account
seasonal fluctuations of excretion and are based on excretion data in a small number of lakes; however, the amounts
of dissolved organic carbon derived from algae appear to
be considerable.

Data on the chemical composition of algal extracellular products from short-term ^{14}C primary production experiments are scarce. Watt (1966) reported ^{14}C-labeled glycollate, amino acids, and polysaccharides in different
amounts, depending on the species of algae present. Both
low and high molecular weight compounds were excreted by
phytoplankton in a number of lakes during 1-3 hr incubation
with ^{14}C (Nalewajko and Lean 1972; Nalewajko and Schindler
1976), but high molecular weight compounds were generally
predominant. As will be discussed later, the possibility
exists that as a result of rapid and selective utilization
by heterotrophs, only "net" extracellular products are being detected and even in short-term experiments these substances may differ substantially in quality and quantity
from the original "gross" extracellular products.

EXTRACELLULAR PRODUCTS OF FRESHWATER BACTERIA

The release of very large molecules by fungi and bacteria appears to be accepted as a commonplace phenomenon
and does not attract as much comment as extracellular release in algae. However, much of the information on extracellular products of fungi and bacteria has been obtained
from clincially important (intestinal and pathogenic types)
or from commercially useful species, and very little is
known about the process in freshwater bacterioplankton.

Various antibiotics are perhaps the best-known extracellular products of fungi and bacteria, but their mode of
origin, as extracellular products, is often not stressed;
e.g., in commercial scale production, the growth medium

of *Penicillium notatum* is used as the main source of penicillin. Other examples are the excretion of streptomycin by *Streptomyces griseus*, and of the polypeptide polymyxin by *Bacillus polymyxa*. These antibiotics are produced in highly artificial cultures. Under natural conditions, the fungus *Trichoderma* has been shown to produce extracellular antibiotics in soils (Brian 1951).

Many species of bacteria excrete enzymes; e.g., some of the changes produced by bacteria acting on carbohydrates (in fermentation), and on proteins (in putrefaction), are brought about by extracellular enzymes such as α- and β-amylases and gelatinase (Skerman 1967; Thornley 1967). In fact, many saprophytic and parasitic organisms have the ability to digest, by means of extracellular enzymes, particles of organic matter too large to enter the cell. Saprophytic fungi excrete cellulases and amylases, while fungi parasitic on higher plants excrete pectinase, a complex of enzymes some of which hydrolyze the calcium pectate of the middle lamella. The production of alkaline phosphatases that, along with other phosphatases, may be extremely important in the utilization of organic sources of phosphorus has been shown in various bacteria (Hodgson and Hopton 1968; Jones 1972). The location of the enzyme close to the cell wall favors some release into the medium (Glew and Heath 1971). The presence of free alkaline phosphatases in lakewaters (Reichardt *et al.* 1967; Berman 1970) could be attributed to release by bacteria, although some freshwater algae also release the enzyme (Healey 1973; Nalewajko, unpublished data).

Reports of bacterial excretion of other macromolecules, such as DNA, RNA, proteins, and polysaccharides, are common (Wilkinson 1958; Cripps and Work 1967). However, the strict definition of extracellular products is often not followed and the term "extracellular" is used

597

simply to denote a substance found outside the cell regardless of whether it originated from healthy cells or
by cell lysis, and without distinction as to its state,
i.e., whether particulate or dissolved. The mechanism of
excretion of these macromolecules is not always obvious.
In *Bacillus subtilis*, RNA, DNA, and protein excretion occurred in the absence of cell lysis (Demain *et al*. 1965).
However, in *Staphylococcus aureus*, the accumulation of
"extracellular" RNA and DNA was accompanied by gross cell
lysis, while protein was released by intact healthy cells
(Cripps and Work 1967). Capsule formation and excretion
of polysaccharides by bacteria appear to be widespread.
While intracellular polysaccharides (as cytoplasmic gran-
ules and in solution inside the cell) may amount to as
much as 60% of the bacterial weight, "extracellular" polysaccharide (defined as capsule and loose, i.e., dissolved
slime) may exceed the bacterial biomass (Wilkinson 1958).
Only the latter, dissolved fraction would be referred to
as extracellular in algal literature. Polysaccharide synthesis and excretion are increased by deficiency of nitrogen, sulfur, and phosphorus but are decreased by potassium
deficiency and by shortage of the organic carbon source.
The type of carbon source supplied is also important. For
example, polysaccharide production in *E. coli* is much higher with glucose than with acetate as the substrate. The
quantity of extracellular products released is greater
under anaerobic conditions than under aerobic and different
substances are excreted (Wilkinson 1958).

　　Much of the interest in extracellular polysaccharides
of bacteria stems from their involvement in determining
virulence and in imparting antigenic properties to cell
surfaces (Wilkinson 1958). The possible role of extracellular macromolecules in aggregation of bacterial cells,
yeast cells, and of bacteria and inorganic colloids in

reactions resembling the antigen-antibody systems has also received attention (Harris and Mitchell 1973).

The excretion of lower molecular weight substances also occurs in fungi and bacteria. For example, various alcohols and organic acids accumulate in the medium during fermentation of carbohydrates and other substrates, and acetic acid is produced by *Acetobacter* during aerobic oxidation of ethanol. In general, these substances appear to be end products of metabolism.

Some of the bacterial extracellular products are derived from processes common to many genera, including bacterioplankton. Bacterial types most commonly isolated from lakes include aerobic spore formers, gram-negative rods, and chromogenic organisms (Gorden *et al.* 1969). Capsules are not common; however, microcapsules or sheaths occur in some species, and the solution of polymers, such as polysaccharides, polypeptides, nitrogen-containing sugars, and sugar acids may take place. Several bacterial isolates from the genera *Flavobacterium, Xanthomonas, Pseudomonas,* and *Bacillus* were found to excrete thiamin (Gorden *et al.* 1969) that stimulated the growth of *Chlorella* in a model aquatic system. Extracellular vitamin B_{12} production in a variety of marine bacteria, isolated from water and sediments, could satisfy the B_{12} requirement of marine diatoms in culture (Haines and Guillard 1974). The bacteria were heterotrophic and could utilize diatom excretory products or the remains of dead diatom cells in the production of the vitamin. From the observed effects of bacteria or their filtrates on the growth of other bacteria and on algae, the production of growth-promoting and growth-inhibiting substances may be inferred but the chemical composition of these is far from clear (Trainor 1965; Gorden *et al.* 1969). Extracellular growth substances could be involved in the apparent symbiotic relationships where algal growth is improved in bacterized as opposed to axenic culture (Trainor 1965).

On the basis of numerous observations, extracellular release in microorganisms may be assumed to be a general phenomenon. Yet, compared to the wealth of data on the relative amounts of carbon accumulated in photosynthesis and excreted by algae, similar data on the proportion of organic substrates assimilated, respired, and excreted by heterotrophs are virtually nonexistent. Wetzel and Manny (1972) give data on excretion of labeled dissolved organic compounds by bacteria, but neither the origin of the bacteria nor the substrates utilized is described.

Some recent direct evidence for excretion in bacterioplankton has been obtained using mixed populations isolated from Lake Ontario and grown in Chu 10 medium with glycollate as a carbon source (Nalewajko and Lean 1972). Over a 48-hr period, the bacteria consumed glycollate ^{14}C at the rate of 3.47 µg/min. Most of the glycollate was respired. Sephadex gel fractionation of filtrates indicated that three substances were excreted. The major extracellular product, accounting for about a half of the total carbon excreted, was a high molecular weight substance (MW > 2500). A substance in the molecular weight range 2500-180 was the next most abundant extracellular product, with a low molecular weight substance ($\leq$ 180) accounting for about 15% of the extracellular carbon. Excretion as a percentage of the total carbon utilized by the bacteria was only 0.51%, however, because respiration losses were high, excretion as a percentage of carbon retained by cells was 13.8%.

In pure culture, *Pseudomonas fluorescens*, a species common in lakewater, utilized glucose, fructose, and sucrose. Rates of uptake were 0.84, 0.08, and 0.01 µg/liter/hr, respectively, from concentrations of about 35 µg/liter.

Extracellular products of a polysaccharide nature were released and consisted of a high molecular weight substance
($\geq$ 5000) and a lower molecular weight fraction. With glucose as the substrate, extracellular products were detected
within 1 hr of incubation with [^{14}C]-glucose and amounted
to 3.7% of ^{14}C label in the cells. Of the total [^{14}C]-
glucose taken up, 67.1% was lost in respiration. Expressed as a percentage of the total ^{14}C assimilated instead of
as a percentage of ^{14}C retained in cells, extracellular
release was 1.3%. Both types of extracellular products
accumulated throughout the duration of the experiment.
After 48 hr, extracellular release was 29.1% (13.7% corrected for respiration). The composition of extracellular
substances and unused glucose as revealed by Sephadex gel
filtration is shown in Figure 18.1. Similar extracellular
substances were released by *Pseudomonas* when utilizing
fructose and sucrose but the utilization rates were much
slower and extracellular substances were not detected in
the filtrates until after about 9 and 24 hr, respectively.
Percent extracellular release values after 48 hr were 2.3%
and 1.4% with fructose and sucrose as the respective substrates (Dunstall 1974; Dunstall and Nalewajko 1975).

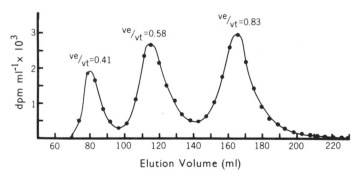

FIG. 18.1. Sephadex gel fractionation of *Pseudomonas* medium after
48 hr growth on ^{14}C-glucose. The peak at V_e/V_t 0.83 represents unused
glucose. Peaks at 0.58 and 0.41 are extracellular products.

The excretion of high molecular weight compounds seems to be a more general phenomenon than excretion of low molecular weight substances, however, the loss of organic matter in short-term experiments at low substrate concentrations does not constitute a significant proportion of the carbon assimilated by the bacteria. If short-term extracellular release by bacteria is indeed small, it need not be taken into account in kinetics experiments of 1-2 hr duration. However, the rapid uptake of small molecules by bacteria combined with some excretion of high molecular weight compounds could modify significantly the composition of DOM, both in contaminated algal cultures and in natural waters.

EXTRACELLULAR PRODUCTS AS A CARBON SOURCE FOR HETEROTROPHS

Among the numerous and varied functions attributed to DOM in general, and to extracellular products in particular (Lucas 1947; Fogg and Westlake 1955; Saunders 1957; Fogg 1966; Whittaker and Feeny 1971), is the potential significance as a carbon source for heterotrophs. Some recent evidence for the utilization of algal extracellular products will be presented here, with special reference to planktonic organisms both in cultures and under natural conditions.

Reutilization of extracellular substances by the algae themselves may occur, particularly in culture (Fogg and Pattnaik 1966; Watt and Fogg 1966). Theoretically at least, extracellular substances liberated by outward diffusion along a concentration gradient may become reabsorbed if the gradient changes to favor inward diffusion into the cells. Several reports of glycollate affecting growth of algae (Sen and Fogg 1966; Droop and McGill 1966) and of glycollate uptake by phytoplankton (Watt 1966) were based on experiments carried out at high substrate concentrations that would favor inward diffusion. Glycollate concentrations of several mg/liter may be reached in culture experiments, at high population densities (Watt and Fogg 1966; Smith 1974), but concentrations in lakes are much lower and at times undetectable (Wright 1968; Fogg *et al.* 1969). Active transport systems for glycollate and other simple organic compounds are widespread in aquatic bacteria but apparently rare in algae (Hobbie and Wright 1965), although evidence is accumulating that they are more widespread in the latter group than initially suggested (Hellebust and Guillard 1967; Hellebust 1971; Lord and Merrett 1971; Sheath and Hellebust 1974; Saunders 1972b). In one algal species, however, the active transport system for glucose was characterized by a half-saturation constant of 5 mg/ liter (Bennett and Hobbie 1972), which makes it unlikely to be effective at the low concentrations encountered in lakes.

In general, although utilization of low molecular weight algal extracellular products by the organism itself, and by other algae, is possible (Jones and Stewart 1969) more evidence is available at present to suggest that utilization by bacteria and fungi is more likely to predominate. Utilization by fungi and bacteria of high molecular weight type II algal extracellular products is also likely to occur. For example, Khailov (1968) reported hydrolysis of extracellular polysaccharides of the alga *Prymnesium parvum* by bacteria, and Gocke (1970) found rapid decomposition of extracellular polypeptides, but not of colloidal nitrogen compounds, by bacteria.

BACTERIAL UTILIZATION OF EXTRACELLULAR PRODUCTS BY ALGAE: EVIDENCE FROM CULTURES

Much evidence has accumulated showing utilization of algal extracellular products in cultures. Vela and Guerra

(1966) showed that several types of bacteria were able to
grow in cultures of *Chlorella pyrenoidosa* at the expense
of algal extracellular products. The growth curves of
Chlorella and a gram-negative bacterium, tentatively identified as a *Flavobacterium* paralleled each other and
reached stationary phase on about the fourth day of growth.
Utilization of oxidizable organic substrates by various
bacteria (including *Bacterium anitratum* and *Mima polymor-
pha)* modified considerably the composition of extracellular
products. A similar selectivity in the utilization of
extracellular products of three species of *Chlorella* by
bacteria has been reported by Maksimova and Pimenova
(1969). A preference was shown for sugars and volatile
organic acids by *Flavobacterium diffusum*, for polysaccharides by *Pseudomonas pyocyanea*, and for keto acids by both
species and by *Bacillus cereus*.

Gocke (1970) showed that bacteria introduced into
axenic *Scenedesmus quadricauda* cultures grew rapidly and
selectively consumed the dissolved amino acids. As a percentage of the total amino acids in the filtrate, serine
and glycine increased, while aspartic and especially glutamic acids decreased, as a result of bacterial activity.
In general, extracellular free amino acids and peptides
were attached more strongly by bacteria than ninhydrin-
positive colloids.

It appears that filtrates from many freshwater algae
may selectively support growth of certain bacteria. Of
eight species of bacteria (all gram negative and all but
one nonmotile) isolated from Lake Ontario, three showed
appreciable growth after 4 days in exponential phase growth
filtrates of *Navicula pelliculosa, Chlorella pyrenoidosa,
Scenedesmus basiliensis,* and *Anabaena flos-aquae*. One
species grew only on *Navicula* and another only on *Anabaena*
filtrate (Nalewajko *et al.* 1976; Dunstall 1974). The

extracellular products of the four species of algae differed significantly. While *Chlorella* excreted glycollate, the other three species excreted a large proportion of the carbon as high molecular weight compounds (Nalewajko and Lean 1972). These are likely to include glycollate and other low molecular weight substances as well as polysaccharides in *Scenedesmus* (Weinmann, in Hellebust 1974) and in *Navicula* (Lewin 1955). Polysaccharides (Moore and Tischer 1964), amides and polypeptides (Fogg 1952) may be present in *Anabaena* filtrates. Carbon from algal extracellular compounds is both assimilated and respired by bacteria (Table 18.1) (Dunstall and Nalewajko 1975). The proportion respired may be very high. For example, 93.3% of the extracellular carbon from *Chlorella* (presumably glycollate) was respired.

As well as modifying the apparent composition of algal extracellular products by selective utilization, bacteria may contribute their own extracellular products, resulting in drastic changes in the composition of the DOM (Nalewajko and Lean 1972). Mixed populations of bacteria isolated from Lake Ontario, when introduced into exponential-phase *Chlorella* cultures growing in the presence of $[^{14}C]$-bicarbonate, rapidly consumed the low molecular weight ^{14}C-labeled extracellular product (presumably glycollate) (Fig. 18.2). The high molecular weight substance predominant in the contaminated but not in the axenic cultures is most likely a bacterial extracellular product.

It is likely that kinetics of extracellular release in nonaxenic cultures of algae are affected by the activity of the bacteria. In axenic *Chlorella pyrenoidosa* a linear increase of $[^{14}C]$DOM in the filtrates was observed. In a parallel experiment, where bacteria capable of utilizing glycollate were added, excretion was also linear over 2 hr, but the rate was significantly lower (Fig. 18.3).

Table 18.1

Uptake and Metabolism of ^{14}C Algal Extracellular Products of Bacteria

| | Sources of extracellular products | | | | | | | |
| | Scenedesmus | | Chlorella | | Anabaena | | Navicula | |
	(a)	(b)	(a)	(b)	(a)	(b)	(a)	(b)
Original carbon (µg/liter)	33.00	8.00	98.00	17.00	24.00	25.00	20.00	
µg carbon in cells	4.82	0.53	3.92	0.73	3.60	2.53	2.36	
µg carbon re-spired	0.49	0.69	0	10.17	2.18	0.50	1.90	
Rate of uptake (first 4 hr) (µg C/liter/hr)	0.82	0.20	0.50	0.14	0.42	0.48	0.52	
Turnover time of DOC pool (hr)	40.00	40.00	196.00	121.00	58.00	52.00	39.00	

[a]Bacterial isolate B, 28-hr incubation.

[b]_Pseudomonas fluorescens_, 25-hr incubation.

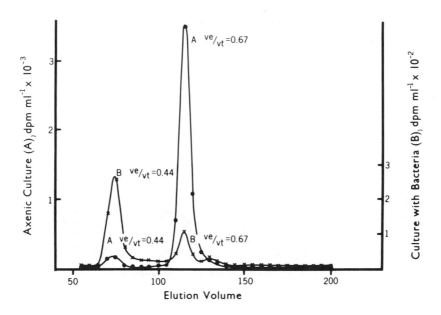

FIG. 18.2. Sephadex gel fractionation of growth medium of 11 day old axenic *Chlorella* (A —— ·) and in a culture of same age 5 days after addition of bacteria (B —— x).

Bacteria consumed the extracellular products at the rate of 8.4 µg C/liter/hr. These data suggest that if bacteria preadapted to algal extracellular products are present in *Chlorella* cultures, utilization of low molecular weight substances is almost simultaneous with excretion. The consumption by bacteria results in much lower PER values in contaminated cultures as compared to axenic ones. Here the values were 5.8% and 8.1%, respectively. Carbon-14 algal excretion products utilized by bacteria may be respired and lost as $^{14}CO_2$ or retained by the cells and estimated in the particulate fraction. Regardless of the proportions respired and retained, algal extracellular release rates will be underestimated. If the proportion retained by bacteria is significant, radioactivity in the

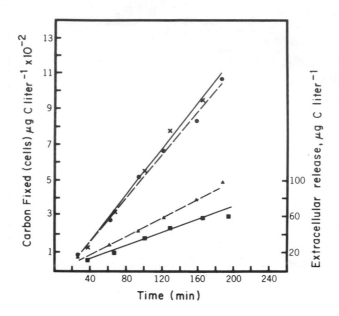

FIG. 18.3. Kinetics of carbon fixation and extracellular release in axenic *Chlorella* and in a culture containing bacteria. Carbon fixed: axenic culture (- - -●), culture with bacteria (—— x). Extracellular release: - - -▲, axenic culture; —— ■ with bacteria.

particulate fraction in the mixed cultures will be higher than in the axenic ones, as observed in this experiment.

Similar data, demonstrating bacterial consumption of glycollate excreted by the marine planktonic diatom *Chaetoceros socialis*, have been obtained by Smith (1974). During a 1-hr experiment, a linear increase in glycollate concentration was observed in axenic cultures, but an equilibrium level was reached within about 20 min in nonaxenic cultures. It is not known whether glycollate is the only substance released by this species but it appears unlikely from reports of extracellular products of *Chaetoceros pelagicus* and other diatoms (Hellebust 1965; Watt 1969;

Watt and Fogg 1966). The equilibrium kinetics could apply to low molecular weight compounds, while high molecular weight compounds accumulate in a linear fashion. The time at which equilibrium is reached probably depends both on the rates of excretion by the alga and on consumption by bacteria. In our experiment, rates of excretion were high and equilibrium was not reached within 3 hr. This probably was a result of low bacterial biomass or metabolic rate.

In cultures of algae excreting both low and high molecular weight substances, the kinetics of extracellular release are likely to be more complex but, again, the presence of bacteria may modify the kinetics significantly. Gocke (1970) reported that in axenic *Scenedesmus quadricauda* cultures excreting amino acids, peptides, and colloidal nitrogen compounds, PER values increased linearly from about 4% to 9% over a period of 35 days. In contaminated cultures, values increased from about 3% to a plateau at about 5% and then declined. The presence of bacteria may also modify the kinetics of algal extracellular release over shorter time periods. Axenic exponential phase *Anabaena flos-aquae* excretes two large molecular weight compounds and one lower molecular weight compound (Nalewajko and Lean 1972) in short-term experiments with ^{14}C. In other experiments (Nalewajko *et al.* 1976), the rate of photosynthesis remained constant for about 8 hr in both axenic cultures and in cultures to which mixed bacteria preadapted to grow on extracellular products of *Anabaena* were added (Fig. 18.4). Photosynthetic rates were somewhat lower in the contaminated culture. Excretion rates were constant for about 160 min and then changed to a higher rate. During the earlier period, the difference in excretion rate between axenic and contaminated cultures was slight, 0.14 and 0.11 µg C/liter/hr, respectively, indicating a consumption rate of 0.03 µg C/liter/hr. In

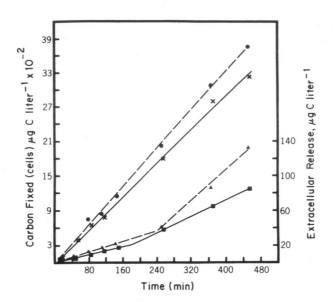

FIG. 18.4. Kinetics of carbon fixation and extracellular release in
axenic *Anabaena* and in a culture containing bacteria. Carbon fixed:
axenic culture (- - -●); culture with bacteria (——— x). Extracellular
release: - - -▲, axenic culture; ——— ■ culture with bacteria.

the subsequent period (ca. 160–480 min), extracellular release in the axenic culture was 0.455 µg C/liter and 0.232
in the contaminated culture; extracellular products therefore were being consumed at a much higher rate (0.233 µg
C/liter/hr) during the second phase. Final PER values
were 3.4% in the axenic and 2.6% in the contaminated cultures. The increased excretion rate after 160 min could
represent the appearance of a new substance, such as a
metabolic end product labeled late in the metabolic sequence.

Other reports of utilization of algal extracellular
products include Walsby (1965), who found that *Penicillium
frequentans*, a frequent contaminant of *Anabaena cylindrica*

cultures, was growing at the expense of the alga. Monahan and Trainor (1970) found that filtrates of *Hormotila blennista* supported the growth of bacteria and stimulated or inhibited two planktonic algae. However, Silvey and Wyatt (1971) found that contaminating bacteria did not thrive in cultures of three *Anabaena* species and concluded that extracellular algal production was not of sufficient quantity and composition to permit extensive bacterial development. Production by the algae of substances inhibitory to the bacteria was not ruled out. Waite (1973) reported that long-term (24-hr) extracellular products from axenic algae were not as readily respired by lakewater populations as water-soluble and water-insoluble algal cell extracts.

UTILIZATION OF ALGAL EXTRACELLULAR PRODUCTS IN SITU

Concurrent observations of the development of heterotrophic populations, and of primary producers, may reveal the probable origin of the DOM needed by the former. If healthy, exponentially growing algal populations are supplying readily assimilable DOM, the heterotrophs will show most active growth at the same time as the primary producers. Such parallel development has been reported in some lakes (Fonden 1969; Schegg 1968). However, instances where bacterial numbers increase after the collapse of algal populations, presumably at the expense of DOM from dead cells, are also frequent (Drabkova 1965; Jones 1971). In fact, Potaenko and Mikheva (1969) have shown that extremely diverse relationships between the algal and bacterial populations may be found. Similarly, Saunders (1971) observed no obvious direct correlations among phytoplankton, DOM, and bacteria. Allochthonous rather than autochthonous DOM may be responsible for supporting high bacterial production rates (Kuznetsov 1968).

Evidence for utilization of phytoplankton extracellular products by bacteria in lakewater is mainly indirect. Bauld and Brock (1974) have demonstrated the transfer of carbon from the benthic blue-green alga *Synechococcus lividus* to the bacterial mat (mainly *Chloroflexis*) underneath via extracellular products. However, in planktonic systems this close spatial arrangement between primary producer and heterotroph may not be common. At least in cultures, healthy algae are seldom used as points for attachment by bacteria, but dying cells may be heavily covered.

Indirect evidence for the importance of glycollate as a carbon source for bacteria has been presented by Wright (1968). In depth profiles in Gravel Lake, the maximum velocity (V_{max} from Michaelis-Menten analysis of uptake kinetics) of glycollate uptake was of the same magnitude as total carbon excreted by primary producers. However, as noted earlier, glycollate is seldom the only extracellular product. On the assumption that no more than one-third of the carbon is excreted as glycollate, the V_{max} would significantly overestimate heterotrophic glycollate uptake.

Saunders and Storch (1971) have shown that bacteria could be extremely dependent on carbon from algal extracellular products. When glucose was supplied at hourly intervals, at rates simulating the observed diurnal patterns of algal excretion, bacterial metabolism was found to oscillate in a similar pattern but with some delay. Such a delay would be expected if substrate (that is, extracellular product) concentrations increased and then decreased during the day, as was demonstrated in Frains Lake. Although glucose is not a common algal excretion product and there is little evidence that low molecular weight substances accumulate to the same extent as in these experiments, the possibility that bacterial metabolic periodicity

may relate to that of phytoplankton is extremely interesting. Positive correlations observed between bacteria and
algae in depth profiles in some lakes (Sorokin 1968; Schmidt
1969) could be interpreted as resulting at least partly
from this dependence of bacteria on a light-produced DOC
pool of algal extracellular products.

A test of the hypothesis that extracellular products
are rapidly removed by bacteria has been carried out by
Anderson and Zeutschel (1970). In their experiments, addition of 50 or 100 mg/liter penicillin and streptomycin
did not affect primary production or excretion rates in
natural seawater. Further work with more subtle control
of the bacterioplankton is needed to investigate this problem. At present, the possibility remains that measured
rates of extracellular release are net values, detecting
only the substances not immediately utilized by bacteria
or removed by absorption (Jones and Stewart 1969; Hobbie
1971; Saunders 1972a). Theoretically, it should be possible to estimate gross or total excretion rates by simultaneous measurements of extracellular release and heterotrophic uptake of substances known to be excreted by the
particular phytoplankton assemblage. Instead of using
such parameters as V_{max}, natural concentrations of these
substances (S_n) should be measured and uptake rates at
these concentrations determined.

A preliminary attempt to estimate gross excretion has
been made in two lakes, L. 239 and L. 227, in the Experimental Lakes Area, northwestern Ontario (Nalewajko and
Schindler 1976). Natural concentrations of individual organic substances in general are extremely low in freshwaters, seldom exceeding a few µg/liter. The lack of suitable chemical methods for measuring such low concentrations, combined with the need to estimate bacterial heterotrophy, was a stimulus to the development of several

613

indirect means of measuring S_n. Of these, the dilution
bioassay (Allen 1968) proved unreliable in the estimation
of glucose and acetate in L. 227 and L. 239. However, the
tracer bioassay (Rigler 1966) developed to measure orthophosphate concentrations was adapted to organic substances.
The values obtained by this method are maximal and not
necessarily actual concentrations. Glucose and acetate
concentrations appeared to be very low, not exceeding 0.5
µg/liter. Turnover times, calculated as the reciprocal
of the rate constant of uptake at natural concentrations,
ranged from 1.8 to 7.4 hr. In L. 227 (August 10, 1973)
glucose and acetate uptake rates, corrected for respiration, were 0.6 and 0.05 µg C/liter, respectively. It was
not possible to estimate natural glycollate concentrations;
however, at 1.6 µg/liter of glycollate, uptake rates, corrected for respiration, were 0.11 µg C/liter/hr. Extracellular release was 1.92 µg C/liter/hr and amounted to
2.7% of the total carbon fixed. Gross extracellular release, estimated as the sum of the measured net release
rate and the rates of glucose, acetate, and glycollate
was therefore at least 2.68 µg C/liter/hr or 3.5% of the
total carbon fixed.

In L. 239 (August 7, 1973) uptake of acetate was 0.012
µg C/liter/hr at a natural substrate concentration of 0.2
µg/liter. No estimates of natural glycollate concentrations were made; however, at an assumed concentration of
0.9 µg/liter of glycollate, the rate of uptake was 0.03
µg C/liter/hr, corrected for respiration. Net extracellular release was 0.53 µg C/liter/hr. Net and gross percentage extracellular release values were 13.1% and 13.9%,
respectively. If other organic compounds had been tested,
total (gross) excretion values undoubtedly would have been
higher. Ideally, uptake rates of compounds known to be
excreted by phytoplankton should be measured. However,

this was not possible in this preliminary study. More
data are needed on the extracellular products of common
planktonic algae in axenic culture before the compounds
to be tested can be selected and their concentrations measured.

At present, then, it is not possible to estimate what
proportion of phytoplankton excretion products consists
of small molecules readily available to heterotrophs. Hy-
pothetically, preferential utilization of low molecular
weight extracellular products by heterotrophs could account
for the low amounts of these substances observed even in
short-term primary production experiments. Analysis of
filtrates from 1-4 hr ^{14}C photosynthesis experiments in
a variety of lakes, and at various seasons, revealed a
predominance of large molecules (Watt 1966) and colloids
(Nalewajko and Lean 1972; Nalewajko and Schindler 1976).
Ninhydrin-positive colloids excreted by *Scenedesmus quadricauda* are resistant to bacterial decomposition (Gocke 1970).
Accumulation of colloids in mixed algal-bacterial cultures
during growth indicates that they are not a readily util-
izable carbon source (Nalewajko and Lean 1972). In lakewater, some binding of low molecular weight phosphorylated
extracellular substances to colloids may occur, with a
simultaneous release of phosphate (Lean 1973). A similar
phosphorus mediated polycondensation appears to take place
in axenic cultures of algae (Lean and Nalewajko 1976); however, it is not possible to estimate the relative amounts
of colloids excreted directly by algae and formed as a result of the polycondensation. In a sense, heterotrophs
may be competing for low molecular weight excretion products with abiotic polycondensation and other processes,
which remove DOC into a less available, colloidal form.
High molecular weight organic substances also may be bound
to each other (Harris and Mitchell 1973) to colloids or

to detritus. As such, they may be included into food chains (Khailov and Finenko 1968) or may be removed from lakewater by sedimentation.

The sources of organic carbon available to bacteria are not only varied but may change seasonally. The relative significance of these sources would therefore be very difficult to assess. Algal extracellular products, according to Saunders (1972c), are a less important source of DOC than organic detritus. This conclusion is based on data from a small eutrophic lake (Frains Lake), where in May and June rates of DOC release by artificial organic detritus exceeded excretion from living phytoplankton by two to six times. Since, in addition, organic detritus ranged from five to 10 times the phytoplankton biomass, it may represent the larger reservoir of DOC. However, at times when phytoplankton exceeds detritus biomass, as, for example, during blooms, the direct DOC path from living algae to bacteria could be relatively more important.

ACKNOWLEDGMENTS

I would like to express my thanks to Mr. T. Dunstall for helpful discussions and to Dr. B. Parker and Dr. G. W. Saunders for a critical reading of the chapter.

REFERENCES

Allen, H. L. 1968. Acetate in fresh water: natural substrate concentrations determined by dilution bioassay. *Ecology 49*, 346-349.

Anderson, G. C., and Zeutschel, R. P. 1970. Release of dissolved organic matter by marine phytoplankton in coastal and offshore areas of the Northeast Pacific Ocean. *Limnol. Oceanog. 15*, 402-407.

Bauld, J., and Brock, T. D. 1974. Algal excretion and bacterial assimilation in hot spring algal mats. *J. Phycol. 10*, 101-106.

Bennett, M. E., and Hobbie, J. E. 1972. The uptake of glucose by *Chlamydomonas* sp. *J. Phycol. 8*, 392-398.

Berman, T. 1970. Alkaline phosphatases and phosphorus availability in Lake Kinneret. *Limnol. Oceanog. 15*, 663-674.

Brian, P. W. 1951. Antibiotics produced by fungi. *Bot. Rev. 17*, 357-430.

Cripps, R. E., and Work, E. 1967. The accumulation of extracellular macromolecules by *Straphylococcus aureus* grown in the presence of sodium chloride and glucose. *J. Gen. Microbiol. 49*, 127-137.

Demain, A. L., Burg, R. W., and Hendlin, D. 1965. Excretion and degradation of ribonucleic acid by *Bacillus subtilis*. *J. Bacteriol. 89*, 640-646.

Drabkova, V. G. 1965. Dynamics of the bacterial number, generation time and production of bacteria in the water of red lake. *Mikrobiologija 34*, 933-938.

Droop, M. R., and McGill, S. 1966. The carbon nutrition of some algae: the inability to utilize glycollic acid for growth. *J. Mar. Biol. Assoc. U.K. 46*, 679-684.

Dunstall, T. G. 1974. Extracellular release in freshwater bacteria: heterotrophic relationships with algae. M.S. Thesis, University of Toronto.

Dunstall, T. G., and Nalewajko, C. 1975. Extracellular release in planktonic bacteria. *Ver. Int. Verein. Limnol. 19*, 2643-2649.

Fogg, G. E. 1952. The production of extracellular nitrogenous substances by a blue-green alga. *Proc. Roy.
Soc. B, 139*, 372-397.

Fogg, G. E. 1958. Extracellular products of phytoplankton
and the estimation of primary production. *Rapp. Cons.
Explor. Mer. 144*, 56-60.

Fogg, G. E. 1966. The extracellular products of algae.
Oceanog. Mar. Biol. Ann. Rev. 4, 195-212.

Fogg, G. E. 1971. Extracellular products of algae in
freshwater. *Arch. Hydrobiol. Beih. 5*, 1-25.

Fogg, G. E., Eagle, D. J., and Kinson, M. E. 1969. The
occurrence of glycollic acid in natural waters. *Ver.
Int. Verein. Limnol. 17*, 480-484.

Fogg, G. E., and Horne, A. J. 1970. The physiology of
antarctic freshwater algae. In *Antarctic Ecology*,
M. W. Holdgate (Ed.). Vol. II. Academic, London and
New York, pp. 632-638.

Fogg, G. E., Nalewajko, C., and Watt, W. D. 1965. Extracellular products of phytoplankton photosynthesis.
Proc. Roy. Soc. Lond. B, 162, 517-534.

Fogg, G. E., and Pattnaik, H. 1966. The release of extracellular nitrogenous products by *Westiellopsis pro-
lifica* Janet. *Phykos 5*, 58-67.

Fogg, G. E., and Westlake, D. E. 1955. The importance
of extracellular products of algae in freshwater. *Ver.
Int. Verein. Limnol. 12*, 219-232.

Fonden, R. 1969. Heterotrophic bacteria in Lake Malaren
and Lake Hjalmaren. *Oikos 20*, 344-372.

Glew, R. H., and Heath, E. C. 1971. Studies on the extracellular alkaline phosphatase of *Micrococcus sodonensis*.
I. Isolation and characterization. *J. Biol. Chem.
246*, 1556-1565.

Gocke, K. 1970. Untersuchungen über Abgabe und Aufnahme
von Aminosäuren und Polypeptiden durch Planktonorganis-
men. *Arch. Hydrobiol. 67*, 285-367.

Gorden, R. W., Beyers, R. J., Odum, E. P., and Eagon, R. G.
1969. Studies of a simple laboratory microecosystem:
bacterial activities in a heterotrophic succession.
Ecology 50, 86-100.

Haines, K. C., and Guillard, R. R. L. 1974. Growth of vitamin B_{12}—requiring marine diatoms in mixed laboratory cultures with vitamin B_{12}—producing marine bacteria. *J. Phycol. 10*, 245-252.

Harris, R. H., and Mitchell, R. 1973. The role of polymers in microbial aggregation. *Ann. Rev. Microbiol. 27*, 27-50.

Healey, P. 1973. Characteristics of phosphorus deficiency in *Anabaena. J. Phycol. 9*, 383-394.

Hellebust, J. A. 1965. Excretion of some organic compounds by marine phytoplankton. *Limnol. Oceanog. 10*, 192-206.

Hellebust, J. A. 1971. Kinetics of glucose transport and growth of *Cyclotella cryptica. J. Phycol. 1*, 1-4.

Hellebust, J. A. 1974. Extracellular products. In *Algal Physiology and Biochemistry*, W. D. P. Stewart (Ed.). B. H. Blackwell, Oxford, England, pp. 838-863.

Hellebust, J. A., and Guillard, R. L. 1967. Uptake specificity for organic substrates by the marine diatom *Melosira nummuloides. J. Phycol. 3*, 132-136.

Hobbie, J. E. 1971. Heterotrophic bacteria in aquatic ecosystems: Some results of studies with organic radio-isotopes. In *The Structure and Function of Freshwater Microbial Communities*, J. Cairns, Jr. (Ed.). Res. Div. Monogr. 3, Virginia Polytechnic Institute and State University, Blacksburg, Va., pp. 181-194.

Hobbie, J. E., and Wright, R. T. 1965. Competition between planktonic bacteria and algae for organic solutes. *Mem. Inst. Ital. Idiobiol. 18* Suppl., 175-185.

Hodgson, G. W., and Hopton, J. W. 1968. Phosphatase activity in streamborne bacteria. *J. Appl. Bacteriol. 31*(i).

Jones, J. G. 1971. Studies on freshwater bacteria: factors which influence the population and its activity. *J. Ecol. 59*, 593-613.

Jones, J. G. 1972. Studies on freshwater micro-organisms: phosphatase activity in lakes of differing degrees of eutrophication. *J. Ecol. 60*, 777-791.

Jones, K., and Stewart, W. D. P. 1969. Nitrogen turnover in marine and brackish habitats. IV. Uptake of the extracellular products of the nitrogen-fixing alga *Colothrix scopulorum*. *J. Mar. Biol. Assoc. U.K. 49*, 701-716.

Khailov, K. M. 1968. Extracellular microbial hydrolysis of polysaccharides dissolved in sea water. *Microbiology 37*, 424-427.

Khailov, K. M., and Finenko, A. A. 1968. Organic macromolecular compounds dissolved in seawater and their inclusion into food chains. *Proc. Symp. Mar. Food Chains*, Denmark, June 18.

Kuznetsov, S. I. 1968. Recent studies on the role of microorganisms in the cycling of substances in lakes. *Limnol. Oceanog. 13*, 211-224.

Lean, D. R. S. 1973. Movement of phosphorus between its biologically important forms in lake water. *J. Fish. Res. Bd. Can. 30*, 1525-1536.

Lean, D. R. S., and Nalewajko, C. 1976. Phosphorus uptake and excretion by freshwater algae. *J. Fish. Res. Bd. Can. 33*(6):1312-1323.

Lewin, J. C. 1955. The capsule of the diatom *Navicula pelliculosa*. *J. Gen. Microbiol. 13*, 162-169.

Lewin, R. A. 1956. Extracellular polysaccharides of green algae. *Can. J. Microbiol. 2*, 665-672.

Lord, M. J., and Merrett, M. J. 1971. The growth of *Chlorella pyrenoidosa* on glycollate. *J. Exp. Bot. 22*, 60-69.

Lucas, G. E. 1947. The ecological effects of external metabolites. *Biol. Rev. 22*, 270-295.

Maksimova, I. V., and Pimenova, M. N. 1969. Influence of concomitant microflora on accumulation of organic compounds in the medium during non-sterile culturing of *Chlorella*. *Microbiology 38*, 509-513.

Monahan, I. J., and Trainor, F. R. 1970. Stimulatory properties of filtrates from the green algae *Hormotila blennista*. I. Description. *J. Phycol. 6*, 263-269.

Moore, B. G., and Tischer, R. G. 1964. Extracellular polysaccharides of algae: effects of life support systems. *Science 164*, 586-587.

Nalewajko, C., and Dunstall, T. G., and Shear, H. 1976. Kinetics of extracellular release in axenic algae and in mixed algal bacteria cultures: significance in estimation of total (gross) phytoplankton excretion rates. *J. Phycol. 12*, 1-5.

Nalewajko, C., and Lean, D. R. S. 1972. Growth and excretion in planktonic algae and bacteria. *J. Phycol. 8*, 361-366.

Nalewajko, C., and Schindler, D. W. 1976. Primary production, extracellular release, and heterotrophy in two lakes in the ELA, northwestern Ontario. *J. Fish. Res. Bd. Can. 33*, 219-226.

Potaenko, Yu. S., and Mikheva, T. M. 1969. Relationships between bacteria and phytoplankton. *Microbiology 38*, 603-607.

Pritchard, G. G., Griffin, W. J., and Whittingham, C. P. 1962. The effect of carbon dioxide concentration, light intensity and isonicotinyl hydrazide on the photosynthetic production of glycolic acid by *Chlorella*. *J. Exp. Bot. 13*, 176-184.

Reichardt, W., Overbeck, J., and Steubing, L. 1967. Free dissolved enzymes in lake waters. *Nature 216*, 1345-1347.

Rigler, F. 1966. Radiobiological analysis of inorganic phosphorus in lake water. *Ver. Int. Verein. Limnol. 16*, 465-470.

Rodhe, W. 1969. Crystallization of eutrophication concepts in Northern Europe. In *Eutrophication: Causes Consequences, Correctives*, G. A. Rohlich (Ed.). Publ. 1700, Div. Biol. Agri., Natl. Acad. Sci. Natl. Res. Council, Washington, D. C. pp. 50-64.

Samuel, S., Shah, N. M., and Fogg, G. E. 1971. Liberation of extracellular products of photosynthesis by tropical phytoplankton. *J. Mar. Biol. Assoc. U.K. 51*, 793-798.

Saunders, G. W. 1957. Interrelations of dissolved organic matter and phytoplankton. *Bot. Rev. 23*, 389-409.

Saunders, G. W. 1971. Carbon flow in the aquatic system. In *The Structure and Function of Freshwater Microbial Communities*, J. Cairns, Jr. (Ed.). Res. Div. Monograph *3*, Virginia Polytechnic Institute and State University, Blacksburg, Va.

Saunders, G. W. Jr. 1972a. The kinetics of extracellular release of soluble organic matter by plankton. *Ver. Int. Verein. Limnol. 18*, 140-146.

Saunders, G. W. 1972b. Potential heterotrophy in a natural population of *Oscillatoria Agardhii* var. *isothrix Skuja. Limnol. Oceanog. 17*, 704-711.

Saunders, G. W. 1972c. The transformation of artificial detritus in lake water. *Mem. Inst. Ital. Idrobiol. 29* Suppl., 261-288.

Saunders, G. W., and Storch, T. A. 1971. Coupled oscillatory control mechanism in a planktonic system. *Nature, New Biol. 230*, 58-60.

Schegg, E. 1968. Relation between plankton development and bacteria in Lake Lucerne and Rotsee. *Schwaz. Z. Hydrol. 30*, 289-296.

Schmidt, G. W. 1969. Vertical distribution of bacteria and algae in a tropical pond. *Int. Rev. Ges. Hydrobiol. Hydrogr. 54*, 791-797.

Sen, Nomita, and Fogg, G. E. 1966. Effects of glycollate on the growth of a planktonic *Chlorella. J. Exp. Bot. 17*, 417-425.

Sheath, G., and Hellebust, J. A. 1974. Glucose transport system and growth characteristics of *Bracteoccus minor. J. Phycol. 10*, 34-41.

Silvey, J. K. G., and Wyatt, J. T. 1971. The interrelationships between freshwater bacteria, algae and actinomycetes in Southwestern Reservoirs. In *The Structure and Function of Freshwater Microbial Communities*, J. Cairns, Jr. (Ed.). Res. Div. Monograph *3*, Virginia Polytechnic Institute and State University, Blacksburg, Va. pp. 249-275.

Skerman, V. B. D. 1967. *A Guide to the Identification of the Genera of Bacteria*. Williams and Wilkins Co., Baltimore.

Smith, W. V. Jr. 1974. The extracellular release of glycolic acid by a marine diatom. *J. Phycol. 10*, 30-33.

Sorokin, Y. I. 1968. Primary production and microbiological processes in Lake Gek-Gel. *Mikrobiologiya 37*, 289-296.

Thomas, J. P. 1971. Release of dissolved organic matter from natural populations of marine phytoplankton. *Mar. Biol. 11*, 311-323.

Thornley, M. J. 1967. A taxonomic study of *Acinetobacter* and related genera. *J. Gen. Microbiol. 49*, 211-257.

Tolbert, N. E., and Zill, L. P. 1956. Excretion of glycollic acid by algae during photosynthesis. *J. Biol. Chem. 222*, 895-906.

Trainor, F. R. 1965. A study of unialgal cultures of *Scenedesmus* incubated in nature and in the laboratory. *Can. J. Bot. 43*, 701-705.

Vela, G. R., and Guerra, C. N. 1966. On the nature of mixed cultures of *Chlorella pyrenoidosa* TX 71105 and various bacteria. *J. Gen. Microbiol. 42*, 123-131.

Waite, D. T. 1973. Some relationships between primary and secondary production in Sunfish Lake, Ontario. Ph.D. Thesis, University of Waterloo, Ontario.

Walsby, A. E. 1965. Biochemical studies on the extracellular polypeptides of *Anabaena cylindrica* Lemm. (Abstract). *Brit. Phycol. Bull. 2*, 514-515.

Walsby, A. E. 1968. Mucilage secretion and the movement of blue-green algae. *Protoplasma 65*, 223-238.

Watt, W. D. 1966. Release of dissolved organic material from the cells of phytoplankton populations. *Proc. Roy. Soc. Lond. B, 164*, 521-551.

Watt, W. D. 1969. Extracellular release of organic matter from two freshwater diatoms. *Ann. Bot. 33*, 427-437.

Watt, W. D., and Fogg, G. E. 1966. The kinetics of extracellular glycollate production by *Chlorella pyrenoidosa*. *J. Exp. Bot. 17*, 117-134.

Wetzel, R. G., and Manny, B. A. 1972. Secretion of dissolved organic carbon and nitrogen by aquatic macrophytes. *Ver. Int. Verein. Limnol. 18*, 162-170.

Whaley, W. G., Danwalder, M., and Kephart, J. E. 1972. Golgi apparatus: influence on cell surfaces. *Science 175*, 596-599.

Whittaker, R. H., and Feeny, P. P. 1971. Allelochemics: chemical interactions between species. *Science 171*, 757-770.

Wilkinson, J. F. 1958. The extracellular polysaccharides of bacteria. *Bact. Rev. 22*, 46-73.

Wright, R. T. 1968. Glycolic acid uptake by planktonic bacteria. Symp. on organic matter in natural waters. Institute of Marine Sciences College, Alaska, September 2-4.

Chapter 19

THE FRESHWATER TO AIR TRANSFER OF
MICROORGANISMS AND ORGANIC MATTER

Duncan C. Blanchard

Atmospheric Sciences Research Center
State University of New York at Albany
Albany, New York

and

Bruce C. Parker

Biology Department
Virginia Polytechnic Institute and State University
Blacksburg, Virginia

CONTENTS

ABSTRACT

Natural and man-induced production of gas bubbles in marine and freshwater ecosystems selectively adsorb inorganic and organic (dissolved and particulate) matter and microorganisms during their upward path through water. Upon reaching the water surface and bursting, a portion of the matter and microorganisms adsorbed by the bubble or concentrated at the surface by other mechanisms is ejected into the atmosphere. The potential importance of these phenomena, now only just being recognized for marine and freshwater ecosystems, include the following:

1. Selective concentration of substances, such as nutrients and toxic materials, as well as microorganisms, in thin films at the air-water interfaces;

2. Different metabolic processes and rates occurring at the surfaces of aquatic ecosystems;

3. Water to air transfer of large quantities of concentrated matter and microorganisms, including pesticides, heavy metals, radioactive wastes, pathogenic organisms, nutrients, etc., which will have significant impact on terrestrial environments; and

4. Influence on meteorology, such as the chemical and microbiological composition of precipitation or rates and types of nucleation of ice in supercooled clouds.

INTRODUCTION

Microorganisms and organic and inorganic matter are continuously transferred from the surface of freshwater

environments into the overlying atmosphere by natural processes that only now are beginning to be appreciated. The
chief objective of this paper is a brief review of the
present information supporting the freshwater to air transfer of microorganisms and matter, and to show that this
pathway in the biosphere probably is far more important
than has been realized previously. We shall make no claim
at treating all of the mechanisms for selective concentration and transport of aquatic microorganisms and organic
matter to the air-water interface and their transfer into
the atmosphere. Furthermore, much of our information comes
from the far more extensive literature concerning the
marine environment. Its extrapolation to freshwater must
be treated cautiously.

Microlayers and Adsubble Processes

The surface microlayer with which we are concerned
is that extremely thin layer of water at the air-water
interface in which materials can be found in concentrations
far higher than that in the bulk water just beneath. The
thickness of this microlayer depends upon the material in
question, but generally it is less than 100 μm and usually
<10. The thinnest microlayers are found when surface-
active substances are spread upon the water, but even here
variations in microlayer thickness of over two orders of
magnitude are common (MacIntyre 1974). For example, a
surface-active monolayer composed of lipids or fatty acids
contains molecules that extend only about 20 Å into the
water, while many protein and glycoprotein monolayers extend to a depth of about 10,000 Å. Thicker microlayers
are found with bacteria, algae, and other plankton. These
can extend to 10 μm or more below the surface.

The microlayer can be sampled in a number of ways but
two methods are in common use. One involves a screen in

which the microlayer and some of the underlying water is removed and held between the screen mesh (Garrett 1965). The other method utilizes a drum-type sampler (Harvey 1966) that rolls through the water and lifts off a thin surface film. However, Hatcher and Parker (1974b), who have used both samplers on the same microlayer, point out that since each device collects a different thickness of the surface layer, one should use caution in calculating from the data the concentrations of material in the micro-layer.

Based almost totally on a review and analysis of marine microlayer studies, Parker and Barsom (1970) pointed out the probable great ecological importance of freshwater microlayers. They proposed that, as with the ocean, not only were microlayers often the sites of excessively high concentrations of microorganisms (neuston or pleuston; Hutchinson 1967), but the microbial community structure and function and the chemical composition of microlayers also were likely to be qualitatively and quantitatively quite different from that of the submicrolayer water. Parker and Barsom (1970) speculated that chlorinated hydrocarbon pesticides might be concentrated in microlayers and that this might involve a major mechanism for movement of these and other microlayer constituents throughout the biosphere; this point subsequently was confirmed (Seba and Corcoran 1969).

Hatcher and Parker conducted year-round studies aimed at characterizing the seasonal chemical and microbiological nature of microlayers relative to the bulk subsurface water for several freshwater ecosystems. Four microlayer samlers were compared for their collection accuracies and efficiencies (Hatcher and Parker 1974b). They found:

1. The communities of microorganisms in the microlayers commonly contained different genera and

nearly always in different proportions to the
submicrolayer water (Hatcher and Parker 1974a).

2. Bacteria, algae, and fungi often reached concentrations one to several orders of magnitude that
of the water from 10-cm depth (Hatcher and Parker
1974c; Parker and Hatcher 1974).

3. Microlayers frequently were enriched in organic
and inorganic matter, including certain heavy
metals (Hatcher and Parker 1974a, 1974c).

These findings generally resembled those for marine microlayers. Hatcher (unpublished) also conducted preliminary
studies of low molecular weight organic acids with gas-
liquid chromatography that showed the frequent occurrence
in microlayers of a variety of organic acids at concentrations a few to many times that of the subsurface water.

One of the several mechanisms discussed by Hatcher
and Parker (1974a), apparently important for inducing
higher concentrations of microorganisms, organic, and some
inorganic matter in freshwater microlayers, is adsorptive
bubble separation processes, abbreviated "adsubble process" by Lemlich (1966, 1972). Briefly, when a bubble of
gas rises through water, a variety of microorganisms and
organic and inorganic substances tends to accumulate about
the bubble surface. Figure 19.1 illustrates how this process operates with surface-active organic material. Adsubble processes are selective and act to fractionate the
particulates and solutes in water. If all or part of the
adsorbed particulates and solutes reach and are deposited
in the microlayer, then clearly the microlayer composition
will be qualitatively and quantitatively different from
the submicrolayer. Adsubble processes have been shown to
occur in the marine environment by a number of workers
(e.g., Blanchard 1975; VanGrieken *et al*. 1974). Of freshwater, considerably less is known. However, Rubin's

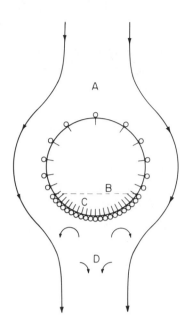

FIG. 19.1. Adsorption of dissolved surface-active material onto the surface of an air bubble rising through the water. (A) The upstream region; (B) rough dividing line between compressed and noncompressed monolayer; (C) compressed monolayer; (D) bubble wake. Adapted from Lemlich (1972) with permission of the author.

(1968) studies of microflotation showed that pH and certain surfactants enhanced the foam separation of *Aerobacter aerogenes* and *Chlorella ellipsoidea* better than *E. coli*, and *Chlamydomonas reinhardti* was more efficiently removed than *Chlorella*. Finally, studies of bacterial aerosols at an aerated waste oxidation pond have revealed that most of the coliform bacteria were in the microlayer (Parker, unpublished). Probably this is the result of the excessively high rate of aeration and bubbling with continuous adsubble processes occurring.

Bubbles and the Transfer of Material to the Atmosphere

Bubbles not only serve as a scavenging mechanisms to transport materials from the bulk to the surface of the water, but they are the agents by which particulate material is ejected into the atmosphere. It has been estimated that each year over the oceans of the world, bubbles are responsible for ejecting about 10^{10} tons of sea salt into the air (Blanchard 1963). This constitutes the major fraction by weight of all the particulate material in the global atmosphere. Although most of the studies on the water to air transfer of material by bubbles were carried out with seawater, the few studies with freshwater indicate that essentially the same mechanism is operating. We will briefly review this work.

When an air bubble (or one composed of any gas) reaches the air-water interface, it will burst either immediately or after a few seconds or minutes. When it does, the bubble collapses rapidly, producing a high-speed jet of water that moves upward from the bottom of the bubble cavity (Woodcock *et al.* 1953) (see Fig. 19.2). The jet, after reaching a height of about one bubble diameter above the water surface, becomes unstable and fragments into several drops. These drops, commonly called jet drops, vary in size but have a diameter roughly one-tenth that of the bubble. The jet drops continue moving upward until their kinetic energy has been expended by doing work against frictional retarding forces and gravity. The drops from a single jet attain heights that may differ by an order of magnitude, but the height attained by the first drop produced by the jet is about 100 times the bubble diameter (Blanchard 1963). Thus, the top jet drop from a 0.1-mm bubble rises about 1 cm, while the same drop from a 2-mm bubble is ejected to a height of nearly 20 cm. For larger bubbles this relation no longer holds and the

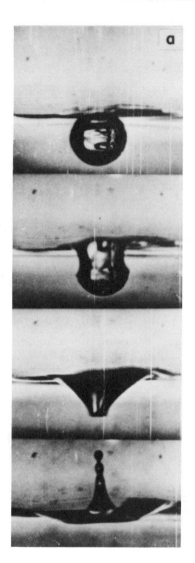

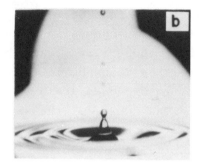

FIG. 19.2. (a) Composite view of high-speed motion pictures illustrating some of the stages in the collapse of a 1.7-mm diameter bubble. The time interval between the top and bottom frames is about 2.3 msec. The angle of view is horizontal through a glass wall. The surface irregularities are caused by a meniscus. (b) Oblique view of the jet and jet drops from a 1-mm diameter bubble.

633

ejection height decreases (see Fig. 19.3). Many of these
jet drops, especially those <50 µm, can be mixed high into
the atmosphere and transported tens or hundreds of kilometers by atmospheric turbulence and convection before
they return once again to water or land.

The source of the kinetic energy for the jet drops
is not to be found in the buoyant force (gravitational
potential energy) of the resting bubble at the interface,
as one might expect, but in the bubble surface free energy
(Blanchard 1963). This energy, the product of bubble surface area and surface tension, powers a circular capillary
wave that moves down the bubble surface when it bursts to
converge at the bottom to produce the jet. The ejection
speeds of the drops from the jet increase with decreasing
bubble size, and can attain astonishingly high values. For
example, the top jet drop from a bubble of 70 µm bursting
in water at a temperature of 4 C is ejected at a speed of
about 80 m/sec (i.e., ca. 290 km/hr or 180 mi/hr)! In spite
of this incredible speed it reaches a maximum height of
only 0.17 cm. This is because of the frictional force of
the air against the drop that produces an initial decelera-
tion of 495,000 g! In the absence of this frictional force
the drop would, by calculation, rise to a height of 335
m.

Since surface tension is the source of the energy of
jet drop ejection it is clear that anything that changes
surface tension can change the entire dynamics (size, number, and ejection height of the jet drops) of bubble bursting. The organic materials on and within freshwaters can
do just that. By being adsorbed to a bubble while it is
rising through the water, or by being transported from
the air-water interface to the bubble surface upon bubble
bursting, they can lower the surface tension of the bubble
and modify their own transport into the atmosphere.

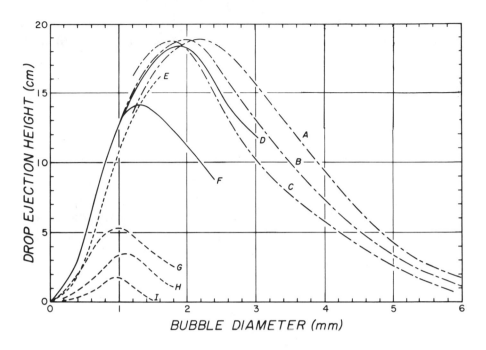

CURVE	DROP POSITION ON JET	WATER TEMP. °C	WATER TYPE	REFERENCE
A	TOP	4	SEA	HAYAMI & TOBA (1958)
B	"	16	"	"
C	"	30	"	
D	"	22-26	"	BLANCHARD & WOODCOCK(1957)
E	"	4	"	PRESENT WORK
F	"	21	DIST.	STUHLMAN('32)
G	2nd	22-26	SEA	BLANCHARD & WOODCOCK('57)
H	3rd	"	"	"
I	4th	"	"	"

FIG. 19.3. Jet drop ejection height as a function of bubble diameter, water temperature, and salinity. See Blanchard (1963) for details.

Typically, a decrease of surface tension will decrease both
the ejection height and size of jet drops (Blanchard and
Syzdek 1972).

In addition to jet drops, film drops are also ejected
into the atmosphere from a bursting bubble. Just before
bursting a thin film of water atop the bubble separates
the air in the bubble from the atmosphere. When this film
ruptures to start the jet drop ejection process, it disintegrates into numerous film drops. These drops are much
smaller than the jet drops, generally less than 10 and
often less than 1 µm. The number of film drops produced
per bubble burst is, among other things, strongly dependent
on bubble size. Bubbles of less than 0.3 mm, bursting in
either seawater or freshwater, do not appear to produce
any film drops at all, but a 2-mm bubble will produce a
maximum of about 100 and a 6-mm bubble a maximum of 1000
film drops (Blanchard 1963; Day 1964).

Organic monolayers on the surface of the water can
modify the film drop production, just as it does the jet
drops. However, there is no agreement on whether the monolayers increase or decrease the production. Both Blanchard
(1963) and Paterson and Spillane (1969) found a decrease,
while Garrett (1968), using a somewhat different experimental technique, found an increase.

The primary mechanism for the production of bubbles
in the sea, and probably also in freshwaters, is the formation of breaking waves or whitecaps. Great quantities of
air are entrained into the water as the waves break. Measurements by Blanchard and Woodcock (1957) in the vicinity
of a breaking wave have shown that at the surface of the
bubble-laden water about 300,000 bubbles are breaking per
square meter per second. Under normal wind conditions
between 3 and 4% of a given body of water is covered with
whitecaps. The bubble spectrum is heavily weighted toward

the small end, with most bubbles being less than 0.5 mm
in diameter.

Precipitation, either in the form of rain or snow,
is a prolific producer of bubbles (Blanchard and Woodcock
1957) and on a local scale is possibly more important than
whitecaps. Precipitation-produced bubbles are very small,
a majority being less than 0.1 mm in diameter. In addition
to breaking waves and precipitation, waterfalls, rapids,
and bow waves from boats can produce bubbles. Since the
relative importance of all these effects in freshwater is
unknown, research in this direction is urgently needed.

We now consider the evidence that bursting bubbles
can eject microorganisms and other materials into the atmosphere in highly concentrated form. Blanchard (1964,
1968, 1975) and Barger and Garrett (1970) found concentrations of surface-active organic material on the marine sea
salt aerosol several thousand times higher than that found
in the sea. On the basis of these findings, Blanchard
(1975) believes that at least 5% of the total productivity
of the oceans becomes airborne each year. Since the sea
is the source of the aerosol and organic monolayers are
found on the sea (Garrett 1967b), it seems certain that
the bubble microtome effect (MacIntyre 1968) is skimming
off the monolayer and ejecting it into the air on jet
drops. Baier (1972) believes that a similar thing happens
in freshwaters and has evidence that suggests that lakes
that become covered with organic monolayers from human
activities can clean themselves in several days by bubbling
action that ejects the material into the atmosphere.

Laboratory studies have shown without a doubt that
surface concentrations of bacteria, similar to those found
at the surfaces of freshwaters (Hatcher and Parker 1974a,
c) and on the sea (Harvey 1966; Sieburth 1971), can be
concentrated in jet drops. Blanchard and Syzdek (1970),

working with a wide range of bubble sizes, found that jet drops from bubbles rising only about a centimeter through a suspension of *Serratia marcescens* contained concentrations of bacteria 10-100 times higher than that in the bulk suspension. Bezdek and Carlucci (1972) obtained the same results with suspensions of marine bacteria in seawater. These findings were extended by Blanchard and Syzdek (1972) with a rotating tank, an experimental device that enables one to use a fixed bubble-generating tip and yet generate bubbles of almost any size desired and at a rate not possible with conventional methods (see Fig. 19.4). With this tank it is easy to generate uniformly sized bubbles at rates up to 30/sec. Thus, in a matter of minutes one may obtain the many thousands of small jet drops that are often required for a proper analysis of bacteria content.

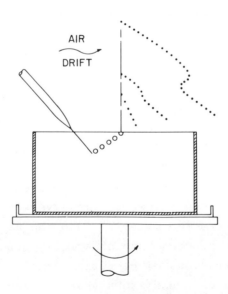

FIG. 19.4. The rotating tank, useful for a great variety of studies of the water to air transfer of materials.

More important than surface concentrations of bacteria in explaining the high concentrations in jet drops appears to be the concentrations of bacteria that develop on the surface of the bubble by scavenging or adsorption. With the aid of a bubble aging tube, another experimental device within which a rising bubble can be prevented from reaching the surface by placing it in a downward directed stream of water, Blanchard and Syzdek (1972) were able to "suspend" or age an air bubble for hundreds of seconds, and thus observe the rate of increase of bacteria concentrations in the jet drops as a function of bubble age or distance it rose through the water. They found that bubbles rising through the water for 20 sec or less collected sufficient *S. marcescens* to enable the jet drops to contain bacterial concentrations over 1000 times that in the bulk suspension. For times in excess of 20 sec, the drop bacterial concentrations did not change appreciably, indicating that bacterial concentrations on the bubble surface had reached a steady-state or saturated condition.

Recently Blanchard and Syzdek (1974) have looked more closely at what happens during the first few seconds (representing not more than 30 cm of distance traveled) of bubble motion through *S. marcescens* suspensions. They found that concentrations of bacteria in the jet drops increased in direct proportion to the bubble rise distance. For example, in an experiment with bubbles of about 1.5 mm diameter the jet drop concentration factor (the ratio of the number of bacteria per unit volume in the jet drops to that in the bulk suspension) increased from about 20 to 120 as the bubble rise path was increased from 1 to 27 cm. This rapid increase in concentration factor over such a short bubble rise path might seem suprising in view of the fact that both Blanchard and Syzdek (1974) and Carlucci and Bezdek (1972) have found that the collection efficiency of a moving bubble for bacteria is very low, of the order of 0.1% or less. However, this low collection efficiency seems to be countered by an extremely high transfer efficiency. The efficiency by which bacteria collected by the bubble are transferred to the much smaller jet drops appears to be more than 50% (Blanchard and Syzdek 1974).

An illustration of how bubbles rising just a few centimeters through natural bacterial populations in freshwaters can produce a steady increase in the concentration of jet drop bacteria is shown in Figures 19.5 and 19.6, taken from the work of Blanchard and Syzdek (1974). They allowed bubbles to rise from a capillary tip placed from 0.5 to 4 cm beneath a sample of water collected from the Hudson River as it passed through Albany. It is clear from Figure 19.5 that the number of viable bacteria in the top jet drop increased steadily with increasing bubble rise distance. Figure 19.6 shows that the concentration factor after only 4 cm of bubble scavenging is about 35.

Judging from colony morphology, over 10 species of bacteria were recovered in the jet drops. Identification was attempted for only six species. Three species, *Bacillus cereus, Aeromonas hydrophila,* and *Acinetobacter lwoffi* contributed about 95% of all the bacteria. Among the remaining 5% were *Pseudomonas putrefaciens* and bacteria of the genera *Flavobacterium* and *Lactobacillus*. None of these is known to be pathogenic to man.

As mentioned earlier, surface-active organic material has been found concentrated on the marine aerosol, but the details on how this happened are not known. To our knowledge there have been only two investigations (Blanchard 1963; Bezdek and Carlucci 1974) on the transfer to jet drops of organic monolayers on the air-water interface, and none on a similar transfer of the monolayers adsorbed onto a bubble while it is rising to the interface.

640

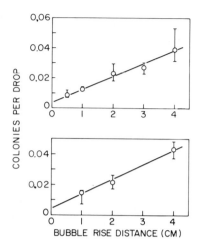

FIG. 19.5. Effect of bubble scavenging in Hudson River water in increasing bacteria count in jet drops. Top: June 21, 1973; drop diameter, 45 μm. Bottom: June 26, 1973; drop diameter, 47 μm.

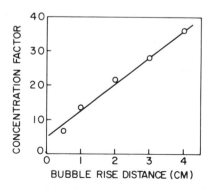

FIG. 19.6. Effect of bubble scavenging in Hudson River water in increasing jet drop bacterial concentration factor. June 21, 1973.

However, circumstantial evidence suggests that such transfer exists. This is reviewed in a recent paper by Blanchard (1975).

We noted earlier that Rubin (1968) presented experimental evidence that some species of bacteria were not
scavenged by bubbles, except under critical conditions
of surfactant concentration and pH. Blanchard and Syzdek
(unpublished) recent found a white variant in their *Serratia marcescens* cultures that, in contrast to the typical
red form, was not concentrated by the bubbles. Thus, the
concentration factor of the white variant in the jet drops
was about unity. While these results remained to be explained, they reinforce the idea that bubbles scavenge
microorganisms and matter selectively, which constitutes
one explanation for the often striking microbiological and
chemical differences observed between microlayers and submicrolayer water.

Bubble concentration and transfer mechanisms in freshwater exist for other than bacteria. Employing essentially
the same rotating tank apparatus (Fig. 19.4) of Blanchard
and Syzdek (1972), and a solution of crystal violet dye,
Parker (in press) examined jet drops of about 70 μm diameter produced by bubbles originating at about 5-cm depth.
The top jet drops were captured on Millipore membranes.
Calculations based on jet drop size and numbers and color-
imetric analyses of crystal violet dye concentrations in
the bulk solution and Millipore membrane-collected jet
drops revealed an approximately 60-fold concentration of
the dye in the jet drops. Similar experiments have been
run to assess the degree to which algae may be concentrated
in jet drops (Parker, unpublished). Following collection
of top jet drops on gelatin-coated glass slides, *Chlorella*
cells were counted in craters produced by the impaction
of the jet drops using phase contrast microscopy. Since

the crater diameters are related to the jet drop diameter, the approximate jet drop volumes could be calculated. Data from two experiments showed that *Chlorella* cell concentrations of $1.2 \times 10^6/cm^2$ and $1.8 \times 10^5/cm^2$ in the bulk suspension were further concentrated in the jet drops 13.4 and 269 times that of the bulk suspension, respectively. This tendency for an inverse relation between concentration factor and bulk cell count is similar to that found for bacteria (Blanchard and Syzdek 1972; Bezdek and Carlucci 1972).

Schlichting (1974) conducted laboratory experiments in which several algal cultures were aerated with bubbles of various sizes and agar plates used to capture airborne algae for culture counts. He also cultured algae ejected by bursting bubbles 0.5-1.0 cm diameter from store aquaria. The size and numbers of airborne droplets apparently were not determined so that calculations of the concentration factors for the algae in the drops cannot be made. However, since Schlichting found no increase in the airborne algae when the bubble rise path was increased, it appears that the concentration factor was near unity; i.e., little or no enrichment took place.

SIGNIFICANCE OF ADSUBBLE PROCESSES AND BUBBLE-BURST PRODUCTION OF JET AND/OR FILM DROPS

The probable importance of adsubble processes and bubble-burst production of jet and film drops is too far reaching to receive elaborate discussion here. We shall attempt, therefore, to summarize briefly a number of significances these processes can have in the biosphere.

Importance of Natural Freshwater Microlayers

The data available suggest that microlayers enriched with microorganisms and organic and inorganic matter occur

in the vast majority of freshwater ecosystems (Goldacre
1949; Hatcher and Parker 1974a,b,c; Parker and Hatcher
1974). On the basis of the frequent orders of magnitude
higher concentrations of organisms and matter in these
microlayers, it follows logically that microlayer metabolic
processes and other reactions occurring in microlayers
will be unlike that (those) in the submicrolayer water.
Gallagher (1975), in studying salt marsh plankton and
neuston (i.e., algae in the microlayer), found that 37%
of the plankton net photosynthesis and 21% of the respiration occurred in the microlayer. These values are striking
when it is realized that the microlayer is perhaps 100 μm
thick, while the mean depth of the salt marsh was 23 cm,
or about 2300 times the bulk of the microlayer. The interesting discussion by Nalewajko (this volume) treating
the effects of light, dissolved O_2, and dissolved CO_2 concentrations on the production of extracellular products
by algae, pose interesting questions concerning the metabolism of microlayer algae. In short, the evidence
available on freshwater microlayers support the assertion
made by Parker and Barsom (1970) that " ... the interaction
of the microlayer with both air and subsurface water may
be of sufficient ecological importance as to be a major
contributing factor in currently unexplained problems of
air and water pollution." A specific example supporting
this statement is the finding by Hatcher and Parker (unpublished) that natural waters possessing permissible concentrations of coliform organisms frequently had numbers
of coliforms concentrated in the microlayer unacceptable
from a public health standpoint.

Although it is likely that much of the nonliving material in the microlayer arrived there by molecular and
eddy diffusion, there is no doubt that bubbles can play
a significant role in the upward transport of these

materials. In seawater Baylor *et al*. (1962) have found
that inorganic phosphate is easily carried to the surface,
and recently Wallace and Duce (1974) have found that bubbles in seawater efficiently carry particulate organic
carbon and heavy metals to the surface.

Not only might bubbles transport materials to the
surface but it was suggested by Sutcliffe *et al*. (1963),
who found organic particles in the foam generated by the
bubbles, that the bubbles are able to convert dissolved
organic material in the sea to particulate particles of a
size to be utilized by filter feeders. It is possible that
bubbles need not reach the surface of the sea (or freshwater) to produce particulate organic carbon. Most of the
bubbles produced by breaking waves and precipitation are
less than 200 μm and, because of surface curvature effects
(Blanchard and Woodcock 1957), tend to go into solution
rapidly even though the water may be saturated with air.
Many of these bubbles dissolve before they reach the surface and conceivably could compress adsorbed organic monolayers into particles (Blanchard 1975). However, one
should be aware that there has been a great deal of controversy over the question of whether bubbles can indeed
convert dissolved to particulate organic carbon (Riley
1970).

Organic monolayers on water surfaces can modify the
air-water transfer of heat and water vapor. They appear
to do this in an indirect way by providing a semirigid interface that, in turn, eliminates small-scale eddy diffusion just beneath the surface, leaving the much slower
molecular diffusion to transport heat between the surface
and the water beneath (Jarvis 1962).

The wave structure can be modified by organic monolayers. The natural slicks or smooth areas that one can
see on most any body of water are caused by the capillary

wave-damping properties of the surface monolayer. It is
quite remarkable that the maximum damping effect is produced by a monolayer that reduces the surface tension by
only 1 dyne/cm from the normal clean water value of about
73 dynes/cm (Garrett 1967a). Barger *et al.* (1970) have
found that by eliminating the capillary waves, the ampli-
tudes of the large wind-driven gravity waves are reduced.

During the past decade concern has been given to the
surface concentrations inadvertantly produced by oil spills
in rivers, harbors, and on the sea (Garrett and Barger
1972). With the increasing use of the oceans and freshwaters as a depository for sewage wastes, concern should
be directed to the surface monolayers and associated microorganisms. Just downwind of the sewage dump site in the
New York Bight, Hardy and Baylor (unpublished) found giant
surface slicks several kilometers across with surface tension reductions of over 20 dyne/cm. If much of this material becomes airborne via bubble action, a potential
health hazard may be present.

Freshwater microlayers not only exist on permanent
lakes, ponds, and rivers, but whenever it rains, or possibly when dew is formed, microlayers are formed on leaves
and other terrestrial litter (Bandoni and Koske 1974).
Monolayer-forming substances naturally produced, rapidly
transport numerous microorganisms both vertically and horizontally over the surfaces of these materials.

Significance of the Aerosol Produced
by Bursting Bubbles

There is an amazing diversity of ways in which jet
and film drops from bursting bubbles play a role either
in processes that take place in the atmosphere or on land
and other surfaces where the droplets may be impacted. In
the marine atmosphere the bubble-induced hygroscopic sea

salt particles (Woodcock 1953), probably originating from
jet drops, are thought to play a major role in the initiation of rain (Woodcock et al. 1971). Film drops may provide a significant fraction of the cloud condensation nuclei (CCN), those submicroscopic particles without which
water vapor cannot condense to form cloud droplets (Blanchard 1969). Although the area covered by freshwaters is
far less than that of the sea, it is possible that large
bodies of freshwater, such as the Great Lakes, may contribute appreciable numbers of CCN to become involved in
cloud-forming processes over or near the lakes.

One of the major problems in atmospheric science is
an understanding of the nucleation of ice in supercooled
clouds. Over the years, numerous investigations have been
carried out with no clear agreement on the major sources
of the ice nuclei. Recently Schnell and Vali (1973) and
Schnell et al. (1973) found that bacterial action on
leaves produced particles that acted as good ice nuclea-
tors. More recently Schnell (1974) indicated that the
bacterium itself was a nucleator. This suggests that bacteria in jet and film drops from bursting bubbles might
play a role in snow-forming processes in clouds. It must
be stressed that this is highly conjectural. It is unlikely that significant numbers of ice nuclei are provided
by jet and film drops from freshwaters, except possibly
in local regions as described above. It remains to be
shown that bacteria in jet drops, and the other organic
materials almost certainly to be in the droplets, are as
effective as those observed in laboratory experiments. If
bacteria, which are known to be producers of cobalamines
and other B vitamins, act as freezing nuclei, these vitamins might be extracted in the precipitation (Parker and
Wachtel, this volume).

Another surprising aspect of the production of jet
drops of interest to atmospheric scientists is that they

carry a positive charge when produced in seawater and a
negative charge when produced in diluted seawater of a
salinity comparable to that of freshwaters. The flux of
positive charge produced by jet drops in the atmosphere
over the sea is a significant fraction of the total flow
of current between the atmosphere and the earth (Blanchard
1963, 1966). When the wind is blowing hard and there is
much bubbling, the flux of negative charge from at least
one body of freshwater, Lake Superior, provides a space
charge that is sufficient to reverse the normal atmospheric
potential gradient in the lowest few meters of the atmosphere from positive to negative (Gathman and Hoppel,
1970). Those who believe that biological processes can
be modified by the magnitude and sign of the atmospheric
potential gradient and the space charge should keep these
bubble-induced electrical effects in mind (Salt 1961;
Blanchard 1961; Reiter 1973; Krueger *et al.* 1974).

The production and dispersal of aeroplankton is another subject that addresses the probable importance of bursting bubbles, jet, and film drops. Maynard (1968a) showed
that freshwater algae living in aquatic forms could be
collected in the atmosphere about 40 miles downwind. She
also collected marine algae in the atmosphere over the
ocean (Maynard 1968b). Schlichting (1964) collected freshwater algae from air blowing off the Great Lakes. From
data presented by Abe (1971) on the concentrations of a
number of species of diatoms in both sea foam and in the
water beneath, we calculate that the concentration in the
surface foam depended on the species, and ranged from about
100 to 10,000 times that in the bulk seawater. It seems
clear that airborne droplets from the foam would show similar enrichments.

Direct measurements that implicate the water to air
transfer of microorganisms by bursting bubbles were made

by Hatcher and Parker (unpublished). They used a specially designed polystyrene container in which petri dishes of culture media can be suspended at different heights over natural water surfaces with the sterile agar facing down. A 30 min exposure with dishes at 6 cm height above the water of a Virginia lake known to possess anaerobic muds with continuous gas production revealed numerous viable bacteria and one green alga that had been captured as aerosols on the agar media.

If bursting bubbles with jet and film drops are important natural processes in freshwater ecosystems, then it follows that such processes may be altered by man. Notable among the obvious man-induced changes is the proliferation of dam construction and construction of farm ponds that was significantly increased the total surface area of freshwater in many localities. Also, the advance of eutrophication that may have increased certain types of gas production in aquatic ecosystems may have changed the rates of water to air transfer of microorganisms and other matter.

A disturbing example of the inadvertant influence of man in producing changes in the water to air transfer of material is the Viareggio phenomenon, observed in Italy along the shore south of Genoa (Cornwell 1971). Rivers carrying vast quantities of pollutants into the sea give rise to large areas of film-covered waters. The films are ejected into the air by bursting bubbles and carried ashore to coat the needles of pine trees. Interference with transpiration results, and thousands of these trees now lie dead or are dying. This is especially tragic since it appears that some vegetation along ocean shores has adapted to the marine aerosol (Boyce 1954) and depends upon airborne nutrients from the sea for its survival (Wilson 1959; Art et al. 1974). One wonders whether a similar

thing may be happening along the shores downwind of polluted areas of the Great Lakes.

The possibilities of the water to air transfer of
microbial aerosols, at least from sewage treatment plants,
was recognized long ago by Fair and Wells (1934), who noted
that " ... the newer processes, including trickling filters, activated sludge, preaeration tanks, and aerated
skimming tanks, afford the medium of evaporating droplets,"
and " ... there appears to be no means of ready escape of
bacteria from sewage into the atmosphere unless the sewage
is broken up into fine droplets that evaporate before set-
tling back ...."

Although Fair and Wells recognized that a microbial
aerosol could escape from sewage plants, neither they nor
anyone else recognized that, through the agency of bubbles,
microorganisms in the water could be highly concentrated
in jet and film drops. Woodcock, in 1955, was apparently
the first to realize the importance of this effect. He
suggested that his observations of the water to air transfer of the aerosol from red tide waters (Woodcock 1948)
that was irritating to the human respiratory tract might
be explained by this mechanism. He further suggested that
microorganisms in any body of water might become airborne
via bubble concentration effects. Since that time others,
including Parker and Barsom (1970) and Goff *et al.* (1973),
have recognized the possibility that bubbles, under certain conditions, might inject high concentrations of human
pathogens into the atmosphere in jet and film drops.

These considerations are far from academic for it is
known that sewage treatment plants, cooling towers, and
tanneries, for example, produce microbial aerosols that
can travel great distances (Spendlove 1974) and produce
viral disease (Wellock 1960). To this we might add the
polluted rivers and lakes, for the data of Figures 19.5

and 19.6 indicate the potential that exists for the efficient transport of pathogens to the atmosphere. Many of
the bubble- and splash-produced particles are in the range
of 1-5 μm where maximal alveolar penetration occurs (Hatch
and Gross 1964).

Although the ability of a bursting bubble to produce
a highly concentrated microbial aerosol has yet to be directly implicated in the aerial transmission of disease,
some interesting speculations have been made by Gruft *et
al.* (unpublished). They point out the extremely high in-
cidence of sensitivity in people living in the coastal
regions of the southeastern United States to the Battey
group (*Mycobacterium intracellulare*) of atypical bacteria
and suggest that the organism may be in the sea salt aerosol that is carried over the land by onshore winds. Their
laboratory experiments have shown that this mycobacterium
can live in seawater at the temperatures normally found,
and that bubbles rising only about 25 cm through saline
water containing the mycobacterium collect sufficient numbers to produce bacterial concentration factors of several
hundred in the jet drops.

Finally, we would like to call attention to the possibilities of interaction between bursting bubbles and
radioactive materials. VanGrieken *et al.* (1974) have conducted studies of adsubble processes and jet drop production of aerosols of several radioisotopes of heavy metals.
Zinc-65 was enriched up to 50 times that in the bulk solution. The relevance of this finding to the possible future
release of radioactive wastes by coastal, offshore, or
other nuclear power installations near aquatic environments
should not be underestimated.

In this brief review we have had to leave out many
of the studies in this fast-growing discipline of adsubble
processes in freshwaters and the sea, and the role of

bursting bubbles in concentrating materials in jet and
film drops. The interested reader will find much detail
in the papers presented at the first two international
symposia on the Chemistry of the Sea-Air Particulate Exchange Processes. Many of these papers appear in a special
issue of the *Journal of Geophysical Research*, 77 (No. 27),
September 20, 1972, and an issue of *Journal de Recherches
Atmosphériques*, 8(3-4), 1974. Those readers interested
in the fate of microbial aerosols should see Dimmick and
Akers' (1969) book.

CONCLUSION

The bursting of bubbles at the surface of freshwaters
provides the main mechanism for the transport of microorganisms and organic and inorganic matter into the atmosphere. These materials tend to concentrate in thin microlayers both at the surface of the water and the surface
of the bubbles. When the bubbles burst, these surfaces
are skimmed off and ejected into the atmosphere in the
form of jet and film drops. This microtome effect of the
bursting bubbles produces a concentration of materials in
the airborne droplets that may be two or more orders of
magnitude higher than that in the bulk water. Since the
ecological consequences of this effect can be significant,
it is a matter of some urgency that before man further
pollutes his freshwaters he understands the details of
this important transfer mechanism between the water and
the air.

ACKNOWLEDGMENTS

For one of us (Duncan C. Blanchard) this work was
sponsored by the Atmospheric Sciences Section, National
Science Foundation, NSF Grant GA-23413.

REFERENCES

Abe, T., and Fukuchi, N. 1971. In situ formation of stable sea foam and its transport. Proceedings of the IEEE Engineering in the Ocean Environment Conference, pp. 171-174.

Art, H. W., Bormann, F. H., Voight, G. K., and Woodwell, C. M. 1974. Barrier island forest ecosystems: Role of meteorologic nutrient inputs. *Science 184*, 60-62.

Baier, R. E. 1972. Organic films on natural waters: Their retrieval, identification, and modes of elimination. *J. Geophys. Res. 77*, 5062-5075.

Bandoni, R. J., and Koske, R. E. 1974. Monolayers and microbial dispersal. *Science 183*, 1079-1081.

Barger, W. R., and Garrett, W. D. 1970. Surface-active organic material in the marine atmosphere. *J. Geophys. Res. 75*, 4561-4566.

Barger, W. R., Garrett, W. D., Mollo-Christensen, E. L., and Ruggles, K. W. 1970. Effects of an artificial sea slick upon the atmosphere and the ocean. *J. Appl. Meteorol. 9*, 396-400.

Baylor, E. R., Sutcliffe, W. H. Jr., and Hirschfeld, D. S. 1962. Adsorption of phosphates onto bubbles. *Deep-Sea Res. 9*, 120-124.

Bezdek, H. F., and Carlucci, A. F. 1972. Surface concentration of marine bacteria. *Limnol. Oceanog. 17*, 566-569.

Bezdek, H. F., and Carlucci, A. F. 1974. Concentration and removal of liquid microlayers from a seawater surface by bursting bubbles. *Limnol. Oceanog. 19*, 126-132.

Blanchard, D. C. 1961. Electrostatic field and freezing. *Science 133*, 1672.

Blanchard, D. C. 1963. The electrification of the atmosphere by particles from bubbles in the sea. *Prog. Oceanog. 1*, 71-202.

Blanchard, D. C. 1964. Sea-to-air transport of surface active material. *Science 146*, 396-397.

Blanchard, D. C. 1966. Positive space charge from the sea. *J. Atmos. Sci. 23*, 507-515.

Blanchard, D. C. 1968. Surface active organic material on airborne salt particles. Proc. Int. Conf. Cloud Physics, Toronto Am. Meteorol. Soc., Boston, Mass., pp. 24-29.

Blanchard, D. C. 1969. The oceanic production rate of cloud nuclei. *J. Rech. Atmosphér. 4*, 1-6.

Blanchard, D. C. 1975. Bubble scavenging and the water-to-air transfer of organic material in the sea. In *Applied Chemistry at Protein Interfaces*, Robert Baier (Ed.), pp. 360-387.

Blanchard, D. C., and Syzdek, L. 1970. Mechanism for the water-to-air transfer and concentration of bacteria. *Science 170*, 626-628.

Blanchard, D. C., and Syzdek, L. D. 1972. Concentration of bacteria in jet drops from bursting bubbles. *J. Geophys. Res. 77*, 5087-5099.

Blanchard, D. C., and Syzdek, L. D. 1974. Bubble tube: Apparatus for determining rate of collection of bacteria by an air bubble rising in water. *Limnol. Oceanog. 19*, 133-138.

Blanchard, D. C., and Syzdek, L. D. 1974. Importance of bubble scavenging in the water-to-air transfer of organic material and bacteria. *J. Rech. Atmosphér. 8* (3-4), 529-540.

Blanchard, D. C., and Woodcock, A. H. 1957. Bubble formation and modification in the sea and its meteorological significance. *Tellus 9*, 145-158.

Boyce, S. G. 1954. The salt spray community. *Ecol. Monogr. 24*, 29-67.

Carlucci, A. F., and Bezdek, H. F. 1972. On the effectiveness of a bubble for scavenging bacteria from seawater. *J. Geophys. Res. 77*, 6608-6610.

Cornwell, J. 1971. Is the Mediterranean Dying? *New York Times Magazine,* February 21.

Day, J. A. 1964. Production of droplets and salt nuclei by the bursting of air bubble films. *Quart. J. Roy. Meteorol. Soc. 90*, 72-78.

Dimmick, R. L., and Akers, A. (Eds.). 1969. *An Introduction to Experimental Aerobiology*. Wiley, New York, 494 pp.

Fair, G. M., and Wells, W. F. 1934. Measurement of atmospheric pollution and contamination by sewage treatment works. Proc. 19th Annual Mtg. New Jersey Sewage Works Assoc., pp. 20-27.

Gallagher, J. L. 1975. The significance of the surface film in salt marsh plankton metabolism. *Limnol. Oceanog. 20*(1):120-123.

Garrett, W. D. 1965. Collection of slick-forming materials from the sea surface. *Limnol. Oceanog. 10*, 602-605.

Garrett, W. D. 1967a. Damping of capillary waves at the air-sea interface by oceanic surface-active material. *J. Mar. Res. 25*, 279-291.

Garrett, W. D. 1967b. The organic chemical composition of the ocean surface. *Deep-Sea Res. 14*, 221-227.

Garrett, W. D. 1968. The influence of monomolecular surface films on the production of condensation nuclei from bubbled sea water. *J. Geophys. Res. 73*, 5145-5150.

Garrett, W. D., and Barger, W. R. 1972. *Control and Confinement of Oil Pollution on Water with Monomolecular Surface Films*. NRL Report 2451, Naval Research Laboratory, Washington, D. C.

Gathman, S. G., and Hoppel, W. A. 1970. Electrification processes over Lake Superior. *J. Geophys. Res. 75*, 1041-1048.

Goff, G. D., Spendlove, J. C., Adams, A. P., and Nicholes, P. S. 1973. Emission of microbial aerosols from sewage treatment plants that use trickling filters. *Health Serv. Rept. 88*, 640-652.

Goldacre, R. J. 1949. Surface films on natural bodies of water. *J. Anim. Ecol. 18*, 36-39.

Harvey, G. W. 1966. Microlayer collection from the sea surface: A new method and initial results. *Limnol. Oceanog. 11*, 608-613.

Hatch, T., and Gross, P. 1964. *Pulmonary Deposition and Retention of Inhaled Aerosols*. Academic, New York.

Hatcher, R. F., and Parker, B. C. 1974a. *Investigations of Freshwater Surface Microlayers*. Virginia Polytechnic and State University, Water Resources Research Center Bulletin 64, 84 pp.

Hatcher, R. F., and Parker, B. C. 1974b. Laboratory comparisons of four surface microlayer samplers. *Limnol. Oceanog. 19,* 162-165.

Hatcher, R. F., and Parker, B. C. 1974c. Microbiological and chemical enrichment of freshwater-surface microlayers relative to the bulk-subsurface water. *Can. J. Microbiol. 20,* 1051-1057.

Hutchinson, G. E. 1967. *A Treatise on Limnology. Vol. II, Introduction to Lake Biology and the Limnoplankton.* Wiley, New York, 1115 pp.

Jarvis, N. L. 1962. The effect of monomolecular films on surface temperature and convective motion at the water-air interface. *J. Coll. Sci. 17,* 512-522.

Krueger, A. P., Reed, E. J., Day, M. B., and Brook, K. A. 1974. Further observations on the effect of air ions on influenza in the mouse. *Int. J. Biometeorol. 18,* 46-56.

Lemlich, R. 1966. Adsubble methods. *Chem. Eng. 73*(21), 7.

Lemlich, R. (Ed.). 1972. *Adsorptive Bubble Separation Techniques.* Academic, New York, 331 pp.

MacIntyre, F. 1968. Bubbles: a boundary-layer "microtome" for micronthick samples of a liquid surface. *J. Phys. Chem. 72,* 589-592.

MacIntyre, F. 1974. The top millimeter of the ocean. *Sci. Am. 230,* 62-77.

Maynard, N. G. 1968a. Aquatic foams as an ecological habitat. *Z. Allg. Mikrobiol. 8,* 119-126.

Maynard, N. G. 1968b. Significance of air-borne algae. *Z. Allg. Mikrobiol. 8,* 225-226.

Parker, B. C. (In press) The water-to-air pathway of the aerobiologic cycle. In *Proceedings of the Soil Microcommunities Conference,* D. Dindal (Ed.). Syracuse University, October 15-17, 1973.

Parker, B. C., and Barsom, G. 1970. Biological and chemical significance of surface microlayers in aquatic ecosystems. *Bioscience 20,* 87-93.

Parker, B. C., and Hatcher, R. F. 1974. Enrichment of surface freshwater microlayers with algae. *J. Phycol. 10,* 185-189.

Paterson, M. P., and Spillane, K. T. 1969. Surface films and the production of sea-salt aerosol. *Quart. J. Roy. Met. Soc. 95*, 526-534.

Reiter, R. 1973. Symposium on Biological Effects of Natural Electric, Magnetic and Electromagnetic Fields. *Int. J. Biometeorol. 17*, 205-309. (Collection of 14 papers presented at Symposium chaired by R. Reiter.)

Riley, G. A. 1970. Particulate organic matter in sea water. *Adv. Mar. Biol. 8*, 1-118.

Rubin, A. J. 1968. Microflotation: Coagulation and foam separation of *Aerobacter aerogenes*. *Biotechnol. Bioeng. 10*, 89-98.

Salt, R. W. 1961. Effect of electrostatic field on freezing of supercooled water and insects. *Science 133*, 458-459.

Schlichting, H. E. Jr. 1964. Meteorological conditions affecting the dispersal of airborne algae and protozoa. *Lloydia 27*, 64-78.

Schlichting, H. E. Jr. 1974. Ejection of microalgae into the air via bursting bubbles. *J. Allerg. Clin. Immunol. 53*, 185-188.

Schnell, R. C. 1974. *Biogenic Sources of Atmospheric Ice Nuclei*. Report #AR111. Department of Atmospheric Resources, University of Wyoming, Laramie. 45 pp.

Schnell, R., Fresh, R., and Vali, G. 1973. Freezing nuclei from decaying vegetation. VIII Nucleation Conf., Leningrad, USSR.

Schnell, R. C., and Vali, G. 1973. Worldwide source of leaf-derived freezing nuclei. *Nature 246*, 212-213.

Seba, D. B., and Corcoran, E. F. 1969. Surface slicks as concentrations of pesticides in the marine environment. *Pestic. Monit. J. 3*, 190-193.

Sieburth, J. McN. 1971. Distribution and activity of oceanic bacteria. *Deep-Sea Res. 18*, 1111-1121.

Spendlove, J. C. 1974. *Developments in Industrial Microbiology. Vol. 15, Industrial, Agriculture, and Municipal Microbial Aerosol Problems*. Amer. Inst. of Biological Sciences, Washington, D. C., pp. 20-27.

Sutcliffe, W. H. Jr., Baylor, E. R., and Menzel, D. W. 1963. Sea surface chemistry and Langmuir circulations. *Deep-Sea Res. 10*, 233-243.

VanGrieken, R. E., Johansson, T. B., and Winchester, J. W. 1974. Trace metal fractionation effects between sea water and aerosols from bubble bursting. *J. Rech. Atmosphér. 8*(3-4), 611-621.

Wallace, G. T., and Duce, R. A. 1974. Concentration of particulate trace metals and particulate organic carbon in marine surface waters by a bubble flotation mechanism. Unpublished manuscript, Graduate School of Oceanography, Univ. of Rhode Island, Kingston, R.I.

Wellock, C. E. 1960. Epidemiology of Q fever in the urban East Bay Area. *Calif. Health 18*(10), 72-76.

Wilson, A. T. 1959. Surface of the ocean as a source of air-borne nitrogenous material and other plant nutrients. *Nature 184*, 99-101.

Woodcock, A. H. 1948. Note concerning human respiratory irritation associated with high concentrations of plankton and mass mortality of marine organisms. *J. Mar. Res. 7*, 56-62.

Woodcock, A. H. 1953. Salt nuclei in marine air as a function of altitude and wind force. *J. Meteorol. 10*, 362-371.

Woodcock, A. H. 1955. Bursting bubbles and air pollution. *Sewage Ind. Wastes 27*, 1189-1192.

Woodcock, A. H., Duce, R. A., and Moyers, J. L. 1971. Salt particles and raindrops in Hawaii. *J. Atmos. Sci. 28*, 1252-1257.

Woodcock, A. H., Kientzler, C. F., Arons, C. F., and Blanchard, D. C. 1953. Giant condensation nuclei from bursting bubbles. *Nature 172*, 1144.

Chapter 20

SEASONAL DISTRIBUTION OF B VITAMINS IN RAINWATER

I. COBALAMINS, BIOTIN, AND NIACIN
IN ST. LOUIS RAINWATER

Bruce C. Parker

Department of Biology
Virginia Polytechnic Institute and State University
Blacksburg, Virginia

and

Mary Ann Wachtel

Saint Louis University
College of Arts and Sciences
St. Louis, Missouri

CONTENTS

II. BIOTIN, NIACIN, AND THIAMIN
IN BLACKSBURG RAINWATER

Bruce C. Parker

ABSTRACT

Bioassays of rainwater for dissolved cobalamins, biotin, and niacin during 15 months in the St. Louis area show that appreciable quantities of these vitamins may reach terrestrial and aquatic ecosystems via atmospheric precipitation. At least for biotin and niacin, the frequency of occurrence and concentration of vitamins is higher during the growing season relative to the November 1 to March 31 period. Limited evidence suggests airborne soil particles and pollen as sources of these vitamins. In addition, clouds are conceived as hypothetical atmospheric ecosystems containing metabolically active microorganisms associated with cloud condensation nuclei and nutrient-rich water droplets.

INTRODUCTION

For convenience and for the historical record, this chapter constitutes a reprinting of the vitamin assay data on St. Louis rainwater reported in the earlier paper by Parker and Wachtel (1971). To this has been added a second section that summarizes data obtained from analyses of rainwater collected at Virginia Polytechnic Institute and State University, Blacksburg, Virginia. The discussion at the end of the second section addresses the data in total.

Parker (1968) first reported the occurrence of cobalamins (vitamin B_{12}) in St. Louis rainwater. This discovery occurred during preliminary tests in a goldfish pond of a new sampling apparatus called a biodialystat (Parker 1967a,b). On several occasions, the B_{12} concentration in

661

biodialystats rose significantly 12-24 hr following initiation of spring rains, and this rise in B_{12} coincided with a bloom of *Chlamydomonas* in the pond. A few days later, the B_{12} concentration dropped concurrently with decline in the algal bloom.

Parker (1968) confirmed that the bulk of cobalamins added to the pond came from rains instead of runoff, vegetation drip, or in situ production. Vitamin B_{12} was so concentrated in several of these 1967 rains derived from convective storms that one could well imagine an appreciable impact of rainborne cobalamins on other aquatic ecosystems in the St. Louis area. Indeed, if a hypothetical lake containing *Euglena gracilis* and a sufficiency of all nutrients except B_{12} received a rainfall of 1.0 cm containing 20 pg/ml cobalamins, *Euglena* could increase to about 10^6 cells/cm^2 of lake surface, according to Parker's calculation. This would produce a visible bloom. The papers of Provasoli (1958) and Menzel and Spaeth (1962) also bear out the probability that cobalamins may limit natural phytoplankton populations.

Results of this earlier study stimulated the initiation of research to evaluate the seasonal contribution of rainfall in the St. Louis area to the cobalamin reserve and cycle in lakes. This program soon expanded to include biotin and niacin.

METHODS

Soluble B_{12} was assayed with *Euglena gracilis* Z strain using the general procedure of Robbins *et al.* (1950). Biotin and niacin assays utilized the bacterium *Lactobacillus plantarum*, according to the methods outlined in Difco Manual, 9th ed. (1953).

Rainwater was collected both at the senior author's home and at Washington University in open stainless steel

or aluminum containers placed on the roof away from over-
hanging vegetation. Most collections were frozen immediately following collection and prior to filtration through
0.22 μm GS Millipore filters. A total of 106 collections
were made between April 25, 1967, and June 9, 1969. Initially collections were sporadic, but from June 1968 to
June 1969 almost every rain and snow in excess of 0.1 cm
liquid depth was collected for assay. The total rainfall
collected during that period approximates 87 cm, which
approaches the average annual rainfall measured by the
U. S. Weather Bureau in this region. Therefore, we are
confident that our collections have been fairly complete.

Euglena assays for cobalamins utilized hand counts
with a hemacytometer during the first year and a Coulter
electronic particle counter subsequently. The inoculum
was standardized at 600 cells/ml. Assays were run in
triplicate, usually with reasonably low variation among
replicate flasks. Standard curves based on *Euglena* growth
for known concentrations of cyanocobalamin varied only
slightly by our method. As performed in our laboratory,
we consider the assay for B_{12} reliable for concentrations
above 0.20 pg/ml.

Lactobacillus assays for biotin and niacin utilized
turbidity measurements at 540 mμm on a B & L Spectronic
20 colorimeter. Except where noted, results are expressed
as means of two separate assay runs each consisting of two
cultures (i.e., four replicates). We find greater variation with replicate cultures of these bacteria. Thus, by
our method, we consider only levels above 1.0 pg biotin
and 1.0 mμg niacin/ml as reliably measured.

We have used several specific modifications in the
published methods in order to improve the reliability of
our assays:

    1. All Pyrex glassware is pretreated in a muffle

furnace 30 min at 600 C to remove trace organics
adhering to the glass.

2. In biotin and niacin assays, we omit the saline
    solution step, which is used for a dilution of
    the inocula. Instead we substitute an extra
    transfer step using vitamin-free medium.

3. Also, in assaying biotin and niacin turbidimetri-
    cally, we read the turbidity of the first tubes
    in a series both at beginning and end. Because
    the reading of a series of 20 duplicate tubes
    may take 15 min, these exponential-phase bacterial
    cultures frequently show a significant change in
    turbidity in 15 min.

4. Prior to media sterilization, the autoclave is
    scrubbed with 95% ethanol. We have shown, at
    least for vitamin B_{12} (Parker 1968), that residual
    vitamin derived from prior autoclaving of complex
    media can contaminate vitamin-free media at high
    temperature and pressure.

RESULTS

Table 20.1 summarizes the data, including those reported earlier for cobalamins (Parker 1968). The term
"nil" is used here both for values that are negative and
for those too low for accurate detection. Such negative
values were especially common with cobalamin assays. A
range in concentrations for biotin or niacin indicates that
one experiment gave the lower reading while the other gave
the higher.

Note the numerous instances where a series of collections was made during continuous rainfall. Of 15 series
collections, cobalamins varied appreciably during the
course of the rain only in one case; on July 15, 1968, the
first fraction of rain contained 0.20 pg cobalamin/ml,

Table 20.1

Concentrations of Cobalamin, Biotin, and Niacin in Rainfall, St. Louis, Missouri, April 25, 1967 to June 9, 1969

Collection Date	Vitamin concentration/ml (means of all replicates)			Calculated vitamin/cm^2 of land and/or water surface (conc/ml x cm rainfall)		
	Cobalamin (pg)	Biotin (pg)	Niacin (mμg)	Cobalamin (pg)	Biotin (pg)	Niacin (mμg)
4/25/67	20.00			15.20		
6/21/67	0.25			0.78		
6/27/67	20.00			32.00		
7/12/67	Nil			Nil		
8/15-16/67	Nil			Nil		
1/29-30/68	Nil			Nil		
3/18/68	3.80	0-5.0	1.4	1.10	0-1.5	0.4
4/17/68†	9.80			1.10		
4/28-29/68	Nil			Nil		
5/7-8/68†	Nil			Nil		
5/15/68†	Nil			Nil		
5/23/68	Nil			Nil		
5/21/68[a]	Nil			Nil		
5/22-23/68[b]	Nil			Nil		

Table 20.1 (Continued)

Collection Date	Vitamin concentration/ml (means of all replicates)			Calculated vitamin/cm² of land and/or water surface (conc/ml x cm rainfall)		
	Cobalamin (pg)	Biotin (pg)	Niacin (mμg)	Cobalamin (pg)	Biotin (pg)	Niacin (mμg)
5/23/68[a]	Nil			Nil		
5/23-24/68[d]	Nil			Nil		
5/28-29/68	Nil	3.5	2.9	Nil	††	††
5/31/68[a]	Nil			Nil		
5/31/68[b]	Nil			Nil		
5/31-6/1/68[c]	Nil			Nil		
6/11/68	<0.20			<0.20		
6/15/68	<0.20	4.3	0-1.0	<0.30	6.5	0-1.5
6/16/68	<0.20			<0.02		
6/19/68	1.40			0.14		
6/22/68	5.60	9.1	0-4.4	2.80	4.6	0-0.7
6/25/68	<0.20	6.5	0-1.2	<0.14	4.6	0-0.8
7/1/68	<0.20			<0.18		
7/14/68	0.20	3.4	0-1.0	0.75	12.8	0-15.0
7/15/68[a]	<0.20			0.04		
7/15/68[b]	4.50			††		

Date						
7/17/68	<0.20	5.8	Nil	<0.48	13.9	Nil
7/18/68	0.30	5.0*	Nil	0.05	0.8	Nil
7/24/68[a]	0.20	9.1	2.4*	0.06	2.7	0.7
7/24-25/68[b]	<0.20	5.0	Nil	<0.06	1.5	Nil
7/25/68[c]	<0.20	5.5	3.2	<0.20	5.5	3.2
7/25/68[d]	<0.20	2.7	Nil	<0.03	0.4	Nil
7/26/68	<0.20	5.4	Nil	<0.04	1.1	Nil
7/31/68	<0.20	5.9	0-1.2	<0.08	2.4	0-0.5
8/7/68	0.20	19.0*	3.2*	0.05	4.8	0.8
8/14/68	<0.20	7.5	2.2	<0.04	1.5	0.4
8/15/68	0.20	4.2	0-3.8	0.20	4.2	0-3.8
8/17/68	<0.20	6.0	1.8	<0.04	1.2	0.4
8/30-31/68	<0.20	18.0	3.8	<0.12	10.8	2.3
9/15-16/68	<0.20	0-3.6	0-4.4	<0.10	0-1.8	0-2.2
9/17/68	0.20	13.2	0-4.0	0.14	9.2	0-2.8
9/17-18/68 (unfrozen)	<0.20	0-1.0	0-3.6	<0.54	0-2.7	0-9.7
9/17-18/68 (frozen)	<0.20	0-2.0	0-4.0	<0.54	0-5.4	0-10.8
9/19/68	<0.20	3.0	3.0	<0.16	2.4	2.4
9/24/68	<0.20	3.3	3.0	<0.10	1.7	1.5
9/28-29/68		3.5	2.9			
10/5-6/68	<0.20	13.0	3.6	0.06	3.9	1.11

Table 20.1 (Continued)

Collection Date	Vitamin concentration/ml (means of all replicates)			Calculated vitamin/cm² of land and/or water surface (conc/ml x cm rainfall)		
	Cobalamin (pg)	Biotin (pg)	Niacin (mμg)	Cobalamin (pg)	Biotin (pg)	Niacin (mμg)
10/13-14/68	<0.20	3.8	3.0	<0.20	3.8	3.0
10/17/68	4.10	18.0	4.6	2.05	9.0	2.3
11/2-3/68[a]	<0.20	7.4	0-0.6	<0.20	7.4	0-0.6
11/4/68[b]	<0.20	6.2	3.3	<0.40	12.4	6.6
11/7/68	<0.20	0-5.0	0-1.6	<0.20	0-6.5	0-1.6
11/14-15/68[a]	<0.20	0-4.4	Nil	<0.20	0-4.4	Nil
11/15/68[b]	<0.20	Nil	Nil	<0.92	Nil	Nil
11/17/68[a]	...	...	0-0.8	++	++	++
11/17-18/68[b]	<0.20	0-2.0	Nil	<0.10	0-1.0	Nil
11/23/68	0.27	9.0*	1.7	0.03	0.9	0.2
11/26-27/68[a]	<0.20	7.6*	3.6*	<0.16	6.08	2.9
11/28/68[b]	<0.20	2.0	0-1.0	<0.24	2.4	0-1.2
11/30-12/1/68	<0.20	0-0.4	0-0.8	<0.28	0-0.6	0-1.1
12/18/68[a]	0.30	0-18.0	Nil	0.30	0-18.0	Nil
12/18/68[b]	0.88	Nil*	2.0*	0.13	Nil	0.3
12/21-22/68	<0.20	Nil	0-1.4	<0.16	Nil	0-1.1

Date						
12/27/68[a]	<0.20	Nil	0-0.4	<0.64	Nil	1.3
12/27-28/68[b]	<0.20	Nil	0-0.4	<0.22	Nil	0-0.4
12/30/68	<0.20	Nil	0-1.0	<0.20	Nil	0-1.0
1/16/69[a]	<0.20	Nil	0-0.6	<0.22	Nil	0-0.7
1/17/69[b]	<0.20	Nil	0.8	<0.04	Nil	0.2
1/17/69[c]	<0.20	Nil	Nil	<0.30	Nil	Nil
1/21/69	<0.20	Nil	Nil*	<0.03	Nil	Nil
1/22/69	<0.20	Nil	Nil	<0.18	++	Nil
1/23/69	<0.20	0-1.6	Nil	++	Nil	Nil
1/27/69[a]	<0.20	Nil	Nil	<0.10	Nil	Nil
1/28/69[b]	<0.20	Nil	Nil	<0.22	Nil	Nil
1/29/69[c]	<0.20	Nil	Nil	<0.42	Nil	Nil
1/29/69[d]	<0.20	Nil	Nil	<0.10	0.3	Nil
2/5-6/69	0.30	1.1	0-1.6	0.09	0-0.9	0-0.5
2/7-8/69	1.50	0-0.4	Nil	3.30	++	Nil
2/16-17/69 (snow)	8.00	2.1	1.9	++	2.9	++
2/22/69	<0.20	2.1	Nil	<0.28	0-0.4	Nil
2/27-28/69	Nil	0-0.2	Nil	Nil	2.2	Nil
3/7/69	Nil	2.0	1.5	Nil	5.6	1.7
3/23-24/69	Nil	1.7	2.1	Nil	0.7	6.9
3/26/69	Nil	2.3		Nil		
4/4/69 (rain)[a]		2.1	2.4		2.1	2.4
4/4/69 (thunder)[b]	Nil	1.6	2.6	Nil	2.2	3.6

Table 20.1 (Continued)

Collection Date	Vitamin concentration/ml (means of all replicates)			Calculated vitamin/cm² of land and/or water surface (conc/ml x cm rainfall)		
	Cobalamin (pg)	Biotin (pg)	Niacin (mμg)	Cobalamin (pg)	Biotin (pg)	Niacin (mμg)
4/8-9/69	Nil	0-1.6	1.7	Nil	0-4.2	4.4
4/13/69[a]	Nil	Nil	1.9	Nil	Nil	1.9
4/14/69[b]	0.20	Nil	1.8	††	Nil	††
4/14/69[c]	Nil	0-1.2	1.0	††	††	††
4/17-18/69[a]	0.22	2.0	2.8	0.22	2.0	2.8
4/18/69[b]	Nil	1.6	1.9	Nil	1.4	1.7
4/19/69	Nil	2.5	3.6	Nil	1.0	1.4
4/27/69	0.22	2.5	3.6	††	††	††
5/7-8/69	0.20	6.9	14.9	0.08	2.8	6.0
5/13/69	0.20	4.7	4.0	0.28	6.6	5.6
5/21-22/69	<0.20	6.8	3.3	<0.20	6.8	3.3
5/31/69	<0.20	2.2	1.7	<0.22	2.4	1.9
5/31-6/1/69	<0.20	Nil	2.2	<0.14	Nil	1.5
6/1/69	Nil	2.4	2.1	Nil	10.8	9.5
6/8/69[a]	<0.20	2.1	1.6	<0.90	9.5	7.2
6/8-9/69[b]	0.23	2.5	2.0	0.18	2.0	1.6

*One experiment only; insufficient water for repeat.

†Collected in 95% EtOH, later removed by vacuum distillation before assay.

a, b, etc. refers to consecutive collections, regarded here as a single rain; this notation also true for other tables.

††Indicates inability to calculate due to inaccurate rainfall depth measurement.

while the second fraction contained 4.50. Of 12 series collections, biotin exhibited a downward trend in concentration during the course of rainfall in six cases, and no appreciable change for the other cases. Of 13 series collections, niacin showed some variability during the course of rains; however, the downward trend in concentration occurred in several experiments. In general, the concentrations of vitamins exhibit no consistent trend during long-term rains. Furthermore, these data provide only weak support for the idea of an early rainout of vitamins, which might be predicted if the vitamins were derived from airborne dust. Also, in several attempts to correlate B vitamin concentrations with total dry weight of nonfilterable particulates in rains, we have found no consistent pattern.

Three rains (April 17, May 7-8, and May 15, 1968) were collected in a large volume of ethanol to preclude any vitamin contribution from microbial activity between the beginning of rain collection and freezing. For assay, the ethanol was removed by vacuum distillation at reduced temperature prior to filtering. Note that April 17, 1968, rain contained appreciable cobalamin. We have not repeated this ethanol-collection procedure for biotin and niacin.

In Table 20.1 rain occurring on September 17-18, 1968, was divided into two portions after collection. One fraction was frozen in routine fashion; the other was left unfrozen and prepared immediately for bioassay. Although the results of replicate assays of this rainwater showed some variation, there was no indication that freezing prior to filtration caused appreciable change in assayable vitamins.

As noted previously, our experience with the assays for these vitamins suggests that concentrations above 0.20 pg cobalamin/ml, 1.0 pg biotin/ml, and 1.0 mμg niacin/ml

are reliably detected. Table 20.2 lists those rains that
contained one or more vitamins in this detectable range.
This table makes it obvious that a significant level of
biotin or niacin occurred more frequently than cobalamin.
Generally, the lower levels for all three vitamins occurred
during winter months. The only high value during winter
for cobalamin was 8.00 pg/ml, derived from a heavy snow
storm on December 16-17, 1969. Also of interest, the highest values for biotin and niacin during 1969 occurred on
May 7-8, the period when the major quantity of pine pollen
was released in the vicinity of the collection vessels.
This pollen fell with the rain in significant visible
amounts.

Table 20.2 shows that all three vitamins occurred in
significant amounts in only five rains, cobalamins and biotin occurred together six times, and cobalamins occurred
with niacin seven times. In contrast to these relationships, biotin and niacin occurred in significant amounts
concurrently 28 times.

Table 20.3 presents additional calculations derived
from Table 20.1. These calculations show that biotin and
niacin occurred in significant amounts in 78 and 61% of
the rains during the growing season, respectively. Such
frequencies of occurrence are considerably higher than the
value of 19% for cobalamins. Table 20.3 also shows that
the average concentrations of vitamins in rain, especially
during the growing season, are well above the levels at
which the growth of assay organisms are stimulated. The
estimated annual contribution of these vitamins by rains
in the St. Louis area are at best conservative.

DISCUSSION

This research, while leaving many questions unanswered, has proved that rainwater may contribute significant

Table 20.2

Rains with over 0.20 pg Cobalamin, 1.0 pg Biotin,
and 1.0 mμg Niacin/ml

Collection Date	Cobalamin	Biotin	Niacin
1967			
4/25	20.00		
6/21	0.25		
6/27	20.00		
1968			
3/18	3.80		1.4
4/17	9.80		
5/28,29		3.5	2.9
6/15		4.3	
6/19	1.40		
6/22	5.60	9.1	
6/25		6.5	
7/14		3.4	
7/15	4.50		
7/17		5.8	
7/18	0.30		
7/24[a]		9.1	
7/24,25[b]		5.0	
7/25[c]		5.5	3.2
7/25[d]		2.7	
7/26		5.4	
7/31		5.9	
8/14		7.5	2.2
8/15		4.2	
8/17		6.0	1.8
8/30,31		18.0	3.8
9/17		13.2	
9/19		3.0	3.0

Table 20.2 (Continued)

Collection Date	Cobalamin	Biotin	Niacin
9/24		3.3	3.0
9/28,29		3.5	2.9
10/5,6		13.0	3.6
10/13,14		3.8	3.0
10/17	4.10	18.0	4.6
11/2,3[a]		7.4	
11/4[b]		6.2	3.3
11/23	0.27		1.7
11/28		2.0	
12/18[a]	0.30		
12/18[b]	0.88		
1969			
2/5,6	0.30	1.1	
2/7,8	1.50		
2/16,17	8.00	2.1	1.9
2/22		2.1	
3/7		2.0	1.5
3/23,24		1.7	2.1
3/26		2.3	
4/4[a]		2.1	2.4
4/4[b]		1.6	2.6
4/8,9			1.7
4/13[a]			1.9
4/14[b]			1.8
4/17,18[a]	0.22	2.0	2.8
4/18[b]		1.6	1.9
4/19[c]		2.5	3.6
4/27	0.22	2.5	3.6
5/7,8		6.9	14.9
5/13		4.7	4.0

Table 20.2 (Continued)

Collection Date	Cobalamin	Biotin	Niacin
5/21,22		6.8	3.3
5/31, 6/1			2.2
5/31		2.2	1.7
6/1		2.4	2.1
6/8		2.1	1.6
6/8,9	0.23	2.5	2.0

amounts of water-soluble vitamins to terrestrial and aquatic ecosystems. The concept of rain as a vector for inorganics must now be expanded to include trace organic substances. Such substances may act in the aquatic environment as (1) energy substrates, (2) carbon skeletons, (3) accessory growth factors, including vitamins, (4) inhibitors, (5) chelators, etc. Parker (1968) reported that rains, in the St. Louis area, at least, sometimes contained more than 8 mg/liter of dissolved organic matter, a concentration equal to or higher than that found in several small lakes in that region. Thus, rains may increase the concentration of total dissolved organics in lakes as well as modify the in situ organic matter composition. The level of 8 mg/liter total dissolved organic matter in St. Louis rains is about 10^6 times the sum of concentrations of the three vitamins treated in this study. Other workers have detected such trace organics as chlorinated hydrocarbon pesticides and herbicides, terpenoids, and organic oxidants (Went et al. 1967; Parker and Barsom 1970). Therefore, the main bulk of rainborne organic solutes is still unidentified.

In the St. Louis area, the concentrations of dissolved cobalamins in rainwater sometimes exceed that within lakes

Table 20.3

Calculation Summarizing Data from Table 20.1

Calculations	Cobalamin	Biotin	Niacin
Total rain samples assayed	103	81	81
Rains with significant concentrations of vitamins[a]	19 (19%)	45 (55%)	34 (42%)
Rains assayed April 1 to November 1	68	46	46
Rains April 1 to November 1 with significant concentrations of vitamins[a]	12 (18%)	36 (78%)	28 (61%)
Rains assayed November 1 to March 31	35	35	35
Rains November 1 to March 31 with significant concentration of vitamins[a]	7 (20%)	9 (26%)	6 (18%)
Average concentration of vitamin/rain, April 1 to November 1	0.98 pg/ml	4.1 pg/ml	2.0 mμg/ml
Average concentration of vitamin/rain, November 1 to March 31	0.43 pg/ml	0.75 pg/ml	0.42 mμg/ml
Estimated annual contribution of vitamins per cm^2 by St. Louis rains	74	250	108

[a]Significance here refers to 0.20 pg cobalamin, 1.0 pg biotin, and 1.0 mμg niacin/ml, as justified in text.

in that area. We have not done biotin and niacin assays
on St. Louis lakes; however, the concentrations of these
two vitamins in rainwater are orders of magnitude above
the levels reported in lakes and the ocean (Vallentyne
1957; Carlucci and Sibernagel 1967). Consequently, we
conclude on theoretical grounds that the contribution of
these three vitamins by rain may augment and influence
appreciably some aquatic environments.

Our data are too limited for an elaborate evaluation
of the ecological importance of rainborne vitamins to
aquatic ecosystems. First, the responses of assay organisms in axenic-defined media need not be identical to those
of vitamin-requiring members of microbial communities.
Even those organisms used for assay of single vitamins
have varying degrees of sensitivity and specificity for
analogs (Baker and Sobotka 1962; Carlucci and Sibernagel
1967). Second, our data treat only that fraction of the
vitamin content of rain that passes through the 0.22 μm
pores of GS Millipore membranes. So far, no method for
determining sestonic or total vitamin has proved satisfactory in this laboratory (Parker 1969). Even our measured filterable vitamin levels may be inaccurate. We
know, for example, that 45-mm diameter Millipore membranes
absorb up to 25 pg of cobalamins, and this amount is not
elutable with water. Thus our procedure may cause loss
of up to 0.25 pg cobalamins/ml by membrane absorption, and
this phenomenon may account in part for the lower frequency
of occurrence of significant B_{12} concentrations in rainwater. We have not determined the amounts of biotin and
niacin absorption on Millipore membranes. Third, the
rather high frequency of negative values for our *Euglena*
B_{12} assay of rainwater may stem from the coexistence of
antimetabolites for this vitamin. Consequently, our bioassay results for these vitamins represent only rough

approximations of the real concentrations dissolved in
rainwater.

A more thorough evaluation of the quantitative importance of rainwater organics to aquatic systems necessitates
more information on (1) other constituents of rainwater,
(2) their frequency and quantity of distribution, (3) the
sources of such organics, and (4) conditions relating to
their occurrence. We have only begun to explore these
subjects. Our preliminary data suggest little connection
between the total particulates and the concentrations of
these vitamins in rains. Also, several attempts to extract
cobalamins from airborne dust have given nil values, suggesting that either the extraction method failed or dry
airborne dust is not a major source of this vitamin. For
example, calcareous clays absorb B_{12} tenaciously, and their
presence in airborne dust in the St. Louis area may interfere with cobalamin extraction.

A clue to one possible source of B vitamins in rains
comes from the correlation between pine pollen and high biotin and niacin levels during the 1969 spring. Pollen and
spores often comprise a major fraction of airborne dust
and rainwater particulates (McDonald 1962; Pop et al. 1964;
Gregory and Monteith 1967). Also the available pollen
and spore count data for the St. Louis area suggest that
the high values of niacin and biotin during the fall of
1968 coincide somewhat with peaks for weed pollen, while
high niacin during late March through May 1969 correlates
with peaks for tree pollen. The dominant trees in the St.
Louis area are *Pinus, Picea, Salix, Populus, Quercus,* and
Ulmus, all of which produce appreciable airborne pollen
during spring.

In conclusion, we wish to suggest a second possible
origin for vitamins and possibly other organics in rain.
We propose that part of these substances may be synthesized

by microorganisms living within clouds. Once borne into
the atmosphere, microorganisms can remain suspended for
long periods; for example, a 50-μm-sized particle will remain suspended indefinitely in the atmosphere over St.
Louis if it is not further aggregated or brought down by
rain.

Fischer *et al.* (1969) records that continental air
may contain on the average 100 particles/cm^2, ranging in
diameter from 0.2 to 2.0 μm. The values are lower, but
nevertheless significant, for progressively larger particles up to about 50 μm diameter. The composition of these
large cloud-inhabiting particulates has not, to our knowledge, been elucidated. However, Gregory (personal communication) has confirmed the presence of microorganisms
in clouds over Britain. Also the works of Schlichting
(1961), of Brown *et al.* (1964) for algae, and of Zobell
(1942) for bacteria document that viable microorganisms
from the seas as well as land surface can reach high altitudes in the atmosphere.

According to our hypothesis, cloud microorganisms and
organic particulates comprise a small fraction of the condensation nuclei for rain droplets. During the period of
condensation, which may vary up to many days, extensive
microbial activity within unfrozen condensation droplets
of some clouds occurs. The diameters of cloud droplets
measure up to 100 μm, and we have no reason to suspect such
droplets are nutrient poor on the basis of airborne dust
and salt nuclei (Aufm Kampe and Weickmann 1957; Weickmann
1957). Furthermore, cloud moisture should absorb some of
the harmful radiation that might otherwise kill these microorganisms or destroy the vitamins produced.

This hypothesis does not exclude contributions of
organic matter to rain from other sources. Indeed Woodcock
(1955) noted that in addition to "giant" hygroscopic nuclei

arising from bursting air bubbles at the sea surface, a sizable amount of organic matter also entered the atmosphere from sea. Also, the works of Valencia (1967) and Garrett (1968) show that large amounts of nonvolatile organic matter from the sea surface is borne into the atmosphere (see also Parker and Barsom 1970). Wilson (1959) found bacteria, diatoms and fragments of phyto- and zooplankton borne into the atmosphere over New Zealand from bursting air bubbles. Neumann *et al.* (1959) also proposed that the large amounts of organic matter found in the atmosphere might also arise from the sea surface.

Our preliminary observations of airborne dust indicate that some microorganisms (yeasts, algae, bacteria) may remain vegetative and metabolically active within the air. May and Druett (1968) also have demonstrated viability of select microorganisms on threads that presumably simulate airborne moisture conditions. If microorganisms can undergo metabolism within clouds, then these and their organic products may augment the composition of rainwater derived from other mechanisms. Frequently, freezing in clouds at these high altitudes just prior to rainfall, a feature especially common to thunderclouds (Weickmann 1957), may squeeze soluble vitamins from cells during falling rains.

To our knowledge, no one previously has envisioned clouds as living ecosystems. Direct evidence supporting this hypothesis is lacking at this time, as is also the evidence to refute it. We hope to test this hypothesis in the course of examining further the influence of rainborne organic substances on aquatic ecosystems.

ACKNOWLEDGMENTS

We are grateful to the Center for the Biology of Natural Systems, Washington University, for support of this research under PHS grant P10 ES 00139 and to Kay Williams for assistance with some of the vitamin assays.

REFERENCES

Aufm Kampe, H. J., and Weickmann, H. K. 1957. Physics of clouds. *Meteorol. Res. Rev. 3*, 182-225.

Baker, H., and Sobotka, H. 1962. Microbiological assay methods for vitamins. *Adv. Clin. Chem. 5*, 173-235.

Brown, R. M., Larson, D. A., and Bold, H. C. 1964. Airborne algae: Their abundance and heterogeneity. *Science 143*, 583-585.

Carlucci, A. F., and Sibernagel, S. B. 1967. Bioassay of seawater. IV. The determination of dissolved biotin in seawater using ^{14}C uptake by cells of *Amphidinium carteri*. *Can. J. Microbiol. 13*, 979-986.

Difco Manual (authors anonymous). 1953. Difco manual of dehydrated culture media and reagents for microbiological and clinical laboratory procedures. Difco Laboratories, Inc., Detroit, Michigan, 350 pp.

Fischer, W. H., Lodge, J. P. Jr., Pate, J. B., and Cadle, R. D. 1969. Antarctic atmospheric chemistry: Preliminary exploration. *Science 164*, 66-67.

Garrett, W. D. 1968. The influence of monomolecular surface films on the production of condensation nuclei from bubbled seawater. *J. Geophys. Res. 73*, 5145-5150.

Gregory, P. H., and Monteith, J. L. (Eds.). 1967. *Airborne microbes*. A symposium of the Society for General Microbiology held in London, April 1967. Cambridge University Press, New York, 397 pp.

May, K. R., and Druett, H. A. 1968. A microthread technique for studying the viability of microbes in a simulated air-borne state. *J. Gen. Microbiol. 51*, 353-366.

McDonald, J. E. 1962. Collection and washout of air-borne pollens and spores by raindrops. *Science 135*, 435-436.

Menzel, D. W., and Spaeth, J. P. 1962. Occurrence of vitamin B_{12} in the Sargasso Sea. *Limnol. Oceanog. 7*, 151-154.

Neumann, G. H., Fonselius, S., and Wahlman, L. 1959. Measurements on the content of non-volatile organic material in atmospheric precipitation. *Int. J. Air Poll. 2*, 132-141.

Parker, B. C. 1967a. Biodialystat: New sampler for dissolved organic matter. *Limno. Oceanog.* 12, 722-723.

Parker, B. C. 1967b. Influence of method for removal of seston on the dissolved organic matter. *J. Phycol.* 3, 166-173.

Parker, B. C. 1968. Rain as a source of vitamin B_{12}. *Nature* 219, 617-618.

Parker, B. C. 1969. Influence of method for removal of seston on the dissolved organic matter. II. Cobalamins. *J. Phycol.* 5, 124-127.

Parker, B. C., and Barsom, G. 1970. Biological and chemical significance of surface microlayers in aquatic ecosystems. *Bioscience* 20, 87-93.

Parker, B. C., and Wachtel, M. A. 1971. Seasonal distribution of cobalamin, biotin and niacin in rainwater. In *The Structure and Function of Freshwater Microbial Communities*, J. Cairns, Jr. (Ed.)., Virginia Polytechnic Institute and State University Press, Blacksburg, Va., p. 195-207.

Pop, E., Boscain, N., Flavia, R., Diaconeasa, B., and Todoran, A. 1964. Effects of atmospheric precipitations on the pollen and spores concentration from the aeroplankton. *Rev Roumaine Biol. Ser. Bot.* 9, 329-334.

Provasoli, L. 1958. Nutrition and ecology of protozoa and algae. *Ann. Rev. Microbiol.* 12, 279-308.

Robbins, W. J., Hervey, A., and Stebbins, M. E. 1950. Studies of *Euglena* and vitamin B_{12}. *Bull. Torrey Bot. Club* 77, 423-441.

Schlichting, H. E. 1961. Viable species of algae and protozoa in the atmosphere. *Lloydia* 24, 81-88.

Valencia, M. J. 1967. Recycling of pollen from an air-water interface. *Am. J. Sci.* 265, 843-847.

Vallentyne, J. R. 1957. The molecular nature of organic matter in lakes and oceans with lesser reference to sewage and terrestrial soils. *J. Fish. Res. Bd. Can.* 14, 33-82.

Weickmann, H. K. 1957. Physics of precipitation. *Meteorol. Res. Rev.* 3, 226-255.

Went, F. W., Slemmons, D. B., and Mozingo, H. N. 1967.
The organic nature of atmospheric condensation nuclei.
Proc. Nat. Acad. Sci. 58, 69-74.

Wilson, A. T. 1959. Surface of the ocean as a source of
air-borne nitrogenous material and other plant nutrients. *Nature 104*, 99-101.

Woodcock, A. H. 1955. Bursting bubbles and air pollution.
Sewage Ind. Wastes 27, 1189-1192.

Zobell, C. E. 1942. *Microorganisms in Marine Air.* Contr.
No. 157, Scripps Inst. Oceanogr., La Jolla, Calif.

INTRODUCTION

The periodic occurrence of cobalamins, biotin, and niacin in significant concentrations in rainwater collected in St. Louis raised several questions:

1. How widespread a phenomenon was it, or more specifically would a rural area, such as Blacksburg, Virginia, also yield rains sometimes rich in B vitamins?
2. Was thiamin also present?
3. Were there any correlations between the vitamins and inorganic ion concentrations or microorganisms?

Studies were undertaken to answer these questions at Virginia Polytechnic Institute and State University.

MATERIALS AND METHODS

All rainwater was collected in precleaned aluminum pails located on the fifth-floor roof of the northeast end of Derring Hall. Immediately following or during the rain, samples usually were filtered through Whatman GFC fiberglass filters and stored in a freezer until assayed. A few samples that had frozen into ice had to be thawed prior to filtration.

Procedures for vitamin bioassays were essentially identical to those of Parker and Wachtel as outlined in part I of this chapter. *Lactobacillus plantarum* (ATCC 8014), as before, was used for assaying biotin and niacin, while *L. fermenti* (ATCC 9338) was employed for thiamin assays.

Analyses for various inorganic ions employed the Hach DR-EL Portable Engineer's Laboratory (Hach Co., Ames,

685

Iowa). It should be remembered that, because of rains of
varying intensity and duration, only select rains enable
collection of sufficient water to allow all analyses.
Generally priority was given to nitrogen and phosphorus
assays.

Culturing for bacterial plate counts employed as the
chief medium Rain Water Agar (RWA) consisting of 15 g
Difco agar, 1 g Difco yeast extract, and 1 g Difco pro-
teose peptone per liter.

RESULTS

Tables of data in this section that compare with those
in the previous section will be numbered similarly, with
an a following (i.e., Table 20.1a). Table 20.1a, thus,
summarizes data for the three vitamins assayed in Blacksburg rainwater collected from April 1, 1970, to June 15,
1971. Of the 43 rainwater collections assayed for one or
more vitamins, the range of concentrations was generally
lower with less frequent appearance of high concentrations
than occurred in the St. Louis rains (Table 20.1).

The paucity of rains over Blacksburg with significant
(as defined in the text) concentrations of biotin, thiamin,
or niacin is still more evident by comparison of Tables
20.2a and 20.2. Of the 15 rains listed in Table 20.2a
that contained significant concentrations of one or more
soluble B vitamins, seven involved convective (thunder)
storms, four were light rains or drizzles, three were ordinary steady rains, and one (5/?/70) lacked recorded observations. All heavy rains (>10 cm) of the nonconvective
type lacked significant concentrations of vitamins assayed.

Of the several experiments run to collect and enumerate bacteria, fungi, and algae on various agar media, only
two points seem relevant to the vitamin assays for Blacksburg. First, no obvious correlation existed for viable

Table 20.1a

Concentrations of Biotin, Niacin, and Thiamin in Rainwater,
Blacksburg, Virginia, April 1, 1970, to June 15, 1971

Collection Date	Vitamin concentration/ml			Calculated vitamin/cm^2 of land and/or water surface (conc/ml x cm rainfall)		
	Biotin (pg)	Niacin (mμg)	Thiamin (mμg)	Biotin (pg)	Niacin (mμg)	Thiamin (mμg)
4/1/70	2.0-6.0	0-1.2	0-0.6	0.4-1.2	0-0.24	0-0.12
4/2/70	1.0	Nil	Nil	3.0	Nil	Nil
4/26/70	1.0-4.0	0.2-2.0	0-1.0	1.5-6.0	0.3-3.0	0-1.5
4/27/70	2.0	Nil	Nil	------Not measured------		
5/8/70	2.0-3.0	0-0.4	0-0.6	2.4-3.6	0-0.48	0-0.72
5/16/70	0-0.4	Nil	0-0.6	0-0.4	Nil	0-0.6
5/?/70	1.0-1.2	Nil	Nil	------Not measured------		
6/2/70	1.2-6.0	0.4-0.6	0-0.2	2.4-12.0	0.8-1.2	0-0.4
6/13-14/70	0.4-5.0	Nil	0-0.4	2.0-25.0	Nil	0-2.0
6/17/70	0-1.0	1.0-2.0	Nil	0-2.7	2.7-5.4	Nil
6/20-2-/70	5.0	0.4-1.4	Nil	5.0	0.4-1.4	Nil
6/25/70	0.4-5.0	Nil	0-0.2	0.4-5.0	Nil	0-0.2
6/25/70	5.0	Nil	Nil	4.0	Nil	Nil
6/26/70	0-4.0	0-0.8	Nil	0-2.8	0-0.56	Nil

687

Table 20.1a (Continued)

Collection Date	Vitamin concentration/ml			Calculated vitamin/cm² of land and/or water surface (conc/ml x cm rainfall)		
	Biotin (pg)	Niacin (mμg)	Thiamin (mμg)	Biotin (pg)	Niacin (mμg)	Thiamin (mμg)
6/27/70	0-2.0	0-0.8	Nil	0-1.0	0-0.4	Nil
7/3/70	0-5.0	0-0.2	Nil	0-5.5	0-0.22	Nil
7/9/70	0-4.0	0-1.0	Nil	0-4.4	0-1.1	Nil
7/9/70	0-2.4	0.2-0.6	Nil	0-10.1	0.84-2.5	Nil
7/10/70	0-0.2	0-1.0	Nil	0-0.06	0-0.3	Nil
7/14/70	0.8-1.8	0-0.2	Nil	1.84-4.14	0-0.46	Nil
7/20/70	0-5.0	0-1.0	Nil	0-2.5	0-0.5	Nil
7/21/70	0-1.0	0-0.6	Nil	0-1.2	0-0.84	Nil
7/22/70	0.2-2.0	0-0.6	Nil	0.28-2.8	0-0.84	Nil
7/23/70	0-1.2	0-0.2	Nil	0-0.96	0-0.16	Nil
7/27/70	2.0-4.0	1.0-2.0		2.0-4.0	1.0-2.0	
8/4/70	Nil	2.0-6.0		Nil	0.6-1.8	
8/10/70	4.0-6.0	Nil		48.0-72.0	Nil	
8/19/70	2.0-3.0	Nil		2.4-3.6	Nil	
8/20/70	0-0.4	2.2-2.4	2.0-2.4	0-0.16	0.88-0.96	
9/11/70	Nil	2.2-2.6	Nil	Nil	3.1-3.6	Nil

SEASONAL DISTRIBUTION OF B VITAMINS

Date					
9/27/70	0–2.8	0–0.8	0–3.6	0–1.0	
10/12/70	0–4.0	0–0.4	0–6.4		0–0.64
10/21/70	Nil	Nil	Nil		Nil
10/30/70	Nil	0–0.6	Nil		0–2.88
11/2/70	Nil	Nil	Nil		Nil
2/22/71	Nil	0–0.8	Nil		0–1.5
3/2/71	Nil	0–3.0	Nil		0–12.6
4/28/71	Nil	Nil	Nil		Nil
5/3/71	Nil	0–2.0	Nil		0–1.8
5/6/71	Nil	0–1.8	Nil		0–1.26
5/13/71	Nil	0–0.2	Nil		0–0.8
6/15/71	Nil	Nil	Nil		Nil

Table 20.2a

Blacksburg Rains with over 1.0 pg Biotin,

1.0 mμg Niacin, and 1.0 mμg Thiamin

Collection Date	Biotin	Niacin	Thiamin
4/1/70	2.0-6.0		
4/26/70	1.0-4.0		
4/27/70	2.0		
5/8/70	2.0-3.0		
5/?/70	1.0-1.2		
6/2/70	1.2-6.0		
6/17/70		1.0-2.0	
6/20-21/70	5.0		
6/25/70	5.0		
7/27/70	2.0-4.0	1.0-2.0	
8/4/70		2.0-6.0	
8/10/70	4.0-6.0		
8/19/70	2.0-3.0		
8/20/70		2.2-2.4	2.0-2.4
9/11/70		2.2-2.6	

numbers of microorganisms and significant concentrations
of the vitamins. Second, numbers of viable microorganisms,
primarily gram-negative chromogenic bacteria of the types
found in lakes and ponds, were frequently highest in the
initial few millimeters of rain and declined in subsequent
collections of the same rains.

Too few rains were collected during the nongrowing
season (November-March) to make comparisons of the types
made for the St. Louis data. Chemical data that was collected on rainwater remaining after vitamin assays showed
no clear-cut correlations with the vitamin levels. The
chemical data have been condensed and are summarized in

Table 20.4a. As with St. Louis rainwater, apparently no positive correlation exists between vitamin and dust, because on several occasions where dust was observed in the aluminum pails (e.g., 9/27/70), significant concentrations of vitamins were not detected in filtrates.

Tables 20.3a and 20.3 summarize comparatively the results obtained from the two geographic areas. There are no data on cobalamins in Blacksburg rainwater but unique data were collected on occurrence of thiamin, which was not assayed in the St. Louis studies. The calculated frequencies (as percents) that significant concentrations of biotin or niacin occurred in Blacksburg rains (Table 20.3a) are both appreciably lower than for St. Louis rains. The Blacksburg frequencies might have been even lower had an equivalent number of rains in the nongrowing season been collected and assayed. Based on calculations of the average concentration of vitamins in rains, the values for biotin and niacin in Blacksburg rains approximate those for St. Louis. However, these calculations in both cases may be biased by the choice of minimum concentrations believed to be significant for each bioassay method, and the frequency of occurrence of significiant vitamin levels may therefore be far more meaningful. This point is illustrated by the calculated estimated annual contribution of vitamins/cm^2 in Tables 20.3a and 20.3. Thus, Blacksburg rains annually, even based primarily on data from rains collected during the growing season, would contribute no more than one-third to one-half of the annual contribution of biotin and niacin by St. Louis rains.

DISCUSSION

The body of data, when compared, implies rather strongly that lesser quantities of the B vitamins biotin and niacin are contributed by rains in the Blacksburg area

Table 20.3a

Calculations Summarizing Data from Table 20.1a

Calculations	Biotin	Niacin	Thiamin
Total rain samples assayed	42	30	35
Rains with significant concentrations of vitamins[a]	11 (26.1%)	5 (16.6%)	1 (2.8%)
Rains assayed April 1 to November 1	39	30	30
Rains April 1 to November 1 with significant concentrations of vitamins[a]	11 (28.2%)	5 (16.6%)	1 (3.3%)
Rains assayed November 1 to March 31	3	0	5
Rains November 1 to March 31 with significant concentrations of vitamins[a]	0		0
Average concentrations of vitamin/rain, April 1 to November 1	2.5-4.1 pg/ml	2.1-3.0 mμg/ml	2.0-2.4 mμg/ml
Average concentrations of vitamin November 1 to March 31	0		0
Estimated annual contribution of vitamins/cm^2 by Blacksburg rains	75-123 pg	40-57 mμg	6-8 mμg

[a]Significant, as used here and in the text, refers to >1.0 pg biotin, >1.0 mμg niacin, and >1.0 mμg thiamin/ml.

SEASONAL DISTRIBUTION OF B VITAMINS

Table 20.4a

Summary of Chemical Data for Precipitation

Collected in Blacksburg, Virginia (mg/liter)

Chemical Form	Rains Analyzed	Concentration Range	Mean Concentration
NH_4-N	36	0.12-2.38	1.13
NO_3-N	29	1.00-8.50	3.28
NO_2-N	30	0.00-0.03	0.007
Ortho-PO_4	34	0.00-0.39	0.043
Condensed PO_4	26	0.01-0.11	0.026
Cl^-	34	0.5-10.0	3.9
SO_4^{-2}	11	0.0-35.0	7.25
Total Hardness	3	5.0-15.0	10.0
Seston >.22 µm	10	0.0-95.0	26.5

relative to those in the St. Louis area. This finding does
not necessarily reduce the potential ecological importance
of B vitamins in Blacksburg area rains, for it is still
possible that an aquatic ecosystem in this region may be
influenced by the concentrations less frequently appearing
in precipitation. For example, an assay for soluble vitamins, in filtered water from a Blacksburg farm pond collected April 1, 1971, revealed concentrations of from 4 to
6 pg biotin and 4-6 mµg niacin and thiamin; these concentrations are among the highest found in any Blacksburg
rainwater sample, and it is unlikely that such levels of
soluble vitamins can drastically limit populations dependent upon exogenous supplies. In contrast, water from
Mountain Lake, Giles County, Virginia, known to be highly
oligotrophic (Obeng-Asamoa and Parker 1972), collected
April 14, 1971, contained no detectable vitamins. The
implication here, also strengthened by the findings of

Daisley (1969) for the year-round concentrations of soluble
cobalamins in numerous lakes, is that oligotrophic ecosystems may be deficient in vitamins and consequently may be
more readily influenced by small additions in precipitation.

A final point might be added to those discussed by
Parker and Wachtel in the preceding section. The chapter
by Blanchard and Parker (this volume) presents evidence for
mechanisms that selectively concentrate inorganic and organic matter and microorganisms in the upper 10-100 μm of
aquatic ecosystems and that bring about the water to air
transfer of these materials. One experiment on filtered
water from a Blacksburg farm pond, referred to above, revealed the presence of two to three times the concentration of biotin, niacin, and thiamin in the upper 100 or
50 μm of water than at a 10-cm depth. This finding, along
with the details presented by Blanchard and Parker (this
volume), suggests the possibility that vitamins in rainwater might be derived, in part, from microorganisms and/or
soluble vitamins concentrated near the air-water interfaces
of aquatic ecosystems. This idea and that suggesting living microorganisms in clouds (Parker 1970) remain highly
speculative but nevertheless are sufficiently important
to our understanding of movements within our biosphere that
research aimed at testing these theories should be strongly
supported.

ACKNOWLEDGMENTS

I am grateful to Barbara Mross, Roger Hatcher, and
several undergraduate biology students who assisted in one
or more aspects of these Blacksburg rainwater studies.

SEASONAL DISTRIBUTION OF B VITAMINS

REFERENCES

SEASONAL DISTRIBUTION OF B VITAMINS

REFERENCES

Daisley, K. W. 1969. Monthly survey of vitamin B_{12} concentrations in some waters of the English Lake District. *Limnol. Oceanog.* 14, 224–228.

Obeng-Asamoa, E. K., and B. C. Parker. 1972. Seasonal changes in phytoplankton and water chemistry of Mountain Lake, Virginia. *Trans. Am. Micros. Soc.* 91, 363–380.

Parker, B. C. 1970. Life in the sky. *Nat. Hist.* 79, 54–59.

695